Springer Handbook of Auditory Research

For further volumes:
http://www.springer.com/series/2506

Annemarie Surlykke • Paul E. Nachtigall
Richard R. Fay • Arthur N. Popper
Editors

Biosonar

With 78 Illustrations

Springer

Editors
Annemarie Surlykke
Department of Biology
University of Southern Denmark
Odense, Denmark

Richard R. Fay
Marine Biological Laboratory
Woods Hole, MA, USA

Paul E. Nachtigall
Marine Mammal Research Program
Hawaii Institute of Marine Biology
University of Hawaii
Kailua, HI, USA

Arthur N. Popper
Department of Biology
University of Maryland
College Park, MD, USA

ISSN 0947-2657 ISSN 2197-1897 (electronic)
ISBN 978-1-4614-9145-3 ISBN 978-1-4614-9146-0 (eBook)
DOI 10.1007/978-1-4614-9146-0
Springer New York Heidelberg Dordrecht London

Library of Congress Control Number: 2014942517

Printed on acid-free paper

Springer is part of Springer Science+Business Media (www.springer.com)

We dedicate this volume to the principal mentors of Annemarie Surlykke and Paul Nachtigall (and friends of all of the coeditors): Bertel Møhl and Ronald J. Schusterman. These two outstanding critical thinkers served to direct the empirical science of echolocation since their participation in the initial Animal Sonar Systems meetings in Frascati Italy in 1966. Both played key roles in the development of our current knowledge of animal echolocation.

Series Preface

The following preface is the one that we published in Volume 1 of the Springer Handbook of Auditory Research back in 1992. As anyone reading the original preface, or the many users of the series, will note, we have far exceeded our original expectation of eight volumes. Indeed, with books published to date, and those in the pipeline, we are now set for more than 50 volumes in SHAR, and we are still open to new and exciting ideas for additional books.

We are very proud that there seems to be consensus, at least among our friends and colleagues, that SHAR has become an important and influential part of the auditory literature. While we have worked hard to develop and maintain the quality and value of SHAR, the real value of the books is very much because of the numerous authors who have given their time to write outstanding chapters and to our many coeditors who have provided the intellectual leadership to the individual volumes. We have worked with a remarkable and wonderful group of people, many of whom have become great personal friends of both of us. We also continue to work with a spectacular group of editors at Springer. Indeed, several of our past editors have moved on in the publishing world to become senior executives. To our delight, this includes the current president of Springer US, Dr. William Curtis.

But the truth is that the series would and could not be possible without the support of our families, and we want to take this opportunity to dedicate all of the SHAR books, past and future, to them. Our wives, Catherine Fay and Helen Popper, and our children, Michelle Popper Levit, Melissa Popper Levinsohn, Christian Fay, and Amanda Fay, have been immensely patient as we developed and worked on this series. We thank them, and state, without doubt, that this series could not have happened without them. We also dedicate the future of SHAR to our next generation of (potential) auditory researchers—our grandchildren—Ethan and Sophie Levinsohn; Emma Levit; and Nathaniel, Evan, and Stella Fay.

Preface 1992

The Springer Handbook of Auditory Research presents a series of comprehensive and synthetic reviews of the fundamental topics in modern auditory research. The volumes are aimed at all individuals with interests in hearing research including advanced graduate students, postdoctoral researchers, and clinical investigators. The volumes are intended to introduce new investigators to important aspects of hearing science and to help established investigators to better understand the fundamental theories and data in fields of hearing that they may not normally follow closely.

Each volume presents a particular topic comprehensively, and each serves as a synthetic overview and guide to the literature. As such, the chapters present neither exhaustive data reviews nor original research that has not yet appeared in peer-reviewed journals. The volumes focus on topics that have developed a solid data and conceptual foundation rather than on those for which a literature is only beginning to develop. New research areas will be covered on a timely basis in the series as they begin to mature.

Each volume in the series consists of a few substantial chapters on a particular topic. In some cases, the topics will be ones of traditional interest for which there is a substantial body of data and theory, such as auditory neuroanatomy (Vol. 1) and neurophysiology (Vol. 2). Other volumes in the series deal with topics that have begun to mature more recently, such as development, plasticity, and computational models of neural processing. In many cases, the series editors are joined by a coeditor having special expertise in the topic of the volume.

Falmouth, MA, USA Richard R. Fay
College Park, MD, USA Arthur N. Popper

Volume Preface

Bats and odondoctes have evolved the ability to use echolocation to find objects in their environments. Yet, despite substantial differences in their environments, there are also substantial similarities, as well as any number of fascinating differences, in how members of these two groups produce, use, and process biosonar signals. One approach to considering biosonar could have been to have separate chapters for bats and odontocetes. Instead, the editors concluded that it might be far more interesting if each chapter focused on a specific aspect of biosonar, and then asked a team of authors to describe, compare, and contrast biosonar in the two groups. To this end, they invited authors who work on bats and odontocetes to collaborate on the chapters. In virtually all cases the authors found this comparative task a challenge, but also very interesting and useful because they often do not get to interact across species groups.

The first chapter, by Surlykke and Nachtigall, presents an overview of the volume and shows how the material is tied together. Chapter 2, by Fenton, Jenson, Kalko, and Tyack, discusses the sounds produced by bats and whales, and how they are used in behavior. In Chap. 3, Au and Suthers discuss the mechanisms of sound production in the two groups. Echolocation sounds are then considered in Chap. 4 by Wahlberg and Surlykke, with a focus on echolocation behavior of wild animals. Detection of echolocation sounds is a critical part of understanding biosonar, and thus hearing of such signals is described in Chap. 5 by Nachtigall and Schuller. Echolocation behavior, including analysis of targets, is the ultimate goal in biosonar, and so this is the theme of Chap. 6 by Simmons, Houser, and Kloepper. However, while these studies have been carried out primarily in the lab, there has been a recent focus on echolocation by wild odontocetes, and the results of these studies are discussed in Chap. 7 by Johnson. Bats and dolphins actively control the timing and spectral content of their echolocation signals, which in return impact the acoustic information the animals receive, and in Chap. 8 Moss, Chiu, and Moore focus on this adaptive control to understand the perceptual organization of sound in echolocating animals and consequently their experience of an auditory scene. Chapter 9 by Madsen and Surlykke examines the ecology and behavior of bats and odontocetes in orientation and prey detection, again using field studies.

As with many other SHAR volumes, the chapters presented here build on chapters in other volumes. In particular, the current chapters carry forward from two earlier volumes, one *Hearing by Whales and Dolphins* (Vol. 12, edited by Au, Popper, and Fay in 2000) and one on *Hearing by Bats* (Vol. 5, edited by Popper and Fay, in 1995). Sound processing by bats has also been a theme of chapters in *Auditory Computation* (Vol. 6, edited by Hawkins, McMullen, Popper, and Fay in 1996).

Odense, Denmark — Annemarie Surlykke
Honolulu, HI, USA — Paul E. Nachtigall
Falmouth, MA, USA — Richard R. Fay
College Park, MD, USA — Arthur N. Popper

Contents

Contributors

Whitlow W.L. Au Hawaii Institute of Marine Biology, University of Hawaii, Kaneohe, HI, USA

Chen Chiu Department of Psychological and Brain Sciences, Johns Hopkins University, Baltimore, MD, USA

Brock (M.B.) Fenton Department of Biology, Western University, London, ON, Canada

Dorian Houser National Marine Mammal Foundation, San Diego, CA, USA

Frants H. Jensen Department of Biology, Woods Hole Oceanographic Institution, Woods Hole, MA, USA

Mark Johnson Sea Mammal Research Unit, University of St. Andrews, St. Andrews, Scotland

Elisabeth K.V. Kalko† Institute of Experimental Ecology, University of Ulm, Ulm, Germany

Smithsonian Tropical Research Institute, Balboa Ancón, Republica de Panamá

Laura Kloepper Department of Neuroscience, Brown University, Providence, RI, USA

Peter Teglberg Madsen Zoophysiology, Department of Bioscience, Aarhus University, Aarhus C, Denmark

Patrick W. Moore National Marine Mammal Foundation, San Diego, CA, USA

Cynthia F. Moss Department of Psychological and Brain Sciences, Johns Hopkins University, Baltimore, MD, USA

Paul E. Nachtigall Marine Mammal Research Program, Hawaii Institute of Marine Biology, University of Hawaii, Kailua, HI, USA

† Elisabeth K.V. Kalko (Author was deceased at the time of publication).

Gerd Schuller Biocenter LMU, Department Biologie II, Neurobiology, Ludwig-Maximilians-University of Munich, Planegg-Martinsried, Germany

James A. Simmons Department of Neuroscience, Brown University, Providence, RI, USA

Annemarie Surlykke Department of Biology, University of Southern Denmark, Odense M, Denmark

Roderick A. Suthers Department of Biology, School of Medicine, Indiana University, Bloomington, IN, USA

Peter L. Tyack School of Biology, Sea Mammal Research Unit, Scottish Oceans Institute, University of St. Andrews, St. Andrews, Fife, Scotland

Magnus Wahlberg Fjord & Bælt and Department of Biology, University of Southern Denmark, Marine Biological Research Center, Kerteminde, Denmark

Chapter 1
Biosonar of Bats and Toothed Whales: An Overview

Annemarie Surlykke and Paul E. Nachtigall

Keywords Auditory evoked potential hearing measurement • Auditory scene analysis • Automatic gain control • Bertel Møhl • Clicks • Comparative echolocation: hearing • Echolocation • Frequency modulated sweep • Ronald J. Schusterman • Tags • Terminal buzzes

1.1 Why Bats and Toothed Whales Together?

Why assemble a number of chapters from different authors into a volume on the echolocation of bats and whales? Bats fly in a medium in which sound travels one fifth as fast as it does underwater where whales swim, and water is 784 times as dense as air. Despite these basic environmental differences, bats and whales have both independently evolved the ability to locate, identify, track, and catch their fast moving prey by sending out bursts of sound and listening for the echoes that soon follow. Like the differences in the animals' environment, investigators who study the two echolocating animal groups tend to operate in different environments. People who study bats tend to go to bat meetings while people who study whales and dolphins tend to go to meetings based on issues related to the underwater environment. Unfortunately there is not enough communication between the two groups.

There is, however, great value in the examination of the comparative processes of echolocation of both bats and dolphins. Given the independent evolution of echolocation in the two animal groups, it is possible to examine how they independently

A. Surlykke (✉)
Department of Biology, University of Southern Denmark, Odense M 5230, Denmark
e-mail: ams@biology.sdu.dk

P.E. Nachtigall
Marine Mammal Research Program, Hawaii Institute of Marine Biology,
University of Hawaii, P.O. Box 1106, Kailua, HI 96734, USA
e-mail: nachtiga@hawaii.edu

A. Surlykke et al. (eds.), *Biosonar*, Springer Handbook of Auditory Research 51,
DOI 10.1007/978-1-4614-9146-0_1, © Springer-Verlag New York 2014

solved similar problems associated with listening to very quiet echoes after making loud sounds and all the other processes involved in echolocating. Previous SHAR volumes have reviewed hearing of bats and whales (Vol. 5 by Popper and Fay 1995, *Hearing by Bats*, and Vol. 12 by Au et al. 2000, *Hearing by Whales and Dolphins*) and thus dealt with sonar in both groups. However, this volume is dedicated to the comparative examination of the echolocation of the two groups of animals. Each chapter is a joint effort that was initiated by inviting at least two investigators—one who studies bats and the other who studies whales—to write a chapter in their area of expertise. The volume was compiled to compare what has been learned, particularly in the last 10 years, in the study of the echolocation of bats and whales in a number of exciting new topic areas, to show the advancement of knowledge and techniques, and hopefully to point the way forward to new areas of fascinating research. Each chapter was also written to help introduce new investigators to aspects of the science of hearing and echolocation as well as allow established echolocation investigators to understand fundamental ideas, theories, and data on subfields of hearing that the investigators may not normally follow closely.

1.2 An Overview of the Volume

Flying at night and diving deeply in the ocean both provide situations with limited light and opportunities for using sound for foraging and orientation. In Chap. 2, Fenton, Jensen, Kalko, and Tyack demonstrate that sound is widely used by both animal groups and both have developed the ability to listen to high frequencies as a part of echolocation. Echolocating bats produce sonar signals with their larynx (Elemans et al. 2011), whereas all odontocetes appear to make echolocation clicks with their nasal complexes (Cranford 2000). Both bats and odontocetes are long lived and most are social mammals, with indications of extensive learning. Both appear to adapt their echolocation signals to the task. Both groups show head anatomy greatly modified to improve the advantages of sound production for echolocation. Most bats echolocate with tonal signals whereas most odontocetes use some sort of click. The obvious exceptions to those generalizations are the few tongue clicking *Rousettus* species from the nonecholocating bat family Pteropodidae (Suthers and Summers 1980), and the frequency modulated (FM) downsweep tonal producing deep-diving beaked whales (Madsen et al. 2005).

Sounds made by the animals to produce echoes are basic to understanding echolocation. Chapter 3 by Au and Suthers describes the basic mechanisms of specialized sound production in whales and bats. Much of the data were collected in laboratory or laboratory/field settings. Although there was once controversy as to whether the site of sound production in whales is the larynx or nasal passages, new work now convincingly demonstrates that the echolocation sounds are produced in the nasal passage's phonic lips. Some interesting controversy remains as to whether a single or double use of the lips occurs in all odontocete species (Madsen et al. 2010). This controversy will be resolved only with future research.

Bats, on the other hand, vocalize using the larynx. The hypertrophied use of the larynx has resulted in a variety of types of signals adapted to the environment in which the bat hunts. Because bat signals are made by pressurized air exiting across the larynx, most pulses are produced in the expiratory phase of the respiratory cycle. Respiratory muscles have become specialized for echolocation, and flying bats emit most of their echolocation calls during the upstroke of the wings (Koblitz et al. 2010). The anatomy of the actual larynx has become very specialized for high-frequency sound production with a very thin, 6–8 μm, nearly transparent membrane extending from the edge of each vocal fold. Both the nasal passages of the whales, and the larynxes of the bats, produce high-frequency signals. Bats precisely control the frequency of their pulses to optimize them for particular perceptual tasks, and although whales demonstrate frequency shifts of signals with hearing loss, much more needs to be done to understand signal change by individual animals. Signals are emitted in directed beam patterns in front of the animals. So far evidence suggests the beam is more precisely directional in the whales, whereas bats can vary the directionality of the beam depending on the environment and ecology of the species. Both groups generally produce signals with low pulse repetition rates of roughly 5–30 Hz, but increasing dramatically in the "buzz" phase right before prey capture. There have been many more bat species examined than cetacean species, but recent efforts in cetacean field studies reveal more variety of signal types than previously thought.

In Chap. 4, Wahlberg and Surlykke examine the outgoing echolocation signals of whales and bats with an emphasis on those studies that have been conducted in the field. Given that one rarely has an animal in an exact controlled position in the field, an understanding of the physics of measuring sounds in both media is essential. Source levels of directional signals are measured using different types of arrays. The geometric shape, along with the number of receivers, determines the amount and quality of echolocation signals recorded. The sound pressure levels of the signals can be determined using peak to peak (pp) or root mean square (RMS) or its variations. Generally peak to peak greatly overestimates the amount of energy actually emitted by the animal. When measured within the center of the beam, maximum source levels for bats range from 99 to 144 dB re 20 μPa RMS at 10 cm. Large delphinids produce source levels ranging from 180 to 220 dB re 1 μPa at 1 m, with the sperm whale measured with intensities up to 236 dB re 1 μPa RMS. All of these sounds would be perceived by humans, if they could hear very high frequencies as very loud sounds above the threshold of pain. Presumably most ears integrate energy over time and therefore to compare the energy, durations of the signals should be included by integrating the squared pressure over the duration of the signal to give the total energy emitted.

It is also important to describe the directionality of the animal's sonar signal by measuring the width of the signal beam. Although this can be calculated in many ways, it is commonly done by measuring the angle between the points where the intensity has fallen by 6 or 3 dB on either side of the point of maximum intensity. One may also calculate the directionality index (DI), which is the ratio in decibels between on-axis intensity of the directional source and the intensity from an omnidirectional reference source radiating the same total power from the sound source (Jakobsen et al. 2013). Although bats apparently manipulate the directionality of their beams, there appears to be a correlation between size and directionality in whales, with the large sperm whale demonstrating the narrowest beam. Although it would seem that the loudest signal should produce the best echo and be the most functional, there are a number of restraints on the optimal signal level, including target range, the level of clutter and reverberation, and hearing. Generally animals decrease the level of the signals as they approach the target. The sonar equation is a very useful tool to describe the physics of the outgoing signal and the echo return (Møhl 1988). Variation in the outgoing signal is useful and can be described in terms of automatic gain control varying the intensity of the outgoing signal.

The intensity and range of frequencies of the outgoing signals must be matched by the ability to hear the echoes in a functional biosonar system. The hearing of odontocetes has recently been measured, as described by Nachtigall and Schuller in Chap. 5, during echolocation. Gain control is not limited to variation in the outgoing signal but hearing sensitivity also changes greatly depending on the echolocation situation. The hearing of a false killer whale (*Pseudorca crassidens*) may be 20 dB more sensitive during the target absent search phase (Supin et al. 2006). Hearing in bats also dramatically changes for protection from the intense self-generated outgoing echolocation signals. The study of bat hearing during echolocation has differed from that of the study of odontocetes and has been considered in a variety of conditions including: (1) simple passive hearing, (2) passive hearing during echolocating behavior (e.g., within groups or eaves dropping), and (3) listening to their own echoes after active emission. Bats may eavesdrop and use the echoes from calls produced by other bats and may change hearing sensitivity by 40 dB immediately after call production (Kick and Simmons 1984).

Big brown bats (*Eptesicus* fuscus) and bottlenose dolphins (*Tursiops truncatus*) are among the most studied of the echolocators. Both species produce broadband signals with energy between about 20 and 130 kHz. The functional use of the broad energy spectrum by these animals is examined in Chap. 6 by Simmons, Houser, and Kloepper. Dolphins and bats change the characteristics of their outgoing signals as they echolocate. Bats produce trains of FM downsweeps whereas dolphins produce a series of broadband clicks. Each successive echo received reduces the signal-to-noise ratio of the detection threshold and both species direct their beam to focus on the target of interest. Both dolphins and bats change the intervals between the signals in response to acoustic conditions and variation in target distance. Bats forage in three stages: (1) search, (2) approach, and (3) terminal buzz (Schnitzler and Kalko 2001) and new field data clearly show that odontocetes also go through similar phases in natural pursuits (Johnson et al. 2006). Dolphins and bats both normally allow the return of an echo before emitting another click to reduce ambiguity, and recently dolphins have been shown to also use terminal buzzes at the conclusion of foraging chases. Most of the broad spectrum used by both species allows for detection and discrimination of desired targets.

Much of the advancement in odontocete echolocation work in the past 10 years has come about by moving from the laboratory into observations of free-swimming animals. Johnson's chapter reviews the latest updates on tagging methods that have so far mainly been deployed on whales, owing to their more convenient size, but is predicted to soon promote a revolution also in bat echolocation research. Chapter 7 by Johnson gives an unprecedented insight into the methods employed for not only using tags but also processing and interpreting the wealth of data acquired by this technique. Visualization of the echo-stream or "echogram" is particularly exciting in providing a picture of the auditory scene in a natural environment, pointing to the prospects in tag techniques for future studies of auditory scene analysis as described in Chap. 8 by Moss, Chui, and Moore. The chapter takes an engineering approach and focuses on the key sensor modalities available in tags to, hopefully, inspire biologists to do new exciting experiments exploiting this powerful tool to combine echolocation sounds with other biologically relevant parameters, for example, depth, position, movement, respiration, or heart rate, all sampled in the wild from animals negotiating the natural environment and the challenges involved.

For simplicity's sake, much of the study of echolocation has made the assumption that the echolocator was looking at the echo from a single object. Logically, one can quickly surmise that that assumption does not adequately reflect the environment of an echolocating animal, but the understanding of how echoes in the natural environment that come from many objects that make up an acoustic scene is still incomplete (Moss et al. 2011). Because echolocators constantly adapt their outgoing signals based on auditory feedback carried by the previous echoes, studies of the echolocating animal's control over signals and hearing provide a unique window into sonar scene perception in actively echolocating animals. Chapter 8 by Moss, Chiu, and Moore examines what is meant by an acoustic scene, and what has been done recently to interpret bat and whale echolocation performance in terms of auditory scene analysis.

This volume brings together bat and whale research and researchers to further cross-area comparisons and collaborations to approach a more general understanding of ecological, behavioral, and physical constraints for using sound for orientation and prey detection. Chapter 9 by Madsen and Surlykke focuses on the comparison of bats and whales based primarily on large amount of new field data from echolocating odontocetes, collected mainly by newly developed tags, as dealt with in Johnson's chapter, and from bats by array techniques. In both cases the escalation in data collection is promoted by the rapidly decreasing size and increasing sample rate and storage size of electronic equipment. The chapter explains the biological impact of physical differences between water and air, but emphasizes that in spite not only of these differences but also the vast differences in size and prey type between bats and odontocetes, the most striking characteristics of their echolocation are the similarities. These extremely different groups of mammals emit signals that closely resemble one another in terms of frequency range, emission rate, and not the least in terms of the adaptive changes in pulse interval and emitted intensity through the three phases, search, approach, and terminal or "the buzz" that constitute the phases of prey pursuit in both air and water.

1.3 Volume Dedication

We have dedicated this volume to Bertel Møhl and Ronald J. Schusterman. These two outstanding critical thinkers served to direct the empirical science of echolocation since their participation in the initial Animal Sonar Systems meetings in Frascati Italy in 1966. Both played key roles in the development of our current knowledge of animal echolocation.

Bertel Møhl personifies the purpose of this SHAR volume. From the very beginning, his research has focused on the comparative aspect of echolocation. Although his true love is whales, in particular the big ones, he has contributed very significantly to research on bat echolocation as well. All his projects have one common denominator: interesting, unsolved problems, and his work is as far away from the "least publishable unit" as possible. His perspective is formed by a very solid training in classical zoology and biology, paired with a pronounced talent for technical innovation—quite an unusual combination for a biologist and thus one main reason for the number of breakthroughs, and intelligent failures, he has been responsible for. He critically tested the matched filter model for bat echolocation and falsified it in a classical Karl Popper fashion by showing that bats' detection thresholds for their own signals were independent of whether the signals were played backward or forward (Møhl 1986). He pioneered the use of large-scale arrays for studying the sonar of odontocetes and found unprecedented emitted intensities and directionalities of first narwhals (Møhl et al. 1990).

But later his favorite subject became the sperm whale (Møhl et al. 2000), the carrier of the largest nose in the animal kingdom, for which he also suggested the "bent horn" to explain the multipulse nature of sperm whale clicks (Møhl et al. 2003a, b). The array studies served as the basis for a great number of subsequent studies of these and other toothed whales in the field by him and many others. Equally important was the array technique for quantifying acoustic parameters emitted by nocturnal bats in the wild (Surlykke and Kalko 2008).

In both animal groups such studies stimulated by Bertel Møhl's ideas have now provided a vast body of results that illustrate the remarkable adaptations to orienting actively with sound, but also point to the enormous diversity in evolution of (acoustic) solutions that allow for life under conditions where light is limited or missing.

Another example of Bertel Møhl's innovative spirit is his pioneering use of on-animal tags. He envisioned and designed an on-board recorder for sound, pressure, and other physical parameters that was equipped with satellite transmitter for upload of data. The tag (which was fairly large compared to today's models) was intended for attachment on the tooth of male narwhals. However, the narwhals were not cooperative, but the experiment was later completed (on sperm whales) in collaboration with Peter T. Madsen and others (Madsen et al. 2002), and has since been succeeded by an overwhelming number of spectacular tag-studies not only by Madsen in collaboration with Mark Johnson, but also many others.

Bertel Møhl has never done experiments just to add to the "stamp collection," but has always had in mind the scientific question as well as the possibility that the hypothesis might be wrong. For these and many other reasons it is hard to exaggerate the inspiration and importance of his work and attitude for his students, colleagues, and friends.

Ron Schusterman started his career by completing his doctorate in experimental psychology under the direction of Winthrop Kellogg, who published the first evidence that cetaceans echolocated (Kellogg et al. 1953). Ron was invited to study echolocation in California sea lions as the co-developer of the BioSonar laboratory at Stanford Research Institute. Despite his best efforts, and the beliefs of many, there was no evidence that sea lions echolocated and his strong experimental designs proved this (Schusterman 1967).

In fact, Schusterman became famous for what he disproved. The experimental designs that he developed, however, became the basis for many excellent experiments examining echolocation in the dolphins and whales that did echolocate. His experimental psychophysics on the learning and sensory systems of California sea lions formed the basis of much of what we know of the species (e.g., Schusterman 1968). His classic presentations and subsequent chapters in the Animal Biosonar Systems books provided the basis of much of the methodology for cetacean echolocation experimentation (Schusterman 1980).

The experience of relying on the empirical evidence and moving toward the simplest explanation of the data marked Ron's career. His analysis of animal language learning moved the thinking of most academics toward the logical simpler explanations (Schusterman and Kastak 1995). His teaching and direction toward sound experimental design, and parsimony in explanation of results, echo through every chapter examining echolocation in cetaceans in this book. He was an outstanding mentor, a respected and honored scholar, and a very dear friend.

1.4 Conclusions

Echolocation is a unique sensory process for study because it requires the animal to actively produce signals in order to detect fast moving prey in its environment. The study of the outgoing signals allows an objective look at the behavior required to actively sense an environment. Changes in both signal production and hearing show a dynamic adjusting sensory system that has allowed successful mammalian foraging in diverse environments. That success has, in turn, provided a path for a fascinating hypertrophy of auditory and neural structures. The study of echolocation serves, therefore, to provide a fascinating roadmap to the evolution of sensory function and behavior.

References

Au, W. W. L., Popper, A. N., & Fay, R. R., Eds. (2000). *Hearing by whales and dolphins*. New York. Springer-Verlag.

Cranford, T. W. (2000). In search of impulse sound sources in odontocetes. In W. W. L. Au, A. N. Popper, & R. R. Fay (Eds.), *Hearing by whales and dolphins*. New York: Springer-Verlag.

Elemans, C. P. H., Mead, A. F., Jakobsen, L., & Ratcliffe, J. M. (2011). Superfast muscles set maximum call rate in echolocating bats. *Science*, 333, 1885–1888.

Jakobsen, L., Brinkløv, S., & Surlykke, A. (2013). Intensity and directionality of bat echolocation signals. *Frontiers in Physiology*, 4, Article 89. doi: 10.3389/fphys.2013.00089.

Johnson, M., Madsen, P. T., Zimmer, W. M. X., Aguilar de Soto, N., & Tyack, P. L. (2006). Foraging Blainville's beaked whales (*Mesoplodon densirostris*) produce distinct click types matched to different phases of echolocation. *Journal of Experimental Biology*, 209, 5038–5050.

Kellogg, W. N., Kohler, N. R., & Morris, H. N. (1953). Porpoise sounds as sonar signals. *Science*, 117, 239–243

Kick, S. A., & Simmons, J. A. (1984). Automatic gain control in the bats sonar receiver and the neuroethology of echolocation. *Journal of Neuroscience*, 4, 2725–2737.

Koblitz, J. C., Stilz, P., & Schnitzler, H. U. (2010). Source levels of echolocation signals vary in correlation with wingbeat cycle in landing big brown bats (Eptesicus fuscus). *Journal of Experimental Biology*, 213, 3263–3268.

Madsen, P. T., Payne, R. S., Kristiansen, N. U., Wahlberg, M., & Møhl, B. (2002). Sperm whale sound production studied with ultrasound time/depth-recording tags. *Journal of Exerimental Biology*, 205, 1899–1906.

Madsen, P. T., Johnson, M., Aguilar de Soto, N., Zimmer, W. M. X., & Tyack, P. (2005). Biosonar performance of foraging beaked whales (*Mesoplodon densirostris*). *Journal of Experimental Biology*, 208, 181–194.

Madsen, P. T., Wisniewska, D., & Beedholm, K. (2010). Single source sound production and dynamic beam formation in echolocating harbour porpoises (*Phocoena phocoena*). *Journal of Experimental Biology*, 213, 3105–3110.

Møhl, B. (1986). Detection by a pipistrelle bat of normal and reversed replica of its sonar pulses. *Acustica*, 61, 75–82.

Møhl, B. (1988). Target detection by echolocating bats. In P. E. Nachtigall & P. W. B. Moor (Eds.), *Animal sonar* (pp. 435–450). New York: Plenum Press.

Møhl, B., Surlykke, A., & Miller, L. A. (1990). High intensity narwhal clicks. In J. A. Thomas & R. A. Kastelein (Eds.), *Sensory abilities of cetaceans* (pp. 295–303). New York: Plenum Press.

Møhl, B., Wahlberg, M., Madsen, P. T., Miller, L. A., & Surlykke, A. (2000). Sperm whale clicks: Directionality and source level revisited. *Journal of the Acoustical Society of America*, 107, 638–648.

Møhl, B., Madsen, P. T., Wahlberg, M., Au, W. W. L., Nachtigall, P. E., & Ridgway, S. H. (2003a). Sound transmission in the spermaceti complex of a recently expired sperm whale calf. *Acoustics Research Letters Online*, 4, 19–24.

Møhl, B., Wahlberg, M., Madsen, P. T., Heerfordt, A., & Lund, A. (2003b). The monopulsed nature of sperm whale clicks. *JASA*, 114, 1143–1154.

Moss, C. F., Chiu, C., & Surlykke, A. (2011). Adaptive vocal behavior drives perception by echolocation in bats. *Current Opinion in Neurobiology*, 21, 645–652.

Popper, A. N., & Fay, R. R., Eds. (1995). *Hearing by bats*. New York: Springer-Verlag.

Schnitzler, H.-U., & Kalko, E. K. V. (2001). Echolocation by insect-eating bats. *BioScience*, 51, 557–569.

Schusterman, R. J. (1967). Perception and determinants of underwater vocalization in the California sea lion. In R. G. Busnel (Ed.), *Les systemes sonars animaux: Biologie et bionique* (pp. 535–610). Jouy-en-Josas, France: Laboratoir de Physiologic Acoustique.

Schusterman, R. J. (1968) Experimental laboratory studies of pinniped behavior. In R. J. Harrison, R. C. Hubbard, R. S. Peterson, C. E. Rice, & R. J. Schusterman (Eds.), *The behavior and physiology of pinnipeds*. New York: Appleton-Century-Crofts.

Schusterman, R. J. (1980) Behavioral methodology in echolocation by marine mammals. In R. G. Busnel & J. Fish (Eds.), *Animal sonar systems* (pp. 11–41). New York: Plenum Press.

Schusterman, R. J., & Kastak, D. (1995). There is no substitute for an experimental analysis of marine mammal cognition. *Marine Mammal Science*, 11(2), 263–267.

Supin, A. Y., Nachtigall, P. E., & Breese, M. (2006). Source-to-sensation level ratio of transmitted biosonar pulses in an echolocating false killer whale. *Journal of the Acoustical Society of America*, 120, 518–526.

Surlykke, A., & Kalko, E. K. V. (2008). Echolocating bats cry out loud to detect their prey. *PLoS ONE*, 3(4), e2036(1)–e2036(10).

Suthers, R. A., & Summers, C. A. (1980). Behavioral audiogram and masked thresholds of the megachiropteran echolocating bat, *Rousettus*. *Journal of Comparative Physiology A*, 136, 227–233.

Chapter 2
Sonar Signals of Bats and Toothed Whales

Brock (M.B.) Fenton, Frants H. Jensen,
Elisabeth K.V. Kalko†, and Peter L. Tyack

Keywords Bat • Call design • Call function • Duty cycle • Evolution • Field • Laboratory • Odontocete • Social context • Toothed whale

2.1 Introduction

At first glance, flying bats and diving toothed whales (odontocete cetaceans) (Table 2.1) appear to be polar opposites among mammals. Adult bats range in size from 2 to 1,500, with low weight an advantage for these flying mammals. In contrast, adult toothed whales range from 55 to 50,000 kg. They inhabit a much denser aquatic environment where they are suspended in the water and have fewer constraints on body mass. Bats and toothed whales are extraordinary among mammals in their adaptations for operating in air (bats) or in the open sea (toothed whales).

---

† Elisabeth K.V. Kalko (Author was deceased at the time of publication).

B. (M.B.) Fenton (✉)
Department of Biology, Western University, London, ON, Canada N6A 5B7
e-mail: bfenton@uwo.ca

F.H. Jensen
Department of Biology, Woods Hole Oceanographic Institution, Woods Hole, MA 02543, USA
e-mail: Frants.Jensen@gmail.com

E.K.V. Kalko†
Institute of Experimental Ecology, University of Ulm,
Albert-Einstein Allee 11, Ulm 89069, Germany

Smithsonian Tropical Research Institute, P.O. Box 0843-03092, Balboa Ancón,
Republica de Panamá

P.L. Tyack
School of Biology, Sea Mammal Research Unit, Scottish Oceans Institute,
University of St. Andrews, St. Andrews, Fife KY16 8LB, Scotland
e-mail: plt@st-andrews.ac.uk

A. Surlykke et al. (eds.), *Biosonar*, Springer Handbook of Auditory Research 51,
DOI 10.1007/978-1-4614-9146-0_2, © Springer-Verlag New York 2014

Table 2.1 The diversity of living toothed whales and bats

	Genera	Species
Order Cetartiodactyla		
Suborder Odontoceti		
Superfamily Platanistoidea		
Platanistidae	1	2
Lipotidae	1	1
Pontoporiidae	1	1
Iniidae	1	1
Superfamily Delphinoidea		
Monodontidae	2	2
Phocoenidae	4	6
Delphinidae	17	33
Superfamily Ziphoidea		
Ziphiidae	6	19
Superfamily Physeteroidea		
Physeteridae	2	3
Total	35	68
Order Chiroptera		
Suborder Yinpterochiroptera		
Pteropodidae	42	169
Superfamily Rhinolophoidea		
Craseonycteridae	1	1
Rhinopomatidae	1	4
Megadermatidae	4	5
Rhinolophidae	1	62
Hipposideridae	9	69
Suborder Yangochiroptera		
Superfamily Emballonuroidea		
Emballonuridae	13	49
Nycteridae	1	13
Superfamily Noctilionoidea		
Noctilionidae	1	2
Mormoopidae	2	8
Phyllostomidae	52	154
Mystacinidae	1	1
Superfamily Vespertilionoidea		
Vespertilionidae	43	342
Natalidae	1	5
Furipteridae	2	2
Thyropteridae	1	4
Myzopodidae	1	2
Miniopteridae	1	11
Molossidae	16	86
Totals	193	989

Whales, like dugongs and manatees, are among the few mammal species that do not routinely come on land. Though flying seems quite different from diving, bats and toothed whales orient and move in a three-dimensional setting where restricted visibility at night for bats and in the underwater environment for toothed whales increases the benefit of using sound to orient and forage. Bats and toothed whales supplement sensory input from visual and positional sensors by listening for echoes from brief acoustic pulses they produce. These echolocators use the time difference between pulse and echo and the characteristics of the echo to form an auditory representation of their surroundings. Though a few species of birds also echolocate (Konishi and Knudsen 1979; Griffin and Thompson 1982), bats and toothed whales possess the most highly developed and adaptable biosonar systems in the animal world.

Three-dimensional settings combined with inconsistent and uncertain visibility could have favored exaptation of acoustic signals for use in orientation. Some differences in the details of echolocation behavior and signal structure between toothed whales and bats reflect differences in the densities of the media in which they operate. The greater density of water means that sound moves about five times faster in water than in air, and less energy is lost due to absorption during propagation underwater. Despite the vastly different environments in which they operate, bats and toothed whales share some striking similarities in their echolocation behavior that are outlined in the present chapter. The transition to the present lifestyles of bats and toothed whales required dramatic changes in patterns of locomotion, perhaps coincident with the evolution of echolocation. The combination of agile, three-dimensional maneuvering and the ability to forage independent of light conditions probably set the stage for the diversification and adaptive radiation of both bats and toothed whales.

Knowledge of echolocation in toothed whales and bats is based on relatively few species. It appears that all toothed whales have the capacity for echolocation, but most fruit bats (Pteropodidae) lack this capacity. At least two species of Old World fruit bats, the *Rousettus* species, use tongue clicks as echolocation signals, while all other bats produce echolocation signals with their larynx. The phylogeny of toothed whales and bats is considered toward the end of this chapter.

The approximately 1,259 species of living bats occur in terrestrial families habitats throughout the world, with the exception of polar areas and some remote oceanic islands. Although most species mainly prey on insects, some bats eat plant products (leaves, fruit, nectar, and pollen), or animals ranging from other arthropods to vertebrates such as fish, amphibians, reptiles, rodents, bats, and birds. In addition, three species of Neotropical vampire bats eat blood. As most bats are active at night, they fly and find resources such as food or roosts when lighting is poor and/or unpredictable. Evolution of biosonar is one of the key innovations that permit bats access to a wide range of resources at night. Bats use echolocation to orient in space, and to detect, classify, and locate food. Many features of the faces of bats are associated with echolocation, whether for transmission of signals or reception of echoes. There are several examples of convergent and parallel evolution among bats, including facial and wing features (Fenton 2010).

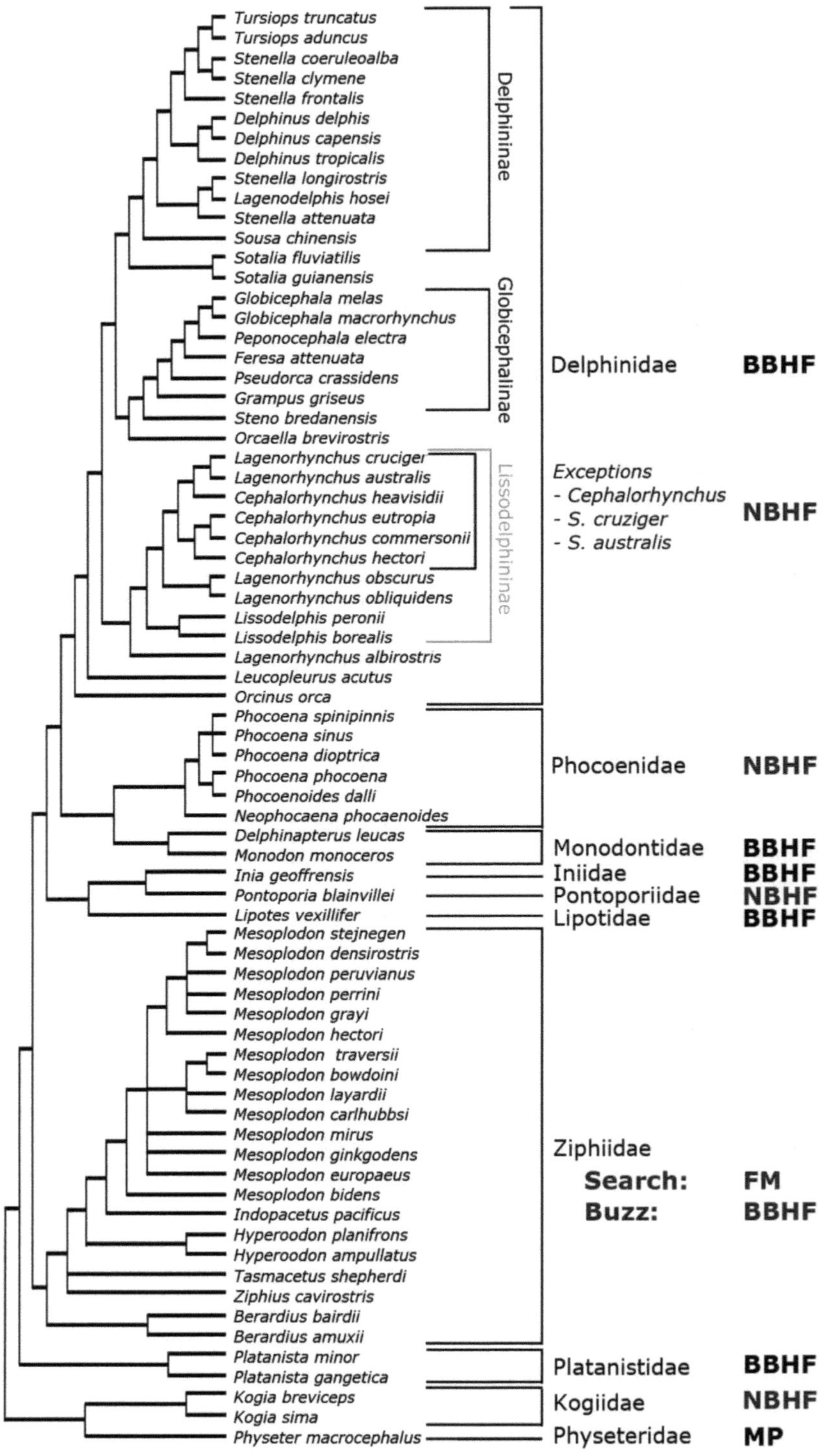

Fig. 2.1 Phylogeny of odontocete cetaceans adapted with permission from McGowen et al. (2009) with echolocation signal type indicated on right side. *BBHF* broadband high-frequency delphinid click, *FM* frequency-modulated beaked whale click, *MP* multipulsed, broadband low-frequency sperm whale click, *NBHF* narrow-band high frequency porpoise-like click. NBHF signals have evolved independently at least four times

The 69 known species of toothed whales comprise some ten different families including the large dolphin (Delphinidae), porpoise (Phocoenidae), and beaked whale (Ziphiidae families) (see Fig. 2.1 for phylogeny). Toothed whales occupy all marine habitats, from rivers and shallow coastal waters to the deep pelagic zones. The polyphyletic river dolphins, including the Amazon River dolphin (*Inia geoffrensis*) and Ganges River dolphin (*Platanista gangetica*), detect and catch freshwater fish in murky water hundreds of kilometers from the ocean. Porpoises mostly feed on small fish in coastal waters near the seafloor, an acoustically cluttered environment. Some populations of the delphinid killer whales (*Orcinus orca*) eat other marine mammals, from seals and porpoises to large baleen whales. Deep diving sperm (*Physeter macrocephalus*) and beaked (ziphiid) whales feed hundreds of meters deep on fish and squid in mid-water or near the seafloor. These different habitats and modes of foraging pose different problems for echolocation, and the shape of the head and jaws of different species of toothed whales has been selected for different modes of echolocation and foraging. River dolphins catch prey with their long thin jaws, a feeding mechanism closest to that of the early ancestors of toothed whales. Species such as killer whales have robust shorter jaws with large teeth, strong enough to grab large prey or to take bites out of large prey such as baleen whales. Many species of toothed whales have a powerful throat musculature that can create a strong suction to suck prey into the mouth (Werth 2006a, b). Variation in the size and shape of the head reflects adaptations for producing directional high-frequency sonar signals and transmitting them into the surrounding seawater.

Bats and toothed whales are long-lived, slowly reproducing animals. Bats tagged in the field can live more than 40 years (Podlutsky et al. 2005). Sperm whales may not become sexually mature until 9 years of age (females) or into their teens (males) (Whitehead 2002). Females may produce only a few offspring in a lifespan that can extend to 80 or more years. There is evidence of social learning in both bat (e.g., Gaudet and Fenton 1984; Page and Ryan 2006; Akre et al. 2011) and toothed whale species (Xitco and Roitblat 1996; Smolker et al. 1997; Krützen et al. 2005). These features set the stage for the evolution of sophisticated social systems, with behavioral flexibility based on social learning. Among nonhuman mammals, evidence for vocal production learning is mostly limited to echolocating bats and cetaceans, a bias that may stem from the three dimensional mobility of these species (Janik and Slater 1997).

The purpose of this chapter is to review the kinds of echolocation signals used by bats and toothed whales, emphasizing how signal design can be related to echolocation performance. While this basically means distinguishing between using clicks and tonal signals of varying frequency modulation for echolocation (Fig. 2.2), it will become clear that echolocating bats and toothed whales are not rigidly constrained by the design of signals they use when echolocating but rather adapt their signals to specific echolocation tasks.

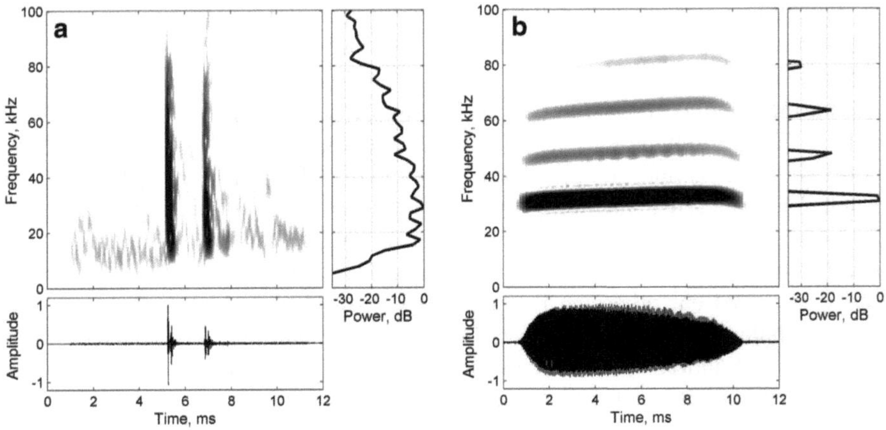

Fig. 2.2 A comparison of two tongue clicks (**a**—*Rousettus aegyptiacus*: Pteropodidae) and a tonal signal (**b**—*Cormura brevirostris*) used by echolocating bats. Both waveforms (*bottom*), spectrograms (*middle*) and Welch power spectrums (*right*) depicted. Spectrograms and power spectrum constructed using 64-point Hanning window with 75 % overlap, and analyzed using 1,024 point FFT (×16 interpolation). (Data from the authors)

2.2 Signal Production

2.2.1 Bats

The mammalian ancestors of bats and cetaceans used the larynx to produce sounds for communication. Most bats also produce their echolocation signals in the larynx by flow-induced oscillation of vocal folds and membranes (Suthers et al. 1972; Suthers 1988; Au & Suthers, Chap. 3). Exceptionally fast muscles allow these bats to call at rates exceeding 160 calls per second (Elemans et al. 2011). However, some species of dog-faced (*Rousettus*) bats echolocate using tongue clicks (Holland et al. 2004). The tongue click echolocation of *Rousettus* is sophisticated because these bats control the directionality and field of view of these echolocation signals (Yovel et al. 2010, 2011a, b). Although most laryngeally echolocating bats emit their calls through open mouths, rhinolophids, hipposiderids, and many phyllostomids emit them through their nostrils (Pedersen 1998), giving rise to fascinating adaptations in facial morphology that serve to shape the transmission of the biosonar beam (Fig. 2.3).

2.2.2 Toothed Whales

Toothed whales have evolved a highly specialized nasal sound production system that differs from sound production in other mammals. The nasal passages have migrated up to a dorsal location, separating from the esophagus and allowing

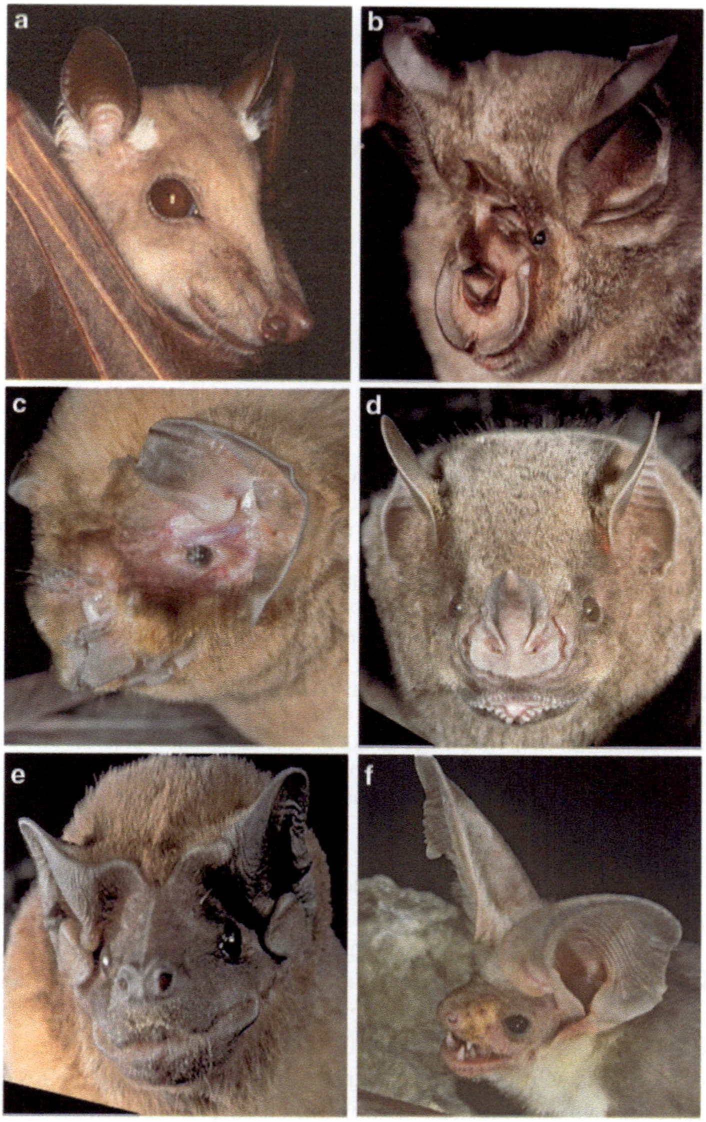

Fig. 2.3 Variation in bat faces illustrating variations in ears and facial structures. Some species have noseleafs; others do not. Tragus is prominent in some species, but not in others. (**a**) *Epomophorus wahlbergi*—nonecholocating pteropodid (Yinpterochiroptera); (**b**) *Rhinolophus hildebrandit*—high duty cycle, laryngeally echolocating rhinolophid (Yinpterochiroptera); (**c**) *Mormoops blainvillii*—low duty cycle laryngeally echolocating mormoopid (Yangochiroptera); (**d**) *Artibeus jamaicensis*—low duty cycle laryngeally echolocating phyllostomid (Yangochiroptera); (**e**) *Molossus molossus*—low duty cycle laryngeally echolocating molossid (Yangochiroptera); (**f**) *Otonycteris hemprichii*—low duty cycle laryngeally echolocating vespertilionid (Yangochiroptera). (Photos by M. B. Fenton)

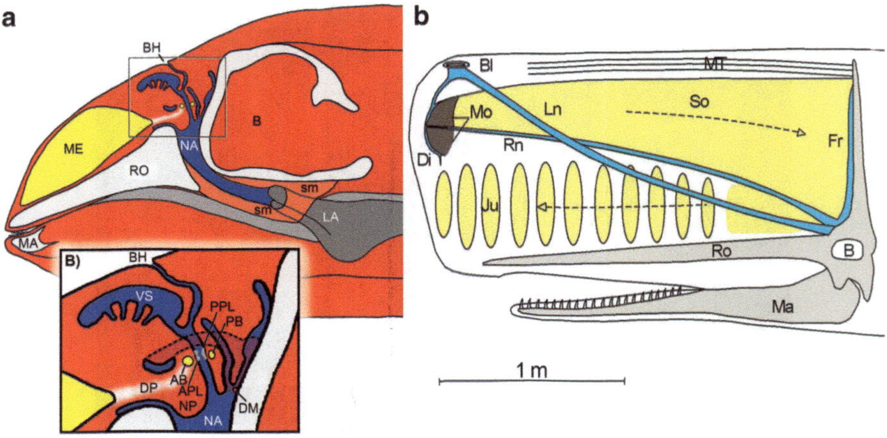

Fig. 2.4 Sound production anatomy of toothed whales. *Left*: Schematic sagittal reconstruction of an adult harbor porpoise (*Phocoena phocoena*) head (adapted with permission from Huggenberger et al. 2009) showing the nasal structures and the position of the larynx (LA). (**a**) Overview. (**b**) Details of *boxed area* in (**a**). *Blue*: Air spaces of the upper respiratory tract; *gray*: digestive system; *light gray*: cartilage and bone of the skull; *yellow*: fat bodies. *AB* anterior *bursa cantantis*, *APL* anterior phonic lip, *B* brain, *BH* blowhole, *DM* diagonal membrane, *DP* low-density pathway, *MA* mandible, *ME* melon, *NA* nasal passage, *NP* nasal plug, *PB* posterior *bursa cantantis*, *PPL* posterior phonic lip, *RO* rostrum, *sm* sphincter muscle of larynx. Note that all toothed whales except sperm whales have two sets of phonic lips (left and right pair of phonic lips), and that only one set is shown here. *Right*: Diagram of the bent-horn model of sound production in sperm whales (adapted with permission from Madsen et al. 2002). *B* brain, *Bl* blow hole, *Di* distal air sac, *Fr* frontal air sac, *Ju* junk, *Ln* left naris, *Ma* mandible, *Mo* monkey lips/museau de singe (equivalent to phonic lips), *MT* muscle/tendon layer, *Rn* right naris, *Ro* rostrum, *So* spermaceti organ

modern cetaceans to breathe at the surface through a blowhole at the top of their head. The upper nasal passages of most toothed whales contain a variety of diverticula as well as a pair of fatty bursae embedded in a pair of connective tissue lips (Fig. 2.4a). This tissue complex, termed the phonic lips, has been identified as the source of echolocation clicks (Cranford et al. 1996; Madsen et al. 2003) and probably also tonal sounds (Madsen et al. 2011). Echolocation signals in odontocete cetaceans are generated by forcing pressurized air through the phonic lips (Ridgway and Carder 1988). Because air can be replenished only at the surface, the blowhole is closed during diving, and air is shunted into vestibular air sacs after passing the phonic lips. This allows toothed whales to recycle air repeatedly for phonation during very long dives by regularly forcing air back into the lower nasal passages (Watkins and Shevill 1972).

Toothed whales have an unusual cranial anatomy. The combination of a concave upper surface of the skull, underlying air sacs, and a fatty melon in the forehead allows these animals to direct most of the energy of their biosonar signals into a highly directional sonar beam and simultaneously improves the coupling of the outgoing sound to seawater (Aroyan et al. 2000). Figure 2.4a illustrates the configuration of phonic lips, skull, air sacs, and melon typical for a harbor porpoise.

The sperm whale has one of the most specialized organs for sound production, devoting about one-third of its total weight and length to soft tissues above the skull (Fig. 2.4b). In the past, biologists hypothesized that the spermaceti organ and the "junk," as these organs are called, functioned in regulation of buoyancy (Clarke 1978) or even as a battering ram (Carrier et al. 2002). Today, the consensus is that these organs function to produce intense echolocation signals (Møhl et al. 2000) that may enable these animals to detect prey at great distances (Madsen et al. 2002, 2007).

In sperm whales, sound energy for echolocation clicks originates from phonic lips in the right nasal passage near the anterior end of the spermaceti organ (Fig. 2.4b; Madsen et al. 2003; Møhl et al. 2003). Most sound energy transmits back through the spermaceti organ, where it reflects from an air sac, called the frontal sac, that sits just dorsal and anterior to the skull. Most of this energy then passes forward through the junk, producing an intense, narrow, forward-directed sonar beam (Møhl et al. 2003; Zimmer et al. 2005a). A small amount of the sound energy continues to reverberate in the spermaceti organ, producing multiple pulses for each sperm whale click, with constant interpulse intervals determined by the dimensions and speed of sound within the spermaceti organ (Gordon 1991; Rhinelander and Dawson 2004).

Physeteroid whales appear to be the only cetaceans with phonic lips only in the right nasal passage (Cranford et al. 1996). Most other toothed whales have two sets of phonic lips that can be operated independently (Cranford et al. 1996, 2011). These two sound generators are nearly symmetric in nonwhistling species, including all porpoises. While both pairs are likely used to produce sound, harbor porpoises (*Phocoena phocoena*) produce echolocation clicks preferentially with the right set of phonic lips (Madsen et al. 2010). In whistling species, such as most delphinids, the two pairs of phonic lips are asymmetric. The right set of phonic lips is larger and may be more specialized for producing high-amplitude echolocation clicks (Cranford 2000). The smaller left set of phonic lips and nasal passages may be more specialized for producing other sounds, such as tonal whistles (Madsen et al. 2011), although both sets of phonic lips can produce clicks (Cranford et al. 2011). As with the syrinx of birds, two independent sound sources allow these animals to simultaneously produce two independent signals such as clicks and whistles (Brill and Harder 1991; Ridgway et al. 2012), or two separate low- and high-frequency components (Hoelzel and Osborne 1986).

2.3 Echoreception

2.3.1 Bats

The pinnae of bats vary considerably in size and shape and they are often mechanically specialized to collect sounds, whether echoes or those generated by prey (Fig. 2.3; Obrist et al. 1993). Features of the heads and faces of some bats, including tragi (Müller 2004) or thickened ear margins (Müller et al. 2006), noseleafs (Hartley and Suthers 1987; Vanderelst et al. 2010), or other facial structures (Zhuang and Mueller 2006) affect the patterns of sounds emitted by or echoes returning to echolocating bats. Ma and Müller (2011) showed that most of the variability in a sample of ears from at least 59 species of bats involved the angle of opening, left–right asymmetry, and changes in the width of the pinnae which could be at the top or the bottom of the structure.

2.3.2 *Toothed Whales*

The ear of terrestrial mammals is adapted for listening in air, which is about 1 % the density of tissue. One of the profound changes in the anatomy of cetaceans relates to adapting the auditory system to function in water, a medium with a density similar to that of tissue. As odontocete cetaceans evolved their hydrodynamic form, pinnae, and the external auditory meatus, which would be less efficient reflectors and conductors of sound in water than in air, were eliminated (Nummela et al. 2004). The ear in terrestrial mammals requires an efficient mechanism to convert airborne sound to fluid-borne sound in the inner ear. Mammals listening underwater face the opposite problem because they need to acoustically isolate the ear. This is achieved by encircling the inner ear within dense tympanic and periotic bones that are isolated in sinuses. The heads of toothed whales contain unusual lipids thought to play a role in sound conduction because their density changes gradually to that of seawater, allowing sound to propagate efficiently from water through lipids to the inner ear (Varanasi et al. 1975). Some of these so-called "acoustic fats" occur near the external auditory meatus where toothed whales show greater sensitivity to low-frequency sound (Bullock et al. 1968; Popov et al. 1990).

The lower jaw of delphinid and phocoenid odontocetes has a hollow section, filled with similar acoustic fats that lead caudally to the rostro-lateral wall of the tympanic bone. Dolphins show particular sensitivity to high-frequency sound when stimulated on the lower jaw (Bullock et al. 1968). Norris (1968) was the first to propose that the pathway for high frequency sound conduction from seawater to the toothed whale ear occurs through a thin section of jaw. Norris hypothesized that sounds are then guided by an acoustic fat channel that leads to the inner ear. With production of intense sounds for echolocation occurring in the skull and melon, the lower jaw pathway appears to offer some isolation from stimulation by the outgoing pulses, and to increase sensitivity by a factor of about 10 (Hemilä et al. 2010). This sound pathway enables excellent sensitivity and directionality in the high-frequency hearing of toothed whales (Au and Moore 1984; Aroyan 2001; Hemilä et al. 2010), which improves echolocation performance.

2.4 Acoustic Structure of Echolocation Signals from Bats and Toothed Whales

2.4.1 Design of Sonar Signals

Our understanding of human sonars and radars has helped to us to understand selection pressures for the evolution of biosonar signals in echolocating animals (Denny 2007). The simplest kind of sonar uses a short broadband click and waits for echoes to return before producing another click. Shorter clicks allow more precise estimation of the arrival time of the echo, and therefore range to the target. Targets differ in how they reflect different frequencies, so broadband signals provide more information for classifying the target. On the other hand, longer tonal calls that provide less information for discrimination may contain more energy within a narrow range of frequencies, providing a longer detection range. The type of signal used by echolocating animals often depends on the ecological constraints of its environment and prey. This has given rise to many variations in signal parameters between species, and even within species solving different echolocation tasks.

2.4.2 Contributions from the Laboratory and the Field

Lazzaro Spallanzani may have been one of the first people to use captive animals to study orientation behavior. His work on the orientation behavior of nocturnal bats and owls set the stage for his discovery that "bats can see with their ears" (Griffin 1958). In 1944, Griffin coined the term "echolocation," setting the stage for a whole new field of research. Griffin et al. (1960) demonstrated how captive little brown bats (*Myotis lucifugus*) used echolocation to detect, track, and close with insects as small as mosquitoes and fruit flies. Since then, literally hundreds of papers have reported results on research with captive bats, expanding our knowledge of call design and echolocation behavior (e.g., Moss et al. 2006; Chiu et al. 2010; Bates and Simmons 2011). Similarly, Norris et al. (1961) demonstrated echolocation in toothed whales by blindfolding bottlenose dolphins, a procedure that could take place only in captivity. The next phase of research on dolphin echolocation involved careful measurement of echolocation signals, hearing capabilities, and target detection and classification capabilities that also required training of captive toothed whales (Au 1993).

The ubiquity of bats and the emergence of commercially available bat detectors, microphones sensitive to ultrasonic frequencies often used by echolocating bats, allowed bat researchers to study the echolocation behavior of bats in the field (e.g., Siemers et al. 2001; Dechmann et al. 2010; Mora et al. 2011). The discovery that different bat species could be identified by their echolocation calls (e.g., Hooper 1964; Ahlen 1981; Fenton and Bell 1981) made it more feasible to study echolocation behavior in the field. An obvious extension of this was monitoring bat activity to assess patterns of movement and habitat use.

The study of echolocation in wild delphinids was similarly facilitated by the development of portable acoustic arrays that allow localization of toothed whales echolocating in the wild (e.g., Møhl et al. 2001; Wahlberg 2002; Au and Benoit-Bird 2003) and acoustic recording tags with sufficient bandwidth and dynamic range to record the full spectrum of echolocation signals (Johnson and Tyack 2003). For some species, these tags have also been able to record echoes from the targets ensonified by toothed whales echolocating in the wild (Johnson et al. 2004), allowing investigators to tap into the biosonar system of the whales themselves to better understand their surroundings and even their interactions with prey (Johnson et al. 2009; Arranz et al. 2011). Call rates derived from tag data can be combined with passive acoustic monitoring of echolocation signals to estimate absolute abundance and density of toothed whales in the wild (Marques et al. 2009), providing a cost-effective method to census animals that are difficult to see.

Now that there are data from several species echolocating in the wild as well as in captivity, clear comparisons can be made between the two situations. Comparisons between source parameters dolphins use in pools, in net pens where trained animals are used to establish maximum range they can achieve by echolocation differ significantly from data obtained in the wild (Au et al. 1974; Wahlberg et al. 2011a). Similarly, big brown bats (*Eptesicus fuscus*) echolocating in the field produce more powerful echolocation signals with greater bandwidth compared to bats echolocating in the lab (Surlykke and Moss 2000). These comparisons demonstrate the necessity of studying animals in both controlled laboratory settings and under more natural situations in the wild.

2.4.3 Echolocation Signals of Bats

The two main types of echolocation signals are clicks and tonal signals (Table 2.2). Most bats produce tonal echolocation signals using their larynx. While many toothed whale species produce clicks, tongue-clicking pteropodids (some species in the genus *Rousettus*; Fig. 2.2) are the only echolocating bats known to use clicks as orientation signals. These clicks are short (50–100 μs), broadband signals similar to those of the clicks of echolocating birds such as oilbirds (*Steatornis caripensis*) and swiftlets (*Collocalia* spp.) (Thomassen and Povel 2006), but dominated by high frequencies. Echolocating *Rousettus* (at least *R. aegyptiacus* and *R. leschenaulti*) often roost in caves and, like the echolocating birds, use echolocation when orienting in their dark roosts. The clicks are often produced in pairs, and as Buchler and Mitz (1980) noted, there is no evidence that clicks are any less effective in echolocation than the tonal signals typical of laryngeally echolocating bats.

Laryngeally echolocating bats typically produce tonal signals that show structured changes in frequency over time (Fig. 2.5). Peak frequencies of bat echolocation signals range from around 8 to >200 kHz, and durations vary from less than 1 ms to more than 50 ms. The echolocation calls of these bats may be broadband or narrowband, or include both of these patterns within a single signal. Traditionally,

Table 2.2 Acoustic features and parameters of echolocation signals of toothed whales

Taxon	Click type	Source level (pp) (dB re 1 μPa at 1 m)	Duration (μs)	ICI (s)	Center frequency (kHz)	−10 dB Bandwidth (kHz)	−3 dB Beam width (degree)	Reference
Physeter macrocephalus	MP	226–236	114	0.5–2	15	4	4	Møhl et al. (2003) Madsen et al. (2002)
Ziphius cavirostris (regular click)	FM upsweep	214	200	0.33–0.50	42	23	12	Zimmer et al. (2005b)
Mesoplodon densirostris (regular click)	FM upsweep	–	271	0.2–0.5	38.3	24.6		Johnson et al. (2006) Madsen et al. (2005a)
Mesoplodon densirostris (buzz click)	BBHF	–	104	0.005–0.020	51.3	54.9		Johnson et al. (2006) Madsen et al. (2005a)
Hyperoodon ampullatus (regular click)	FM upsweep	194–220	276	0.07–0.95	43		19.8	Wahlberg et al. (2011b)
Hyperoodon ampullatus (buzz click)	BBHF	–	–	0.006–0.013				Wahlberg et al. (2011b)
Tursiops truncatus	BBHF	186–228	13–72	0.003–0.526	33–102		9	Wahlberg et al. (2011a) Au et al. (1986)
Tursiops aduncus	BBHF	177–219	8–48	0.003–0.255	45–109	91	8	Wahlberg et al. (2011a)
Grampus griseis	BBHF	202–222	30–75		58–91	66		Madsen et al. (2004)
Pseudorca crassidens	BBHF	201–225	18–55		33–68	63		Madsen et al. (2004)
Phocoena phocoena	NBHF	178–205	57–175	0.03–0.20	130–142	14–46	16	Villadsgaard et al. (2007) Au et al. (1999)

(continued)

Table 2.2 (continued)

Taxon	Click type	Source level (pp) (dB re 1 μPa at 1 m)	Duration (μs)	ICI (s)	Center frequency (kHz)	−10 dB Bandwidth (kHz)	−3 dB Beam width (degree)	Reference
Cephalorhynchus commersonii	NBHF	165–190	52–138	–	123–137	−3 dB: 21		Kyhn et al. (2010)
Lagenorhynchus australis	NBHF	169–196	65–153	–	123–138	−3 dB: 15		Kyhn et al. (2010)
Kogia breviceps	NBHF	175	119	0.04–0.07	129	15		Madsen et al. (2005b)

BBHF broadband high frequency, *FM* frequency-modulated upsweep, *MP* multipulsed echolocation click (but functional component is essentially a very powerful broadband delphinid click at low frequency), *NBHF* narrowband high frequency

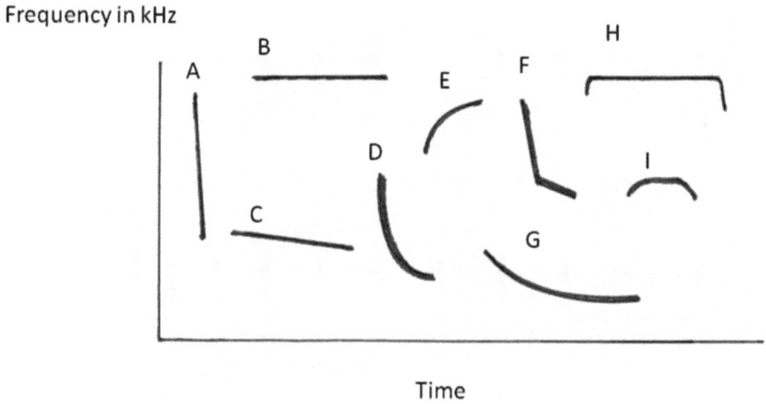

Fig. 2.5 Variations in patterns of frequency change over time in search phase echolocation calls produced by laryngeally echolocating bats. Frequency-modulated (FM) calls can be broadband (**A**, **D**, **F**, **E**) or narrower in bandwidth (**C**, **G**). Constant frequency (CF) calls (**B**) are dominated by one frequency. Often bats combine feature in calls, for example, **H**, which starts with an upward FM sweep, then is CF and ends in a downward FM sweep. Other variations such as **F** and **I** are common. Some molossids produce upwardly sweeping FM signals (**E**)

frequency-modulated (FM; e.g., Fig. 2.5a, d, f) are distinguished from narrowband calls called constant frequency (CF; Fig. 2.5b, h) even if they have bandwidths of several kiloHertz.

In terms of patterns of frequency change over time, laryngeally echolocating bats use a variety of tonal signals, often with characteristic but flexible time/frequency patterns (Fig. 2.5). In general, bat echolocation calls can be characterized as either CF (Fig. 2.5a, h) or FM (Fig. 2.5b–g) calls, or as broadband clicks (*Rousettus*; Fig. 2.2a). Bats often combine CF and FM components in a single call. FM signals range from broadband (steep FM sweeps; Fig. 2.5a, d) to narrowband (shallow FM sweeps; Fig. 2.5c, g). FM sweeps typically, but not always, sweep from high to low frequency. Some researchers describe shallow FM sweeps as quasi-constant frequency (QCF). Simmons and Stein (1980) demonstrated how steep FM signals give better resolution of target detail and position than shallow FM signals but steep FM signals provided less effective range because of the low energy in any one frequency. Bats dynamically adjust the structures of their echolocation calls while hunting, suggesting that they may be able to dynamically modify echolocation signals to suit changing echolocation requirements.

Not all laryngeally echolocating bats produce ultrasonic echolocation signals. Furthermore, the descriptive terms used to describe the calls or their components (e.g., FM, steep FM, shallow FM, CF-FM, quasi-CF, or CF; Fig. 2.5) do not describe the echolocation behavior of the bats but only the typical call structure. The variability in signal design (e.g., Kalko and Schnitzler 1993) is clear when you examine how calls change as bats progress from searching for, to detecting and closing with prey (Fig. 2.6).

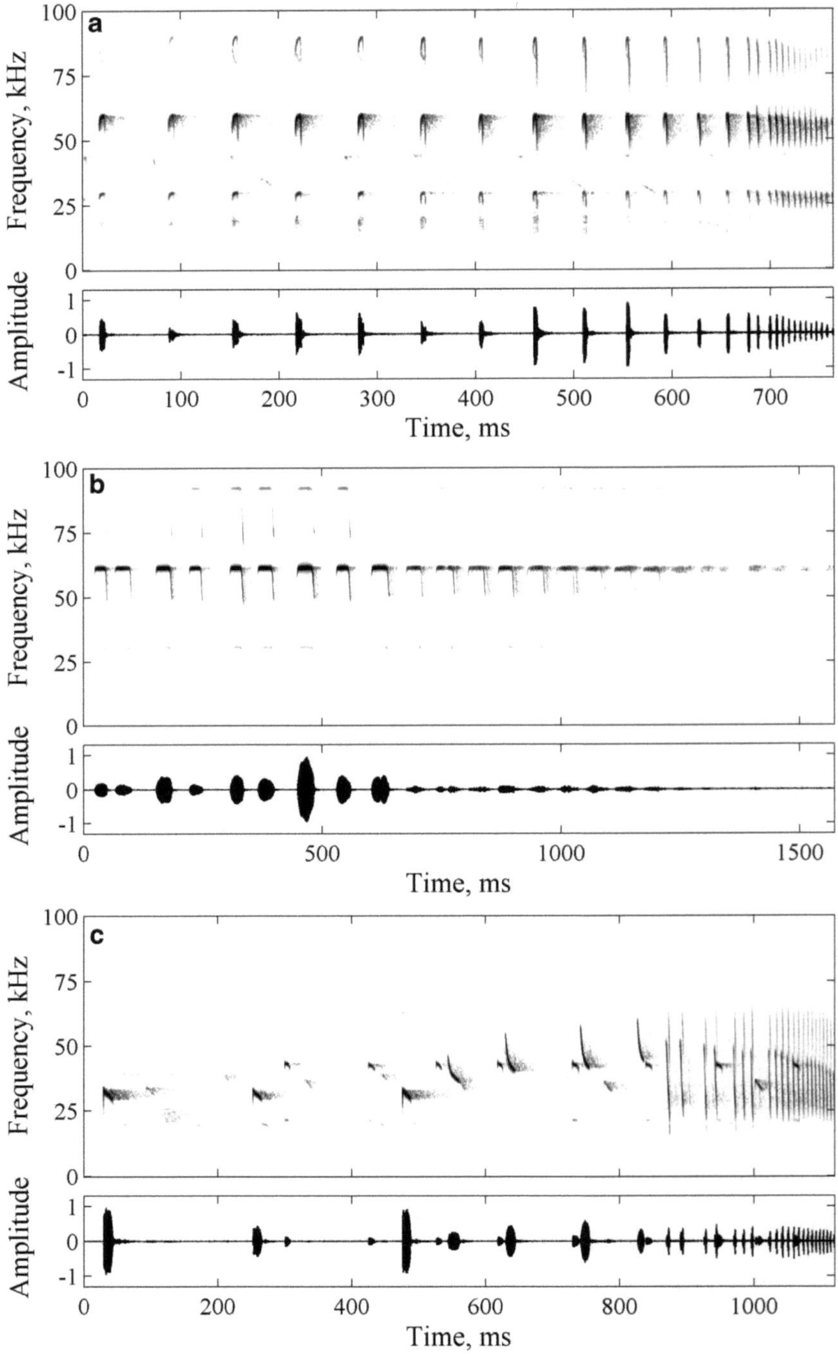

Fig. 2.6 Echolocation behavior during foraging by three species of laryngeally echolocating bats. (**a**) *Pteronotus parnellii*. (**b**) *Saccopteryx bilineata*. (**c**) *Molossus rufus*. Sample rate 200 kHz (**A+C**) to 250 kHz (**b**). Spectrograms constructed using 512-sample Hanning windowed segments with 75 % overlap and power spectrum interpolated 16 times. Changes from search calls (*left*) to buzz calls (*right*) shown for *Saccopteryx bilineata* (**d**) and *Molossus rufus* (**e**). (Data from the authors)

One critical parameter for defining the performance of echolocation is the intensity or level of the echolocation sound (see Wahlberg & Surlykke, Chap. 4). As originally reported by Griffin (1958), the echolocation calls of laryngeally echolocating bats vary considerably in intensity with "whispering bats" producing signals of approximately 60 dB re 20 μPa at 0.1 m and high intensity bats' calls closer to 120 dB re 20 μPa at 0.1 m. More recent work, often from field studies involving arrays of microphones, demonstrates that high-intensity bats can produce calls of approximately 130 dB re 20 μPa at 0.1 m (Holderied et al. 2005; Surlykke and Kalko 2008). Furthermore, some whispering bats can produce more intense calls than previously reported (Mora and Macias 2007; Brinkløv et al. 2009, 2010, 2011).

Higher frequency sounds have shorter wavelengths than lower frequencies, giving the potential for finer discrimination of information about targets. But the atmosphere attenuates more sound energy for higher frequency signals, so that sounds above 30 kHz provide less effective range than sounds at lower frequencies (Lawrence and Simmons 1982). To complicate this trade-off further, it is important that the signal wavelength be smaller than the relative size of the sound production aperture (the mouth or nostrils of an echolocating bat), especially if the signal needs to be directional. Although earlier work suggested that bat echolocation is limited to very short range (5 m for a 19 mm diameter sphere) for detection of insect-sized targets (Kick 1982), more recent work (Surlykke and Kalko 2008) suggests effective ranges of ≥20 m. Effective range helps to put the situation for echolocating bats in perspective. If a bat is flying 5 m s^{-1}, then detecting a moth at 20 m means having about 4 s from detection to contact (depending also on the speed and direction that the moth is traveling in).

In general, echolocating bats and toothed whales adjust the intercall interval (ICI) according to the distance to target, so that ICI exceeds the time it takes for the signal to travel to the target and back again. In this situation, ICIs may reflect distance to target. This relationship is less clear when bats produce more than one call in a short time or for bats that can detect a Doppler-shifted echo while still emitting a call. The Egyptian fruit bat (*Rousettus aegyptiacus*) and some echolocating birds typically emit clicks in pairs (e.g., Holland et al. 2004), and in some situations *E. fuscus* emit echolocation calls in bursts. This "stroboscopic" approach to echolocation involves a brief series of calls produced in rapid succession with adjacent calls sometimes differing in frequency (Hiryu et al. 2010). SENTENCE ON DOPPLER CW? Some beaked whale species do not alter their (see section 2.5.1) as they close on prey, but rather maintain a slow ICI until they rapidly accelerate the click rate as they attempt to capture a prey item within a body length of range (Madsen et al. 2005a, b).

Some bats that produce pairs of echolocation calls change the details of every second call (e.g., Denzinger et al. 2001; Kingston et al. 2003; Mora et al. 2011). Greater white-lined bat (*Saccopteryx bilineata*) alternate calls when foraging, but not when commuting (Ratcliffe et al. 2011). One of the most remarkable species in this regard is Wagner's sac-winged bat (*Cormura brevirostris*) which emits calls in triplets of ascending frequencies, the "doh, ray, me bat" (Jung et al. 2007). Alternating frequencies between echolocation calls may be a mechanism for dealing with clutter, normally defined as echoes generated by objects other than the target of interest (Ratcliffe et al. 2011). Many sheath-tailed bats (Emballonuridae) produce echolocation calls with two or three harmonics (overtones at frequencies equal to a multiple of the fundamental frequency). Here higher frequency harmonics may help to improve discrimination whereas lower frequency components provide greater range because they are less susceptible to atmospheric attenuation (Lawrence and Simmons 1982; Holderied et al. 2005).

Many species of laryngeally echolocating bats often include harmonics in their signals, but their use is not consistent across species or families (Fenton et al. 2011). We noted earlier that the big brown bat uses information in harmonics in its echolocation calls to better distinguish targets from clutter (Bates et al. 2011). Big brown bats more often included harmonics in their echolocation calls when flying in confined situations, perhaps to reduce clutter (Fenton et al. 2011). In most cases, harmonics in echolocation calls will be detected only when the bats fly relatively close to the microphone(s), making it difficult to detect them consistently. While emballonurids and rhinopomatids use harmonics consistently, vespertilionids and molossids apparently do not.

The role of different calls and/or call components can influence a bat's perception of its surroundings and potential targets (Simmons and Stein 1980). Calls at lower frequencies and with longer duration may have a greater effective detection range but at the cost of less precise location of targets. Shorter, broadband FM calls provide less range but better range resolution and prey localization (Simmons and Stein 1980). In other species such as the lesser bulldog bat (*Noctilio albiventris*), narrowband components of signals in the first part of the calls open windows of analysis in the bat's brain, setting the stage for further analysis (e.g., Roverud and Grinnell 1985).

The extent of intraspecific variation in echolocation calls of bats has been well documented (e.g., Kalko et al. 1998; Siemers et al. 2001), and two examples clearly make this point. Holderied et al. (2005) documented the echolocation calls of foraging Botta's serotine bats (*Eptesicus bottae*) in different stages of searching for, detecting, and attacking insects (Fig. 2.7). Mora et al. (2011) provided comparable data for the little goblin bat (*Mormopterus minuta*; Fig. 2.8), a species with impressive variation. Other species show less variation, for example, species of vespertilionids in the subfamilies Kerivoulinae and Murininae (Kingston et al. 2003), species effective at detecting flutter (Lazure and Fenton 2011). There appear to be many examples of parallels and convergences among signal design in echolocating bats (Siemers et al. 2001; Siemers and Schnitzler 2004).

2.4.4 Echolocation Signals of Toothed Whales

The echolocation signals of toothed whales can be grouped into four basic categories (Fig. 2.9; Table 2.2): short broadband high-frequency (BBHF) clicks, narrowband high-frequency (NBHF) clicks, FM upsweeps, and the intense broadband lower frequency clicks of sperm whales, which are made up of multiple pulses

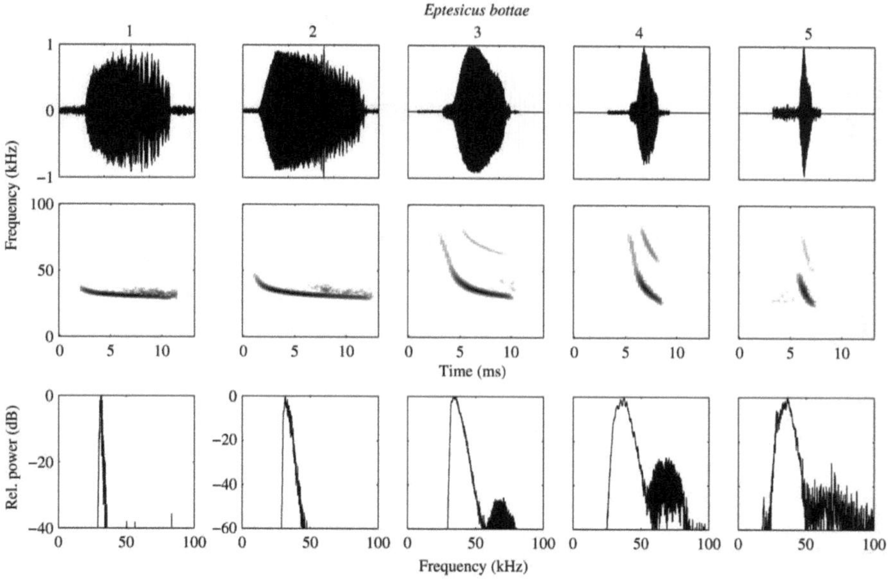

Fig. 2.7 Variations in echolocation calls produced by foraging *Eptesicus bottae*. (After Holderied et al. 2005. Reproduced with permission of *Journal of Experimental Biology*) calls 1 through 5 are progressively closer to contact with prey

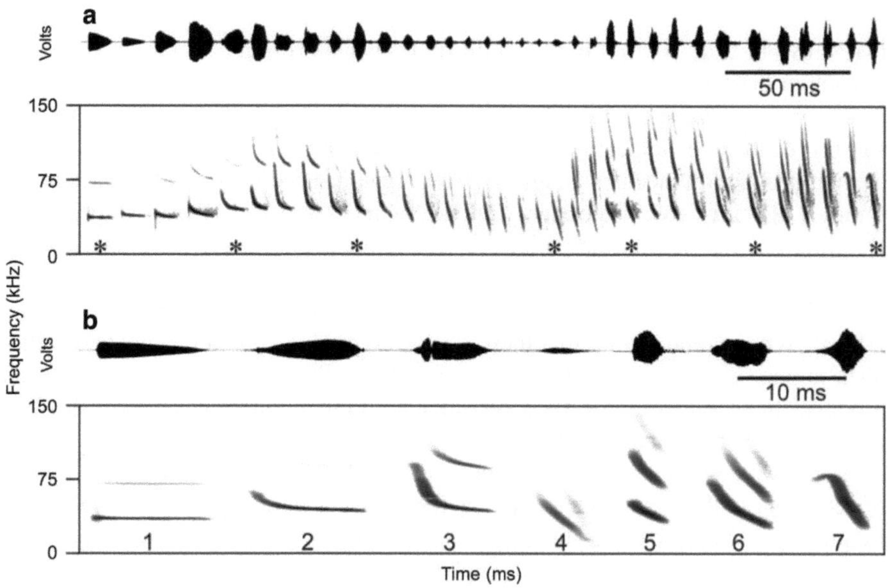

Fig. 2.8 Variation in echolocation calls produced by foraging *Mormopterus minuta*. (After Mora et al. 2011. Reproduced with permission of the editor of *Acta Chiropterologica*)

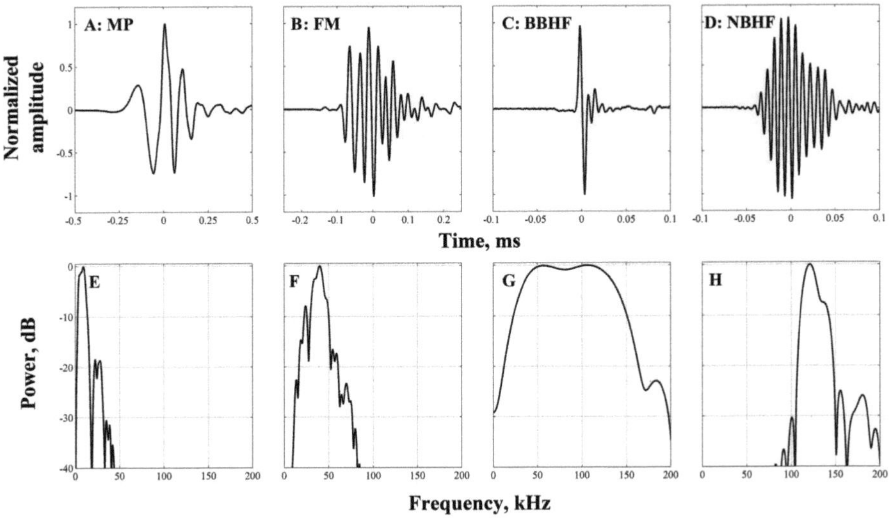

Fig. 2.9 Echolocation clicks of odontocetes can be categorized into four general types of clicks. This is demonstrated here with waveforms (**A–D**) and power spectra (**E–H**) of the typical search clicks of four toothed whales. From *left* to *right*: A broadband, low-frequency pulse from a sperm whale (**A+E**) (*note*: only the functional p1 pulse shown here). A frequency-modulated click from a Cuvier's beaked whale, *Ziphius cavirostris* (**B+F**). A broadband, high-frequency click from a bottlenose dolphin, *Tursiops aduncus* (**C+G**). A narrow-band, frequency-modulated click from a harbor porpoise, *Phocoena phocoena* (**D+H**). (Data from the authors)

(MPs) (Møhl et al. 2000). Figure 2.9 illustrates waveforms and spectra of examples of each of these kinds of click.

The most prevalent type of biosonar signal for toothed whales appears to be a short, broadband echolocation signal that puts as much energy into as short a time as their production mechanism allows. This signal is used by toothed whales in the families Platanistidae, Lipotidae, Iniidae, Monodontidae, and most species of Delphinidae, the oceanic delphinids. These animals can direct the sound energy forward in a narrow beam—for bottlenose dolphins, about half of the sound energy is within 9° of the direction with maximum energy (Au et al. 1986; Wahlberg et al. 2011a). When you consider that these clicks are produced by a pneumatic mechanism moving soft tissue structures, it is remarkable that the bottlenose dolphin can produce clicks louder than 220 dB re 1 μPa at 1 m with durations as short as 10 μs (Au et al. 1974; Wahlberg et al. 2011a).

Sperm whales have an extremely powerful, highly directional sonar signal capable of detecting weak targets such as squid at long range. These clicks have lower center frequencies than those of most other toothed whales. As in air, low-frequency sound is not absorbed as much by travel through seawater, so low-frequency clicks provide better performance than higher frequencies at long ranges. Most reports of sperm whale clicks have not controlled for orientation with respect to the sonar axis, and the signals appear to be multipulsed clicks. However, Møhl et al. (2003) identified on-axis clicks, and reported "monopulsed clicks, lasting 100 μs," a half-power beamwidth of

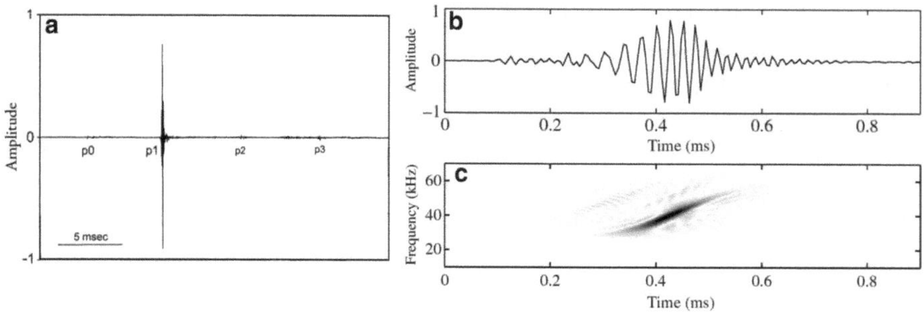

Fig. 2.10 (**a**) On-axis recording of a multipulsed (MP) click from a sperm whale (*Physeter macrocephalus*), with individual pulses marked. (Adapted from Møhl et al. 2003.) The on-axis recording illustrates how nearly all energy is concentrated in the strong, outgoing P1 pulse. (**b**) Waveform and (**c**) time-frequency (Wigner) distribution of the frequency-modulated (FM) search click from a Blainville's beaked whale (*Mesoplodon densirostris*). (Adapted from Johnson et al. 2006)

4°, "with source levels up to 236 dB *re*: 1 μPa (rms), and with centroid frequencies of 15 kHz". The functional part of the echolocation signal for the sperm whale is an extremely directional, broadband delphinid-like signal capable of producing a sound pressure level rivaling that of the most powerful naval sonars.

Sperm whales also produce a less intense, lower frequency (LF), more omnidirectional component of their echolocation click that may be a byproduct of the sound generation system (Zimmer et al. 2005a). However, echoes from the seafloor and sea surface are audible from this LF component of sperm whale clicks. This LF component of the click is audible over ranges of up to 10 km and may function in orientation and in communication, specifically for maintaining contact and/or monitoring the behavior of other members of the group (Tyack 1997). Sperm whales live in stable family groups, which often are out of visual contact so listening for the omnidirectional LF component of clicks could help these animals maintain group cohesion (Fig. 2.10a).

Ziphiid beaked whales are a large but poorly known family of toothed whales. Curiously, beaked whales seem to produce a steep FM click while searching for prey, an upsweep contrasting with the typical downsweeps of bat calls, and over a much shorter duration (Fig. 2.10b, c, from Johnson et al. 2006). The two largest ziphiids, *Hyperoodon ampullatus* (northern bottlenose whale) and *Berardius bairdii* (Baird's beaked whale), produce clicks with center frequencies of about 24 kHz and durations of 350–450 μs (Dawson et al. 1998; Hooker and Whitehead 2002; Wahlberg et al. 2011b). Smaller beaked whale species produce higher frequency clicks with source levels up to about 210 dB re 1 μPa at 1 m (Zimmer et al. 2005b for Cuvier's beaked whale), and little energy below 20 kHz. The clicks of Cuvier's (Zimmer et al. 2005b) and Blainville's (Madsen et al. 2005a) beaked whales have durations (97 % energy) of 200 and 270 μs respectively, and center frequencies around 40 kHz. In these species, the switch from search to approach is marked by an abrupt change from FM clicks to a rapid buzz clicks similar to the BBHF clicks used by delphinids and platanistids (Fig. 2.9) (Johnson et al. 2006).

Beaked whales hunt mid-water or epi-benthic fish and squid in relatively uncluttered environments. The primary source of clutter for foraging beaked whales is echoes from other organisms in the water column, although echoes from the seafloor are also recorded from search clicks (Johnson et al. 2004; Madsen et al. 2005a). The selectivity of foraging beaked whales suggests a capacity for subtle classification. Schnitzler and Kalko (2001) point out that FM clicks are well suited to precise localization of prey, and also for classification of prey based on spectral features. Jones et al. (2008) demonstrated that the echoes of prey selected by Blainville's beaked whale have spectral features that differ from rejected targets, suggesting that spectral features are used in prey selection. Johnson et al. (2006) suggest that the FM clicks of beaked whales may provide a means for increasing the bandwidth of the signal the better to classify prey using spectral cues, while also increasing the energy of the click for detecting weak targets by extending the duration compared to delphinid clicks. The short buzz clicks will then allow rapid temporal updates about prey position during the last stages of capture.

Four groups of toothed whales use narrowband high-frequency (NBHF) clicks, including (1) porpoises (Phocoenidae) (Villadsgaard et al. 2007); (2) six species of nonwhistling dolphins of the genera *Cephalorhynchus* (Kyhn et al. 2009, 2010) and *Lagenorhynchus* (two of the species labeled *Sagmatias* by McGowen et al. (2009)) (Kyhn et al. 2010); (3) the Franciscana river dolphin, *Pontoporiidae* (Melcon et al. 2012); and (4) the pygmy and dwarf sperm whales of the genus *Kogia* (Madsen et al. 2005a). Recorded NBHF clicks have peak frequencies of 130–135 kHz, modest source levels of 165–205 dB re 1 μPa at 1 m, durations of 50–175 μs, and bandwidths of 6–14 kHz (Table 2.2). Most discussions of the evolution of echolocation signals focus on phylogenetic relationships and ecological niches. Porpoises and nonwhistling delphinids have similar body types and, along with *Pontoporia*, are thought to forage near the bottom in shallow cluttered environments. *Kogia* is more closely related to the sperm whale than any other toothed whales. It has similarities in sound production anatomy with the sperm whale, such as just having one set of phonic lips. *Kogia* spp. live in open water and find prey in the deep sea, making their foraging niche radically different from that of the other NBHF species.

2.5 Patterns of Call Production

One of the most important problems facing echolocating animals is ensuring that outgoing signals do not interfere with the detection of returning echoes. Acoustic energy from the outgoing signals will be lost while the signal propagates through the environment, due to absorption of sound energy converted into heat during propagation through the medium (Urick 1995). Furthermore, due to spreading of energy into an increasing area, only a small fraction of the incident energy will be reflected back from a target. Consequently, the outgoing biosonar signals will inevitably be much stronger than the returning echoes. Section 2.5.1 addresses how echolocating animals separate their strong, outgoing signals from the faint, returning echoes containing information of their surroundings and prey.

2.5.1 Duty Cycle

Echolocating animals use two general mechanisms to separate pulse and echo: separation in time and separation in frequency. Most species separate the outgoing pulse and returning echo in time. In engineering terms, these echolocators operate at a low duty cycle by producing short signals separated by much longer periods of silence where returning echoes can be detected (Fig. 2.11a–c). Low duty cycle echolocators cannot broadcast and receive at the same time and this limits the duration of biosonar signals of many low duty cycle bats throughout the process of finding and catching a prey item.

About 160 species of bats (the families Hipposideridae and Rhinolophidae as well as the mormoopid *Pteronotus parnellii*—Parnell's mustached bat) use a different mechanism that separates pulse and echo in frequency by exploiting Doppler shifted echoes. Echolocation signals of these species are typically very long compared to the total time between signals, and in engineering terms, they are produced at high duty cycle (Fig. 2.11d–f). The echolocation calls of these bats are dominated by one frequency and they are typically designated constant frequency (CF) calls. When these signals reflect from a target, the wavelength of the echo (and consequently the frequency) is altered slightly through Doppler shifts generated by the relative movement of bat and target so that the returning echo will be shifted to a higher frequency when the bat and the target are approaching each other. In addition, smaller oscillating Doppler shifts are generated by fluttering movement of insect wings, helping these echolocators to discriminate fluttering prey (Schnitzler and Kalko 2001; Lazure and Fenton 2011). High duty cycle echolocating bats use calls with FM and CF components. These bats change their outgoing CF signals to compensate for Doppler shifts (Doppler shift compensation [DSC]) (Schnitzler 1973), and they have an acoustic fovea, an area of the inner ear that is particularly sensitive to slight changes in echo frequency (Neuweiler et al. 1980). This combination of features does not occur among bats that echolocate at a low duty cycle, even though many LDC species use signals that are CF or narrowband FM.

A high duty cycle approach to echolocation appears to have made bats more effective at detecting, locking onto, and tracking flying prey, especially in situations where vegetation generates clutter (Fenton et al. 2011). Lazure and Fenton (2011) demonstrated that high duty cycle echolocating rhinolophids and hipposiderids responded more often than low duty cycle echolocators to fluttering targets. These fluttering motions create both frequency and amplitude modulations of the echo, which provide cues for detecting and classifying prey (Schnitzler and Flieger 1983).

The high speed of sound in water compared to air, coupled with slower swimming speeds, means that echolocation signals of toothed whales, including NBHF species, are not narrow enough in bandwidth to be Doppler sensitive (Thorpe et al. 1991). Like most bats, all toothed whales are therefore low duty cycle echolocators.

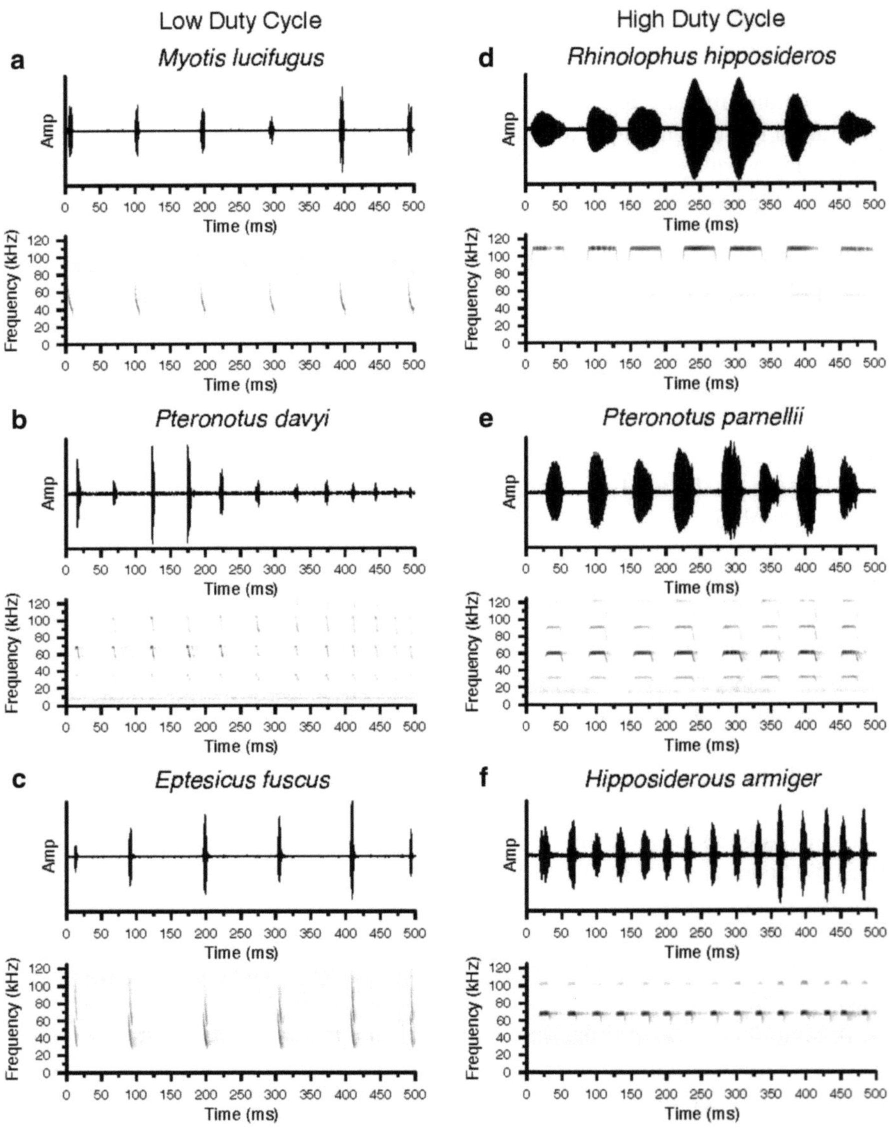

Fig. 2.11 A comparison of patterns of call production by high duty cycle and low duty cycle largyngeally echolocating bats. (Adapted from Fenton et al. 2012. Reproduced with permission of *Journal of Experimental Biology*)

2.5.2 Feeding Buzzes

One of the most striking convergences in the echolocation of bats and toothed whales involves changes in the production of biosonar signals as echolocating animals search for prey, select individual targets, approach those targets, and finally

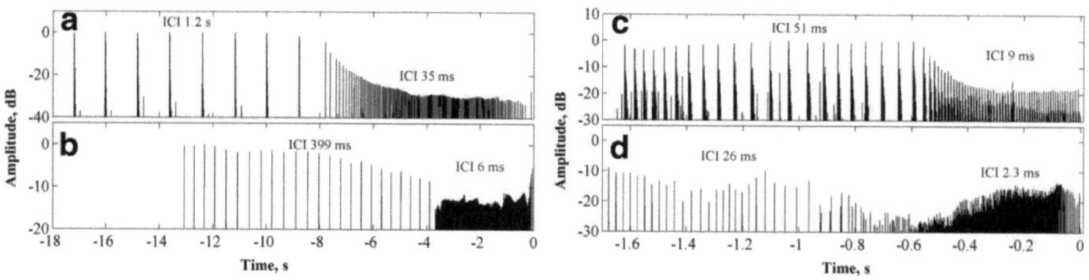

Fig. 2.12 Sequence of echolocation clicks recorded as sperm whales (**a**), Blainville's beaked whales (**b**), Atlantic bottlenose dolphins (**c**), and harbor porpoises (**d**) search for, approach, and attempt to capture prey. (**a–c**) Recorded using acoustic tags placed on the animal; (**d**) recorded from a captive animal with a hydrophone on a target. (Data provided by W. Zimmer, P. Tyack, and S. DeRuiter)

attempt to capture them (Figs. 2.6 and 2.11). Any animal attempting to capture fast-moving prey benefits from rapid updates on the location of its target. While visual predators enjoy nearly instantaneous feedback on the location of their prey, low-duty cycle echolocating animals are constrained by the time it takes for a biosonar signal to travel through the environment to the target and back again. The returning echo of a target must arrive before a new biosonar signal is emitted, meaning that the ICI, and consequently the biosonar update rate, is normally limited by the two-way-travel time through the environment plus a delay as the information is processed in the brain (Au 1993). As low duty cycle echolocating animals approach their prey and the round trip travel time decreases, many species shorten the time between signals to obtain more rapid updates on the location of the prey. In the final stages of prey capture, where a quick response to sudden movements of the prey can mean the difference between capture and escape, a rapid update rate on the position of the prey is even more important. Echolocating animals therefore accelerate echolocation signals into a buzz with very high repetition rates during the terminal phase of prey capture. This was first documented in bats (Fig. 2.6) (Griffin et al. 1960; Kalko 1995) and later in toothed whales (Fig. 2.12; sperm whale: Miller et al. 2004; beaked whale: Madsen et al. 2005a; delphinid pilot whale: Aguilar de Soto et al. 2008; and porpoise: DeRuiter et al. 2009) and likely represents a fundamental component of catching moving prey using biosonar. Echolocating species such as frugivorous bats that do not hunt moving prey items are not reported to buzz, indicating that buzzing is unnecessary for navigation or foraging on stationary prey.

As described for *Eptesicus fuscus*, many bats produce FM signals at low duty cycle in search phase and through attacks on prey (Moss et al. 2011). Search phase signals of *Eptesicus fuscus* are often emitted at a repetition rate of 5–10 calls per second. As they approach the prey, they emit approach calls at repetition rates of 20–80 calls per second, and when they switch into a buzz for final prey capture, calls are emitted at up to 200 calls per second (Kalko 1995; Moss et al. 2011). These high call rates are enabled by superfast muscles in the larynx (Elemans et al. 2011).

In toothed whales, the terminal buzz or capture phase is probably best understood for beaked whales (see also Madsen & Surlykke, Chap. 9), where tags are not only able to record the outgoing clicks but can also record the echoes of prey (Johnson et al. 2004; Madsen et al. 2005a). Beaked whales produce FM clicks at low duty cycle in search and approach mode. The ICI of Blainville's (*Mesoplodon densirostris*) and Cuvier's (*Ziphius cavirostris*) beaked whales does vary during these phases, but does not show a systematic reduction as range to the target reduces. Rather, the beaked whale makes a sudden transition from FM clicks to buzz clicks with shorter duration and a much faster ICI (Madsen et al. 2005a). When Blainville's beaked whales move through a heavily cluttered scene and face the task of selecting suitable prey items, they have relatively high ICIs, typically between 0.2 and 0.4 s (2–5 clicks per second). During this search and selection phase, the whale emits FM clicks with an average length of 270 μs and a −10 dB bandwidth from 26 to 51 kHz (Johnson et al. 2006). During the capture phase, Blainville's beaked whale switches to a buzz with much lower ICIs of 0.01 s at the start of the buzz to 0.003 s at the end of the buzz (100–350 clicks per second). These buzz clicks are short (105 μs) transient echolocation clicks with −10 dB bandwidths from 25 to 80 kHz or higher, and without the FM upsweep pattern characterizing regular search phase echolocation clicks of this species (Johnson et al. 2006). As with bats, the high update rate during buzzing is essential for successful prey capture in the final capture phase, where the spatial relationship between predator and prey may change rapidly over a short time period, especially if prey tries to evade the predator. The similarity of sonar-based foraging between 10-g bats feeding on fluttering insects moving about in air, and 1,000-kg whales feeding on squid and fish at depths exceeding 1 km is remarkable.

2.5.3 Adaptive Changes in Signal Structure

Echolocating animals not only change the repetition rate as they close in on prey items. Modifying other signal parameters can be both essential for an effective biosonar function, and can help shape the information gathered so as to best support the perceptual requirements of different echolocation functions Moss et al. 2011). These changes can include changing the duration, frequency, or intensity of echolocation signals, as well as the biosonar beam pattern (Moss et al. 2011).

Many low duty cycle bats often simultaneously shorten the duration of signals as well as the time between calls to avoid overlap between outgoing pulse and returning echo. This is especially apparent when animals switch into a buzz (Fig. 2.6). For example, the duration of biosonar signals produced by big brown bats ranges from 15 to 20 ms during search, through 2–5 ms during approach phase, and to as little as 0.5–1 ms during the terminal buzz phase (Moss et al. 2011). While the repetition rate normally depends on the range to the target, Aytekin et al. (2011) used a distractor to show that the call duration of the big brown bat is decreased to prevent overlap between the call and an echo from the closest object of interest. However, overlap between call and other objects can be tolerated. In an experiment where a big brown bat was trained to negotiate through the opening of a small net before catching a tethered mealworm, the bat adjusted call duration to avoid overlap between the call and the net echo at first. Once the path through the net opening was planned, the bat increased the duration of its biosonar call to find the tethered mealworm and tolerated overlap between call and net echo even while flying through the net (Surlykke et al. 2009).

Toothed whale signals are much shorter than bat signals, but the way beaked whales switch from long FM signals in the search and approach phase to short, broadband signals during the buzz provides an appealing analogy to the decreased duration of bat calls. Although shorter duration is likely not necessary to avoid pulse-echo overlap, the shorter signal duration may combine with fast repetition rate to facilitate more accurate ranging of close prey items at this stage (Johnson et al. 2006). This can best be envisioned using a recently developed visual representation of the auditory scene of a tagged whale, termed an echogram, which is constructed like the output of a fish finder by stacking traces of echolocation signal and echo energy from the start of each click on top of each other (Johnson et al. 2006; Arranz et al. 2011). In an echogram from a Blainville's beaked whale, the temporal resolution increases drastically when the whale switches to a buzz, allowing for much more detailed tracking of the spatial relationship of predator and prey (Fig. 2.13; Johnson et al. 2004).

During feeding buzzes, most low duty cycle bats reduce call amplitude, frequency, and bandwidth with increasing biosonar repetition rate (Kalko and Schnitzler 1993; Kalko 1995), possibly because the fast-twitch muscles responsible for sound production are not capable of producing loud broadband signals at very high repetition rates. A lower biosonar frequency of terminal buzz calls simultaneously decreases the directionality and may be useful for keeping track of rapidly moving prey at close range (Jakobsen and Surlykke 2010). However, some Asian rainforest bats seem to break this trade-off between call bandwidth and pulse repetition by producing short, high-frequency, broad bandwidth echolocation calls throughout the entire buzz sequence. In these species, durations of signals in the buzz are often somewhat longer than those produced by other species (Schmieder et al. 2010). The high frequency and bandwidth may help these species track and catch insects through their dense and cluttered rainforest environment.

Like most bats, toothed whales also decrease the amplitude of echolocation signals when they enter the buzz phase (Fig. 2.6). Amplitude and frequency are normally positively correlated in toothed whales (Au et al. 1995) meaning that buzz clicks may have a lower frequency and consequently a broader beamwidth. As for bats trying to catch prey items at close range, broadening the beam right before capture may prevent rapid escape behaviors from taking the prey outside the narrow biosonar beam of a toothed whale.

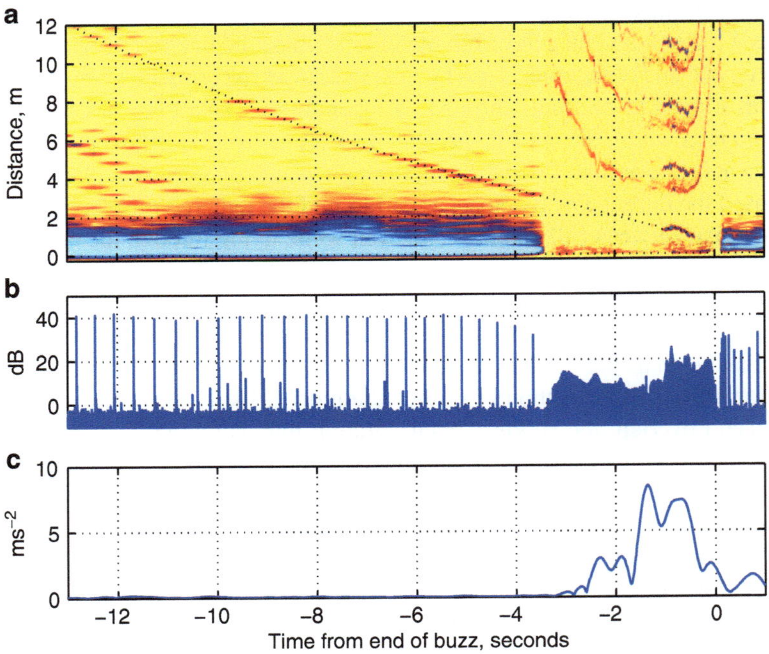

Fig. 2.13 Echolocation of prey by a tagged Blainville's beaked whale, *Mesoplodon densirostris*, as recorded with an acoustic tag. (**a**) Alignment of successive clicks made by the tagged whale at the appropriate time on the *x*-axis. The *y*-axis indicates the time elapsed between the outgoing click and the returning echo expressed as distance to the target assuming a sound speed of 1,500 m s^{-1}. The color scale indicates the energy of the signal from *blue* indicating intense to *yellow* indicating faint. (**b**) The envelope of the recorded signal on the logarithmic scale from which (**a**) was calculated. The high-level clicks produced before –4 s are regular clicks. The clicks in the buzz from –3.4 to 0 s are so rapid that they appear continuous in (**b**). (**c**) The magnitude of the dynamic acceleration of the tagged whale. The peak at the end of the buzz probably indicates movements associated with prey capture. (Reproduced with permission from Johnson et al. 2004)

2.5.4 Time-Varying Gain Control

Man-made sonars have a limited dynamic range where echoes can be processed. Therefore to maintain a steady echo level over increasing distance, sonars use a time-varying gain control mechanism. The gain control (see also Wahlberg & Surlykke, Chap. 4) can be on both the transmitting and receiving side of the biosonar system. In man-made sonars, it is most often located on the receiving side, and achieved by increasing the receiving sensitivity with time from emission of the sonar pulse to compensate for the decreasing echo levels from more distant targets.

Evidence of time-varying gain control in the receiver has been found for bats that reduce sensitivity by tightening muscles attached to middle ear bones prior to emitting an echolocation call, then relax those muscles resulting in a gradual increase in hearing sensitivity over the next 6.4 ms (Henson 1965; Suga and Jen 1975; Kick and Simmons 1984). A similar time-varying gain control has been found in the auditory system of the false killer whale (*Pseudorca crassidens*) where it involves a combination of forward masking and active hearing control (Nachtigall and Supin 2008). The harbor porpoise also adjusts its hearing sensitivity to the received echo level during echolocation (Beedholm et al. 2006; Linnenschmidt et al. 2012). Chapter 5 explores the auditory sensitivity during echolocation in more depth.

A time-varying gain on the transmitter side has been reported for several species of bats (Kobler et al. 1985; Hartley 1992; Hiryu et al. 2007) and toothed whales (Au and Benoit-Bird 2003; Atem et al. 2009). The reported reduction in source level for toothed whales is highly variable, but is best approximated by a 20-dB reduction in level with every tenfold reduction in range (Au 2004). This time-varying gain control has been hypothesized to be a passive consequence of the pneumatic sound generator being unable to generate high source levels when the repetition rate goes up (Au and Benoit-Bird 2003; Beedholm and Miller 2007), at least at closer range (Jensen et al. 2009). This mechanism of adjusting the amplitude of outgoing clicks with ICI and distance is further debated in Chap. 4 by Wahlberg and Surlykke. Some species such as beaked whales do not show a gradual decrease in biosonar source level when approaching prey. Blainville's beaked whales maintain a constant output level and relatively stable repetition rate throughout the approach, and then switch into a buzz with clicks of lower amplitude and higher repetition rate (Madsen et al. 2005a; Johnson et al. 2006). Before beaked whales switch to a buzz and attempt to capture prey, their echo levels may increase by 30 dB (Madsen et al. 2005a). Though toothed whales may have a corresponding decline in acoustic sensitivity, it is equally possible that their biosonar does not depend on a constant echo level to optimize performance. Rather, it is possible that the long, stable repetition rate and source levels during approach allow the whale to maintain a relatively constant clutter or reverberation level while the target echo increases in level, facilitating auditory scene analysis in a complex multitarget environment (see Wahlberg & Surlykke, Chap. 4; Madsen & Surlykke, Chap. 9). The subsequent buzz provides rapid updates required for capturing prey (Madsen et al. 2005a). However, the variation in strategies observed may also indicate that toothed whales are flexible in their abilities to modulate source level, auditory sensitivity, and signal structure to improve target detection and classification in clutter and reverberation as they close on a target.

2.6 Challenges Faced and Solved

Toothed whales operate in a wide range of conditions and environments, with visibility underwater ranging from tens of centimeters to tens of meters in surface waters during daytime, to the deep canyons of the oceans where the only brief glimpse of light is produced by bioluminescent organisms. The diversification of echolocation has radically expanded opportunities for finding and catching prey under very challenging conditions, and allowed toothed whales and bats to diversify in amazing ways. Some species such as porpoises find prey near the bottom where they have to find small prey items between other sound reflecting objects normally termed clutter. In contrast, some deep diving species find prey in mid-water depths with little clutter other than echoes from other organisms. It has been estimated that sperm whales can detect and track squid at 300 m or greater (Madsen et al. 2007). Killer whales hunting salmon can detect them at approximately 100 m (Au et al. 2004), while porpoises detect fish at tens of meters (Villadsgaard et al. 2007; Kyhn et al. 2009, 2010). Beaked whales often pass by hundreds of sound reflecting targets before attacking one as prey to eat (Madsen et al. 2005a), probably discriminating the preferred targets based on characteristic features in the echoes (Jones et al. 2008). Passive listening may in some cases supplement echolocation when hunting. Gannon et al. (2005) indicated that bottlenose dolphins may listen for sounds made by their soniferous prey, and then use echolocation to locate it. Echolocation provides information not only about prey items, but also about the environment. Toothed whales routinely hear echoes from the sea floor (Tyack 1997) and sea surface (Zimmer et al. 2005a) and we can even use echoes recorded on acoustic tags to gather a representation of their environment (Arranz et al. 2011). Jaquet et al. (2001) argue that as sperm whales dive, they adjust their click rates to echolocate on the seafloor, and Verfuss et al. (2009) showed that captive porpoises adjust their click rates to range lock on landmarks. These results show that toothed whales use echolocation for orientation as well as foraging.

The diversity of echolocating bats is even more impressive than toothed whales. Echolocating bats are mainly nocturnal and fly in habitats ranging from open skies to forest thickets (Schnitzler and Kalko 2001). Most of them detect, track, and identify, and then attack flying insects. Other animal-eating bat species listen for prey-generated sounds (from footfalls to mate attraction calls) to detect, identify, and locate their targets (e.g., Aldridge et al. 1990; Holderied et al. 2011). Some flower-visiting phyllostomid bats detect acoustic reflectors on flowers (von Helversen and von Helversen 1999) or leaves (Simon et al. 2011) and then orient themselves to sources of nectar. Some fruit-eating bats may use echolocation to detect their food (e.g., Kalko and Condon 1998; Thies et al. 1998), but olfaction may be more important in the final decision about what to take and eat. In contrast, nectar-feeding and fruit-eating bats rely more on spatial cues than either shapes or scents when foraging (Carter et al. 2010). While the blood-feeding vampire bats (*Desmodus*, *Diaemus*, *Diphylla*) use echolocation to orient in roosts, its role in locating prey appears to be secondary relative to sounds emanating from prey (Groger and Wiegrebe 2006), olfaction (Turner 1975), and heat perception (Gracheva et al. 2011). Using echolocation, some bats can recognize habitat features such as water surfaces (Grief and Siemers 2010) or may be able to find roosts (Ruczynski et al. 2007). Bats such as Indian false vampire bat (*Megaderma lyra*) often echolocate in unfamiliar situations (Ratcliffe et al. 2005), but the same individuals may cease to produce echolocation signals in familiar areas where spatial memory may be sufficient for orientation (Fiedler 1979).

2.6.1 Clutter

Although bat biologists typically consider clutter to be an important challenge to echolocating bats, many bats obviously orient effectively and take prey even in the face of many echoes from background objects (e.g., Siemers and Schnitzler 2000). From a biosonar perspective, discriminating targets from clutter depends on signal parameters such as sound frequency and duration, the dimensions and characteristics of the clutter-generating objects, and their orientation relative to the sound source. Directional hearing and signal transmission further helps limit the masking effects of echoes from surrounding objects. From a locomotory point of view, clutter avoidance is influenced by the echolocation behavior of the bat as well as its size, flight speed, and ability to maneuver (e.g., Aldridge and Rautenbach 1987).

Bats seem to have intricate signal processing mechanisms for echolocating through clutter. To negotiate through a maze of hanging chains, the big brown bat (*Eptesicus fuscus*) emits echolocation calls in pairs to maintain a high echolocation rate, and uses frequency shifts within call pairs to separate the returning echoes (Hiryu et al. 2010). These bats use not only the fundamental frequency of the call, but also higher harmonics that are more directional, in order to reduce the masking effects of echoes from clutter away from the acoustic axis of the sonar beam (Bates and Simmons 2011; Bates et al. 2011).

There have been few laboratory studies of toothed whale sonar performance in clutter (Au and Turl 1983; Au 1992), and it is difficult to quantify clutter in aquatic habitats, especially for highly mobile echolocators. Beaked whales are obviously able to discriminate and select a few tasty treats among lots of potential targets (Jones et al. 2008; Arranz et al. 2011). Even more remarkably, bottlenose dolphins can use their short broadband pulses to detect targets such as small fish completely buried in sand (Rossbach and Herzing 1997). No human sonars have been able to solve this problem in spite of significant investment by naval engineers.

2.6.2 Jamming Avoidance

Large numbers of bats emerging from a roost, particularly through a small opening obviously must deal effectively with the proximity of others and the din of echolocation calls, their own and those of conspecifics. In some cases, for example, Brazilian free-tailed bats (*Tadarida brasiliensis*), calls quite distinct from echolocation signals are produced during emergence (Gillam et al. 2010). Suthers (1967) showed how greater bulldog bats (*Noctilio leporinus*) adjusted their echolocation calls in the presence of conspecifics. Such changes may help bats deal with a confusing acoustic environment containing many other conspecifics echolocating at the same time, and have been proposed to play a role in so-called jamming avoidance (Ulanovsky et al. 2004).

Most toothed whales are social, and one of the dominant sources of interference for their own echolocation could similarly be the echolocation signals of conspecifics. Much of the time when toothed whales forage, they may separate and forage independently unless they are using a social strategy for foraging, such as corralling prey (Similä and Ugarte 1993). For delphinids foraging in groups, it is possible that the directionality of their echolocation signals and hearing as well as differences in click repetition rates between animals may help reduce interference from the echolocation signals of conspecifics. In some human sonars, called bistatic sonars, a receiver can gather information from echoes of signals produced by other platforms. Captive dolphins can be trained to detect targets by listening to echoes from the clicks of other dolphins (Xitco and Roitblat 1996), but whether animals actually exploit this in the wild remains untested.

2.6.3 Communication

The communicative role of echolocation signals has been repeatedly demonstrated in a range of bats. This often involves changing the frequencies dominating the echolocation calls (e.g., Ulanovsky et al. 2004; Gillam et al. 2007), typically over a very short time frame (Gillam and McCracken 2007). Details of echolocation calls can permit recognition of individuals (Voight-Heucke et al. 2010). Möhres (1966) reported that captive greater horseshoe bats (*Rhinolophus ferrumequinum*) recognized roost mates by their echolocation calls and the topic remains one of active investigation (Barclay 1982; Jones and Siemers 2011). Some bats, notably molossids, also produced "social buzzes," periods of high pulse production during social interactions (Swartz et al. 2007; Bayefsky-Anand et al. 2008). Kingston and Rossiter (2004) showed how changes in the harmonic content of echolocation calls could influence both prey detection and the role of echolocation signals in communication. Dechmann et al. (2009, 2010) suggested that monitoring the echolocation calls of conspecifics was a key to group hunting and sociality in molossid bats.

Many toothed whale species also use click-based signals for communication, and it is possible that species learn about conspecifics by eavesdropping on echolocation. Sperm whales use rhythmic patterns of clicks, called codas, for communication (Watkins and Schevill 1977; Weilgart and Whitehead 1997). Coda clicks are similar to echolocation clicks, but have a slower decline in intensity of successive pulses within the click, and may be less directional than echolocation clicks (Madsen et al. 2002). Porpoises also use stereotyped repeated patterns of echolocation clicks in specific behavioral contexts, apparently to communicate (Amundin 1991; Nakamura et al. 1998; Clausen et al. 2010). Clausen et al. (2010) report that porpoises produced buzzes with particularly high click repetition rates during aggressive interactions. Dawson (1991) argues that *Cephalorhynchus hectorii* (Hector's dolphin, which produces NBHF echolocation clicks similar in structure to those of porpoises), also use patterns of these clicks for communication as well as for echolocation. Blainville's beaked whales (*Mesoplodon densirostris*) sometimes produce "rasps" or stereotyped bursts of their FM search clicks as they begin their deep foraging dives, and Aguilar de Soto et al. (2011) have argued that these sounds have a function in communication. While porpoises or Hector's dolphins do not produce lower-frequency sounds such as whistles, and may have few other options for communicating other than adapting echolocation signals, clicks may also be important for communication in whistling species.

2.6.4 Predator–Prey Interactions

Both echolocating bats and toothed whales encounter prey (at least insects, marine mammals and some fish) capable of detecting echolocation signals (e.g., Roeder 1967; Fullard 1987; Popper et al. 2004). Detection of echolocation signals usually triggers defensive behavior ranging from negative phonotaxis (Roeder 1967; Wilson et al. 2008, 2011) to production of counter signals opening new areas for mimicry (Barber and Conner 2007; Ratcliffe and Nydam 2008). The clicks of some arctiid moths jam the echolocation of some bats (e.g., Corcoran et al. 2009). Goertlitz et al. (2010) described "stealth" echolocation behavior in foraging barbastelles (*Barbastella barbastella*) and Barrett-Lennard et al. (1996) and Deecke et al. (2005) described how killer whales feeding on other marine mammals produce more stealthy echolocation signals than killer whales feeding on fish species that do not detect ultrasound. Mammal-eating killer whales hunting seals or other toothed whales with very sensitive underwater hearing, tend to avoid echolocating and calling during hunting to reduce the risk of alerting their prey (Barrett-Lennard et al. 1996; Deecke et al. 2005). Harbor seals listen for the acoustic signals of killer whales to detect and avoid them, and respond more strongly to the calls of the mammal-eating transient killer whales than to familiar fish-eating resident killer whales (Deecke et al. 2002).

2.7 Phylogeny and Diversification of Echolocation in Bats and Toothed Whales

An analysis using a molecular supermatrix of families of living mammals combined with a relaxed molecular clock indicates that modern mammals diversified after the Cretaceous–Paleogene mass extinction (Meredith et al. 2011). The data suggest that bats diversified earlier than toothed whales perhaps by 20 million years, although both had been derived by about 65.5 million years ago (Meredith et al. 2011). Both groups belong to Laurasiatheria, derived from ancestors that had diversified in the Cretaceous. During the Cretaceous the presence of birds and pterosaurs could have limited opportunities for bats, while various lineages of marine reptiles could have had the same effect on the evolution of whales (Nummela et al. 2004).

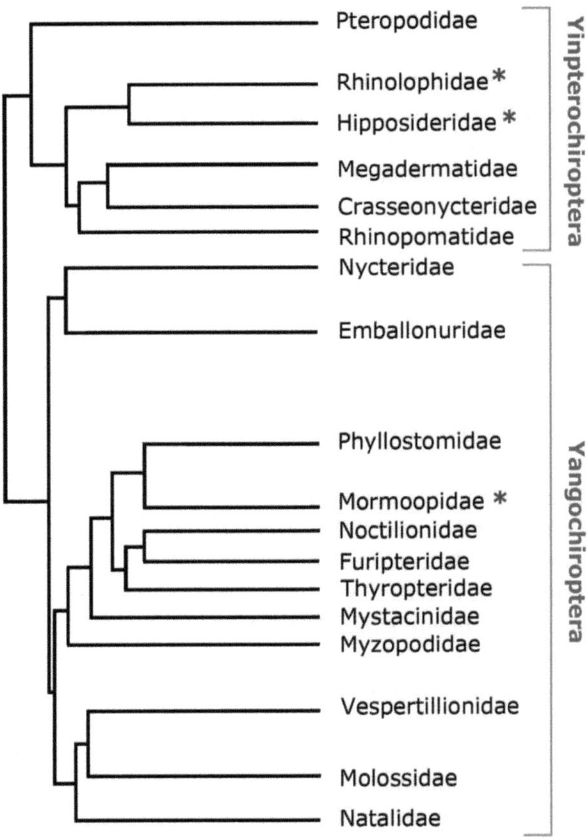

Fig. 2.14 Time-corrected phylogeny of modern families of bats. (Adapted from Teeling et al. 2005.) The *asterisk* indicates families that use high duty cycle echolocation. Pteropodidae contains the genus *Rousettus*, with at least two species that produce echolocation clicks using their tongues

Diversification of cetaceans beyond an initial radiation appears to have been associated with a fundamental change in sound reception, adapting an ear designed for sound reception in air to one that works in water (Nummela et al. 2004). This change was achieved via the mandible to an acoustic fat channel to the inner ear. The adaptation of the cetacean ear for sound reception in water would have been key to the development of echolocation. In bats, contact between stylohyal and tympanic bones in laryngeal echolocators served as a possible reafferent mechanism for registering outgoing pulses in the brain (Veselka et al. 2010). This contact may be a comparable key to the development of echolocation in bats. Liu et al. (2010) report convergence among bats and toothed whales in the *Prestin* gene with modifications that may improve high-frequency performance of the hair cells in the inner ear, which convert sound energy into neural impulses.

Theories about the evolutionary relationships among bats have changed dramatically in the last 10 years (Teeling et al. 2005). This is reflected by a change in subordinal classification from Megachiroptera and Microchiroptera to Yinpterochiroptera and Yangochiroptera (Fig. 2.14). The arrangement of families in Yinpterochiroptera and Yangochiroptera differs substantially from that in Megachiroptera (family Pteropodidae) and Microchiroptera (all other living bats). Details of the pectoral girdle demonstrate that the earliest known fossil bats (*Onychonycteris finneyi*) could fly, but whether or not this species could echolocate remains a topic of discussion (Simmons et al. 2008; Veselka et al. 2010).

If laryngeal echolocation was ancestral, it evolved once in bats, was subsequently lost in the Pteropodidae, but echolocation was regained in a different form as tongue clicks in some species of *Rousettus*. If laryngeal echolocation is a derived feature, then it may have evolved at least twice in bats (Fig. 2.14). High duty cycle echolocation has evolved at least twice in bats (Teeling 2009), at least once in the Yinpterochiroptera (Hipposideridae, Rhinolophidae), and once in the Yangochiroptera (Mormoopidae—*Pteronotus parnellii*) (Fig. 2.14).

McGowen et al. (2009) analyzed molecular data to provide a comprehensive phylogeny of cetaceans with estimates of divergence dates. We reproduce their phylogenetic tree, indicating biosonar signals typical of each taxon (see Fig. 2.1). More classical taxonomy includes many features of the cranial anatomy of toothed whales that are driven by adaptations for sound production. Fahlke et al. (2011) point out that the asymmetry of the skull coevolved in ancient cetaceans with a "complex of traits linked to directional hearing (such as pan-bone thinning of the lower jaws, mandibular fat pads, and isolation of the ear region." These characters evolved before the split between baleen (mysticete) and toothed (odontocete) cetaceans, and may have enabled directional hearing underwater. Fahlke et al. (2011) argue that toothed whales built upon these adaptations, evolving phonic lips in the nasal passages for high-frequency sound production, nasal sacs in the upper respiratory tract that function to recycle air for sound production, and that couple with skull shape and specialized fats in the melon to form a directional acoustic beam. This system for producing echolocation sounds appears to have originated once in cetacean evolution but subsequently diversified into several forms, with physeterids the most anatomically specialized (Au 1993; Møhl et al. 2003; Madsen et al. 2005b).

There are some constraints in the physical relationships between size of a sound source, frequency of sound produced, and directionality, but within these constraints, toothed whales show flexibility in adapting their echolocation systems to different tasks and conditions. In an evolutionary sense, this is shown in the convergence of NBHF clicks in four evolutionarily distinct groups of toothed whales with quite different sound production anatomies: (1) Phocoenid porpoises, (2) non-whistling delphinids of the genus *Cephalorhynchus* and several *Lagenorhynchus* species (called *Sagmatias* by McGowen et al. 2009), (3) the Franciscana or La Plata river dolphin, *Pontoporia blainvillei*, and (4) physeterids of the genus *Kogia*.

There is little evidence for deciding which of the other forms of click in toothed whales is closest to ancestral. However, broadband echolocation clicks are intuitively the simplest, and are shared not only by some of the more ancestral toothed whales such as Ganges River dolphins (Jensen et al. 2013) but also by the other basal lineages, the sperm whales (the outgoing p1 pulse) and the ziphiidae (buzz clicks). If we assume that the BBHF clicks represent clicks closest to the ancestral form, then it appears that the four NBHF taxa may have independently evolved NBHF clicks. The coastal porpoises and nonwhistling delphinids (such as *Cephalorhynchus*) have similar body types, and along with the La Plata dolphin, they are thought to have similar foraging niches, feeding near the bottom in shallow cluttered coastal or estuarine environments. This similarity is consistent with the hypothesis of convergence of similar signals for species in similar ecological niches. By contrast, the pygmy and dwarf sperm whales in the genus *Kogia* feed deep in the open ocean. Though this represents a very different ecological niche, might there still be similarities in how all four taxa use NBHF signals for echolocation?

Madsen et al. (2005b) argue that toothed whales using NBHF signals exploit a low noise window at about 130 kHz in the ocean. To do so, these animals must have an auditory system that averages energy over smaller bandwidths than would be typical for mammals at these frequencies. Narrower filters have been measured in *Kogia* (Ridgway and Carder 2001) and inferred anatomically in *Kogia* for the high frequencies of their clicks (Ketten 2000). Popov et al. (2006) report that phocoenid porpoises have constant bandwidth filters, and Lemonds et al. (2011) argue that most delphinids do as well for frequencies between 30 and 120 kHz, meaning that these taxa have a higher relative spectral resolution at higher frequencies than is typical for mammalian hearing. These observations suggest that NBHF clicks have lower capabilities for classifying targets and determining range, but that they may increase probability of detecting prey by concentrating the signal in a narrow frequency band and reducing the impact of ambient noise that would mask echoes.

Acoustic crypsis may be another important factor in the evolution of NBHF signals. The very high frequencies used by NBHF animals attenuate rapidly as they pass through seawater, limiting the range of echolocation, but also reducing the likelihood of being overheard by a predator. One of the predominant marine mammal predators, the killer whale (*Orcinus orca*) is itself a toothed whale with acute ultrasonic hearing. However, its auditory sensitivity may decrease quickly above 100 kHz (Szymanski et al. 1999). Morisaka and Connor (2007) proposed that the independent convergent evolution of NBHF clicks in four phylogenetically separate species, combined with the simultaneous nonwhistling nature of these animals, is a mechanism to decrease the risk of being detected by killer whales. The signals and behavior of other toothed whales may also have been shaped to avoid predation. For example, beaked whales produce sound deep enough that they are not easily detected at the surface (Aguilar de Soto et al. 2011). This cryptic behavior of seldom making a sound when swimming in water shallower than 200 m may be seen as a behavioral adaptation to avoid detection by killer whales. These observations suggest considerable flexibility in the evolution of sonar and communication signals as toothed whales balance requirements to orient, forage, and avoid detection by predators. The role of echolocation in the predatory behavior of bat-eating bats (either hunter or quarry) remains to be assessed in the field.

2.8 Summary

This chapter has introduced echolocation in bats and toothed whales, including the role that it appears to have played in the origin and diversification of these two groups. Mechanisms of sound production and reception are compared in the two groups along with different patterns of call production. Echolocating toothed whales and bats operate in a range of situations, from open spaces to cluttered ones. There is a high diversity of signal types and different degrees of variability within and between bat and toothed whale species and families. Signal variability is mainly linked to ecological factors (e.g., the high degree of convergence in signal types across bat families that forage in similar habitat types). Three of the four toothed whale taxa that have converged on narrowband high frequency signals have similar foraging niches. The fourth NBHF taxon has a different foraging niche, but may have evolved NBHF signals for similar functions as the other three, suggesting the need for further study into how these species use their signals in the wild. Phylogeny also plays an important role in shaping parts of signal design, for example, the consistent use of the second harmonic in emballonurids versus the first harmonic in molossids. However, the variability of signal types within some families suggests that phylogeny does not always constrain signal structure. Bats that appear to consistently hunt flying insects in relatively open spaces (e.g., molossids and some vespertilionids), show considerable variability in call design, while those hunting in more cluttered surroundings and often relying on acoustic cues from prey (e.g., phyllostomids, nycterids, and megadermatids) show less variation in call design, using short, multiharmonic FM calls. Sperm whales and other toothed whales also show low variability in echolocation signals despite many species having a highly variable diet.

By separating pulse and echo in time (low duty cycle) most echolocators (all toothed whales and most bats) reduce the problem of outgoing pulses interfering with detection of echoes. Some bats (rhinolophids, hipposiderids, and the mormoopid *Pteronotus parnellii*) avoid this problem by separating pulse and echo in frequency. There is more information in the calls than we have covered in our analysis of echolocation. In both toothed whales and bats, there is a continuum of function between echolocation and social signals, with echolocation signals simultaneously conveying information to the individual that produced the sound and to eavesdropping conspecifics. Many bat and toothed whale species broadcast calls with differences in frequency content that may relate to body size or individual variation in vocal anatomy, and can provide cues for both species and individual identification for nearby animals. Signals normally associated with echolocation may also have evolved a more specific communication function especially during short range interactions including aggressive encounters. Both bats and toothed whales deal with some prey that can detect echolocation calls; when foraging on these prey, bats and toothed whales may modify their echolocation to reduce the risk that the prey can take evasive action.

Comparative data from laboratory and field studies demonstrate the flexibility of bat and toothed whale biosonar. Some aspects of biosonar signals can be related to phylogeny, but the convergent evolution of similar sonar signals across phylogenetic borders and the high variability in calls even within a species, suggest that phylogeny may not constrain echolocation signals as much as previous work has suggested. This summary also emphasizes the synergy of data originating from the lab, from the field, or from behavioral experiments conducted under semi-natural conditions.

References

Aguilar de Soto N., Johnson, M. P., Madsen, P. T., Díaz, F., Domínguez, I., Tyack, P., & Brito, A. (2008). Deep foraging sprints in short finned pilot whales off Tenerife (Canary Islands). *Journal of Animal Ecology*, 77, 936–947.

Aguilar de Soto, N., Madsen, P. T., Tyack, P., Arranz, P., Marrero, J., Fais, A., Revelli, E., & Johnson, M. (2011). No shallow talk: Cryptic strategy in the vocal communication of Blainville's beaked whales. *Marine Mammal Science*, 28 (2), E75–E92.

Ahlen, L. (1981). Identification of Scandinavian bats by their sounds. *Sveriges Lantbruksuniversitet, Institutionen för Viltekologi*, 6, Uppsala.

Akre, K. L., Farris, H. E., Lea, A. M., Page, R. A., & Ryan, M. J. (2011). Signal perception in frogs and bats and the evolution of mating signals. *Science*, 333, 751–752.

Aldridge, H. D. N. J., & Rautenbach, I. L. (1987). Morphology, echolocation and resource partitioning in insectivorous bats. *Journal of Animal Ecology*, 56, 763–778.

Aldridge, H. D. J. N., Obrist, M., Merriam, H. G., & Fenton, M. B. (1990). Roosting, vocalizations, and foraging by the African bat, *Nycteris thebaica. Journal of Mammalogy*, 71, 242–246.

Amundin, M. (1991). *Sound production in odontocetes with emphasis on the harbor porpoise*, (Phocoena phocoena). PhD dissertation, University of Stockholm.

Aroyan J. L. (2001). Three-dimensional modeling of hearing in *Delphinus delphis. Journal of the Acoustical Society of America*, 110, 3305–3318.

Aroyan, J. L., McDonald, M. A., Webb, S. C., Hildebrand, J. A., Clark, D., Laitman, J. T., & Reidenberg, J. S. (2000). Acoustic models of sound production and propagation. In R. R. Fay & A. N. Popper (Eds.), *Hearing by whales and dolphins* (pp. 409–469). New York: Springer-Verlag.

Arranz, P., Aguilar de Soto, N., Madsen, P. T., Brito, A., Bordes, F., & Johnson, M. P. (2011). Following a foraging fish-finder: Diel habitat use of Blainville's beaked whales revealed by echolocation. PLoS ONE 6(12): e28353.

Atem, A. C. G., Rasmussen, M. H., Wahlberg, M., Petersen, H. C., & Miller, L. A. (2009). Changes in click Source Levels with distance to targets: Studies of free-ranging white-beaked dolphins *Lagenorhynchus albirostris* and captive harbour porpoises *Phocoena phocoena. Bioacoustics*, 19, 49–65.

Au, W. W. L. (1992). Application of the reverberation-limited form of the sonar equation to dolphin echolocation. *Journal of the Acoustical Society of America* 92, 1822–1826.

Au, W. W. L. (1993). *The sonar of dolphins*. New York: Springer-Verlag.

Au, W. W. L. (2004). Echolocation signals of wild dolphins. *Acoustical Physics*, 50, 454–462.

Au, W. W. L., & Turl, C. W. (1983). Target detection in reverberation by an echolocating Atlantic bottlenose dolphin (*Tursiops truncatus*). *Journal of the Acoustical Society of America*, 73, 1676–1681.

Au, W. W. L., & Moore, P. W. B. (1984). Receiving beam patterns and directivity indices of the Atlantic bottlenose dolphin. *Journal of the Acoustical Society of America*, 75, 255–262.

Au, W. W. L., & Benoit-Bird, K. J. (2003). Automatic gain control in the echolocation system of dolphins. *Nature*, 423, 861–863.

Au, W. W. L., Floyd, R. W., Penner, R. H., & Murchison, A. E. (1974). Measurement of echolocation signals of the Atlantic bottlenose dolphin, *Tursiops truncatus* Montagu, in open waters. *Journal of the Acoustical Society of America*, 56, 1280–1290.

Au, W. W. L., Moore, P. W. B., & Pawloski, D. (1986). Echolocation transmitting beam of the Atlantic bottlenose dolphin. *Journal of the Acoustical Society of America*, 80, 688–694.

Au, W. W. L., Pawloski, D., Nachtigall, P. E., Blonz, M., & Gisner, R. G. (1995). Echolocation signals and transmission beam pattern of a false killer whale (*Pseudorca crassidens*). *Journal of the Acoustical Society of America*, 98, 51–59.

Au, W. W. L., Kastelein, R. A., Ripper, T., & Schooneman, N. M. (1999). Transmission beam pattern and echolocation signals of a harbor porpoise (*Phocoena phocoena*). *Journal of the Acoustical Society of America*, 106, 3699–3705.

Au, W. W. L., Ford, J. K. B., Horne, J. K., & Allman, K. A. N. (2004). Echolocation signals of free-ranging killer whales (*Orcinus orca*) and modeling of foraging Chinook salmon (*Oncorhynchus tshawytscha*). *Journal of the Acoustical Society of America*, 115, 901–909.

Aytekin, M., Mao, B., & Moss, C. F. (2011). Spatial perception and adaptive sonar behavior. *Journal of the Acoustical Society of America*, 128, 3788–3798.

Barber, J. R., & Conner, W. E. (2007). Acoustic mimicry in a predator-prey interaction. *Proceedings of the National Academy of Sciences of the USA*, 104, 9331–9334.

Barclay, R. M. R. (1982). Interindividual use of echolocation calls: Eavesdropping by bats. *Behavioral Ecology and Sociobiology*, 10, 271–275.

Barrett-Lennard, L. G., Ford, J. K. B., & Heise, K. A. (1996). The mixed blessing of echolocation: Differences in sonar use by fish-eating and mammal-eating killer whales. *Animal Behaviour*, 51, 553–565.

Bates, M. E., & Simmons, J. A. (2011). Perception of echo delay is disrupted by small temporal misalignment of echo harmonics in bat sonar. *Journal of Experimental Biology*, 214, 394–401.

Bates, M. E., Simmons, J. A., & Zorikov, T. V. (2011). Bats use harmonic structure to distinguish their targets from background clutter. *Science*, 333, 627–630.

Bayefsky-Anand, S., Skowronski, M. D., Fenton, M. B., Korine, C., & Holderied, M. W. (2008). Variations in the echolocation calls of the European-free-tailed bat (*Tadarida teniotis*, Molossidae). *Journal of Zoology*, 275, 115–123.

Beedholm, K., Miller, L.A., & Blanchet M.A. (2006). Auditory brainstem response in a harbor porpoise show lack of automatic gaine control for simulated Echoes. *Journal of the Acoustical Society of America*, 119(3):EL41–EL46.

Beedholm, K., & Miller, L. (2007). Automatic gain control in harbor porpoises (*Phocoena phocoena*)? Central versus peripheral mechanisms. *Aquatic Mammals*, 33, 69–75.

Brill, L. R., & Harder, P. J. (1991). The effects of attenuating returning echolocation signals at the lower jaw of a dolphin (*Tursiops truncatus*). *Journal of the Acoustical Society of America*, 89, 2851–2857.

Brinkløv, S., Kalko, E. K. V., & Surlykke, A. (2009). Intense echolocation calls from two 'whispering' bats, *Artibeus jamaicensis* and *Macrophyllum macrophyllum* (Phyllostomidae). *Journal of Experimental Biology*, 212, 11–20.

Brinkløv, S., Kalko, E. K. V., & Surlykke, A. (2010). Dynamic adjustment of biosonar intensity to habitat clutter in the bat *Macrophyllum macrophyllum* (Phyllostomidae). *Behavioral Ecology and Sociobiology*, 64, 1867–1871.

Brinkløv, S., Jakobsen, L., Ratcliffe, J. M., Kalko, E. K. V., & Surlykke, A. (2011). Echolocation call intensity and directionality in flying short-tailed fruit bats, *Carollia perspicillata* (Phyllostomidae). *Journal of the Acoustical Society of America*, 129, 427–435.

Buchler, E. R., & Mitz, A. R. (1980). Similarities in design features of orientation sounds used by simpler, nonaquatic echolocators. In R-G. Busnel & J. F. Fish (Eds.), *Animal sonar systems* (pp. 871–874). New York: Plenum Press.

Bullock, T. H., Grinnell, A. D., Ikezono, E., Kameda, K., Katsuki, Y., Nomoto, M., Sato, O., Suga, N., & Yanagisawa, K. (1968). Electrophysiological studies of central auditory mechanisms in cetaceans. *Zeitschrift für vergleichende Physiologie*, 59, 117–156.

Carrier, D. R., Debahn, S. M., & Otterstrom, J. (2002). The face that sunk the Essex: Potential function of the spermaceti organ in aggression. *Journal of Experimental Biology*, 205, 1755–1763.

Carter, G. G., Ratcliffe, J. M., & Galef, B. G. (2010). Flower bats (*Glossphaga soricina*) and fruit bats (*Carollia perspicillata*) rely on spatial cues over shapes and scents when relocating food. *PLoS ONE*, 5(5), e10808.

Chiu, C., Reddy, P. V., Xian, W., Krishnaprasad, P. S., & Moss, C. F. (2010). Effects of competitive prey capture on flight behaviour and sonar beam pattern in paired big brown bats, *Eptesicus fuscus*. *Journal of Experimental Biology*, 213, 3348–3356.

Clarke, M. R. (1978). Buoyancy control as a function of the spermaceti organ in the sperm whale. *Journal of the Marine Biological Association, UK*, 58, 27–71.

Clausen K. T., Wahlberg M., Beedholm K., Deruiter S., & Madsen P. T. (2010). Click communication in harbor porpoises *Phocoena phocoena*. *Bioacoustics*, 20, 1–28.

Corcoran, A. J., Barber, J. R., & Connor, W. E. (2009). Tiger moth jams bat sonar. *Science*, 325, 325–327.

Cranford, T. W. (2000). In search of impulse sound sources in odontocetes. In W. W. L. Au, A. N. Popper, & R. R. Fay (Eds.), *Hearing by whales and dolphins* (pp. 109–155). New York: Springer-Verlag.

Cranford, T. W., Amundin, M., & Norris, K. S. (1996). Functional morphology and homology in the odontocete nasal complex: Implications for sound generation. *Journal of Morphology*, 228, 223–285.

Cranford, T. W., Elsberry, W. R., Van Bonn, W. G., Jeffress, J. A., Chaplin, M. S., Blackwood, D. J., Carder, D. A., Kamolnick T., Todd, M. A., & Ridgway, S. H. (2011). Observation and analysis of sonar signal generation in the bottlenose dolphin (*Tursiops truncatus*): Evidence for two sonar sources. *Journal of Experimental Marine Biology and Ecology*, 407, 81–96.

Dawson, S. M. (1991). Clicks and communication—the behavioural and social contexts of Hector's dolphin vocalisations. *Ethology*, 88, 265–276.

Dawson S., Barlow J., & Ljungblad D. (1998). Sounds recorded from Baird's beaked whale, *Berardius bairdii*. *Marine Mammal Science*, 14, 335–344.

Dechmann, D. K. N., Heucke, S. L., Giuggioli, L., Safi, K., Voight, C. C., & Wikelski, M. (2009). Experimental evidence for group hunting via eavesdropping in echolocating bats. *Proceedings of the Royal Society of London B: Biological Sciences*, 276, 2721–2728.

Dechmann, D. K. N., Kranstauber, B., Gibbs, D., & Wikelski M. (2010). Group hunting – a reason for sociality in molossid bats? *PloS ONE*, 5(2), e9012.

Deecke, V. B., Slater, P. J. B., & Ford, J. K. B. (2002). Selective habituation shapes acoustic-predator recognition in harbour seals. *Nature*, 420, 171–173.

Deecke, V. B., Ford, J. K. B., & Slater, P. J. B. (2005). The vocal behavior of mammal-eating killer whales: Communicating with costly calls. *Animal Behaviour*, 69, 395–405.

Denny, M. (2007). *Blip, ping and buzz: Making sense of radar and sonar*. Baltimore: Johns Hopkins University Press.

Denzinger, A., Siemers, B. M., Schaub, A., & Schnitzler, H-U. (2001). Echolocation by the barbastelle bat *Barbastella barbastellus*. *Journal of Comparative Physiology A: Neuroethology, Sensory, Neural, and Behavioral Physiology*, 187, 521–528.

DeRuiter, S. L., Bahr, A., Blanchet, M.–A., Hansen, S. F., Kristensen, J. H., Madsen, P. T., Tyack, P. L., & Wahlberg, M. (2009). Acoustic behaviour of echolocating porpoises during prey capture. *Journal of Experimental Biology*, 212, 3100–3107.

Elemans, C. P. H., Mead, A. F., Jakobsen, L., & Ratcliffe, J. M. (2011). Superfast muscles set maximum call rate in echolocating bats. *Science*, 333, 1885–1888.

Fahlke, J. M., Gingerich, P. D., Welsh, R. C., & Wood A. R. (2011). Cranial asymmetry in Eocene archaeocete whales and the evolution of directional hearing in water. *Proceedings of the National Academy of Sciences of the USA*, 108, 14545–14548.

Fenton, M. B. (2010). Convergences in the diversification of bats. *Current Zoology*, 56, 454–468.
Fenton, M. B., & Bell, G. P. (1981). Recognition of species of insectivorous bats by their echolocation calls. *Journal of Mammalogy*, 62, 233–243.
Fenton, M. B., Skowronski, M. D., McGuire, L. P., & Faure, P. A. (2011). Variation in the use of harmonics in the calls of laryngeally echolocating bats. *Acta Chiropterologica.*, 13, 169–178.
Fenton, M. B., Faure, P. A., & Ratcliffe, J. M. (2012). Evolution of high duty cycle echolocation in bats. *Journal of Experimental Biology*, 215, 2935–2944.
Fiedler, J. (1979). Prey catching with and without echolocation in the Indian vampire (*Megaderma lyra*). *Behavioral Ecology and Sociobiology*, 6, 155–160.
Fullard, J. H. (1987). Sensory ecology and neuroethology of moths and bats: Interactions in a global perspective. In M. B. Fenton, P. A. Racey, & J. M. V. Rayner (Eds.), *Recent advances in the study of bats* (pp. 244–273). Cambridge, UK: Cambridge University Press.
Gannon, D. P., Barros, N. B., Nowacek, D. P., Read, A. J., Waples, D. M., & Wells, R. S. (2005). Prey detection by bottlenose dolphins, *Tursiops truncatus*: An experimental test of the passive listening hypothesis. *Animal Behaviour*, 69, 709–720.
Gaudet, C. L., & Fenton, M. B. (1984). Observational learning in three species of insectivorous bats (Chiroptera). *Animal Behaviour*, 32, 385–388.
Gillam, E. H., & McCracken, G. F. (2007). Variability in the echolocation of *Tadarida brasiliensis*: Effects of geography and local acoustic environment. *Animal Behaviour*, 74, 277–286.
Gillam, E. H., Ulanovsky, N., & McCracken, G. F. (2007). Rapid jamming avoidance in biosonar. *Proceedings of the Royal Society of London B: Biological Sciences*, 274, 651–660.
Gillam, E. H., Hristov, N. I., Kunz, T. H., & McCracken, G. F. (2010). Echolocation behavior of Brazilian free-tailed bats during dense emergence flights. *Journal of Mammalogy*, 91, 967–975.
Goertlitz, H. R., ter Hofstede, H. M., Zeale, M. R. K., Jones, G., & Holderied, M. W. (2010). An aerial hawking bat uses stealth echolocation to counter moth hearing. *Current Biology*, 20, 1568–1572.
Gordon, J. C. D. (1991). Evaluation of a method for determining the length of sperm whales *Physeter catodon* from their vocalizations. *Journal of Zoology, London*, 224, 301–314.
Gracheva, E. O., Cordero-Morales, J. F., Gonzalez-Carcacia, J. A., Ingolia, N. T., Manno, C., Arangueren, C. I., Weissman, J. S., & Julius, D. (2011). Ganglion-specific splicing of TRPV1 underlies infrared sensation in vampire bats. *Nature*, 476, 88–91.
Grief, S., & Siemers, B. M. (2010). Innate recognition of water bodies in echolocating bats. *Nature Communications*, 1, 107.
Griffin, D. R. (1958). *Listening in the dark*. New Haven, CT: Yale University Press.
Griffin, D. R., & Thompson, D. (1982). Echolocation by cave swiftlets. Behavioral Ecology and Sociobiology, 10, 119–123.
Griffin, D. R., Webster, F. A., & Michael, C. R. (1960). The echolocation of flying insects by bats. Animal Behaviour, 8, 141–154.
Groger, U., & Wiegrebe, L. (2006). Classification of human breathing sounds by the common vampire bat, *Desmodus rotundus*. *BMC Biology*, 4,18.
Hartley, D. J. (1992). Stabilization of perceived echo amplitudes in echolocating bats. I. Echo detection and automatic gain-control in the big brown bat, *Eptesicus fuscus*, and the fishing bat, *Noctilio leporinus*. *Journal of the Acoustical Society of America*, 91, 1120–1132.
Hartley, D. J., & Suthers, R. A. (1987). The sound emission pattern and the acoustical role of the noseleaf in the echolocating bat, *Carollia perspicillata*. *Journal of the Acoustical Society of America*, 82, 1987–1900.
Hemilä, S., Nummela, S., & Reuter, T. (2010). Anatomy and physics of the exceptional sensitivity of dolphin hearing (Odontoceti: Cetacea). *Journal of Comparative Physiology A: Neuroethology, Sensory, Neural, and Behavioral Physiology*, 196, 165–179.
Henson, O. W. (1965). Activity and function of middle-ear muscles in echolocating bats. *Journal of Physiology, London*, 180, 871–887.
Hiryu, S., Hagino, T., Riquimaroux, H., & Watanabe, Y. (2007). Echo-intensity compensation in echolocating bats (*Pipistrellus abramus*) during flight measured by a telemetry microphone. *Journal of the Acoustical Society of America*, 121, 1749–1757.

Hiryu, S., Bates, M. E., Simmons, J. A., & Riquimaroux, H. (2010). FM broadcasting bats shift frequencies to avoid broadcast-echo ambiguity in clutter. *Proceedings of the National Academy of Sciences of the USA*, 107 (15), 7048–7053.

Hoelzel, A. R., & Osborne, R. W. (1986). Killer whale call characteristics: Implications for cooperative foraging strategies. In B. C. Kirkevold & J. S. Lockard (Eds.), *Behavioral biology of killer whales* (pp. 373–403). New York: Alan R. Liss.

Holderied, M. W., Korin, C., Fenton, M. B., Parsons, S., Robinson, S., & Jones, G. (2005). Echolocation call intensity in the aerial hawking bat *Eptesicus bottae* (Vespertilionidae) studied using stereo videogrammetry. *Journal of Experimental Biology*, 208, 1321–1327.

Holderied, M., Korin, C., & Moritz, T. (2011). Hemprich's long-eared bat (*Otonycteris hemprichii*) as a predator of scorpions: whispering echolocation, passive gleaning and prey selection. *Journal of Comparative Physiology A: Neuroethology, Sensory, Neural, and Behavioral Physiology*, 197, 425–433.

Holland, R. A., Waters, D. A., & Rayner, J. M. V. (2004). Echolocation signal structure in the megachiropteran bat *Rousettus aegypticaus* Geoffroy 1810. *Journal of Experimental Biology*, 207, 4261–4369.

Hooker, S. K., & Whitehead, H. (2002). Click characteristics of northern bottlenose whales (*Hyperoodon ampullatus*). *Marine Mammal Science*, 18, 69–80.

Hooper, J. D. H. (1964). Bats and the amateur naturalist. *Studies in Speleology*, 1, 9–15.

Huggenberger, S., Rauschmann, M. A., Vogl, T. J., & Oelschläger H. H.A. (2009). Functional morphology of the nasal complex in the harbor porpoise (*Phocoena phocoena* L.). *Anatomical Record*, 292, 902–920.

Jakobsen L., & Surlykke A. (2010). Vespertilionid bats control the width of their biosonar sound beam dynamically during prey pursuit. *Proceedings of the National Academy of Sciences of the USA*, 107, 13930–13935.

Janik, V. M., & Slater, P. J. B. (1997). Vocal learning in mammals. *Advances in the Study of Behaviour*, 26, 59–99.

Jaquet, N., Dawson, S., & Douglas, L. (2001). Vocal behavior of male sperm whales: Why do they click? *Journal of the Acoustical Society of America*, 109, 2254–2259.

Jensen, F. H., Bejder, L., Wahlberg, M., & Madsen, P. T. (2009). Biosonar adjustments to target range of echolocating bottlenose dolphins (*Tursiops* sp.) in the wild. *Journal of Experimental Biology*, 212, 1078–1086.

Jensen, F. H., Rocco, A., Mansur, R.M., Smith, B.D., Janik, V.M., & Madsen, P.T. (2013). Clicking in shallow rivers: Short-range echolocation of Irrawaddy and Ganges river dolphins in a shallow, acoustically complex habitat. PLoS ONE, 8(4):e59284.

Johnson, M., & Tyack, P. L. (2003). A digital acoustic recording tag for measuring the response of wild marine mammals to sound. *IEEE Journal of Oceanic Engineering*, 28, 3–12.

Johnson, M., Madsen, P. T., Zimmer, W. M. X., Aguilar de Soto, N., & Tyack, P. L. (2004). Beaked whales echolocate on prey. *Proceedings of the Royal Society of London B: Biological Sciences*, 271, S383–S386.

Johnson, M., Madsen, P. T., Zimmer, W. M. X., Aguilar de Soto, N., & Tyack, P. L. (2006). Foraging Blainville's beaked whales (*Mesoplodon densirostris*) produce distinct click types matched to different phases of echolocation. *Journal of Experimental Biology*, 209, 5038–5050.

Johnson, M. P., Aguilar de Soto, N., & Madsen, P. T. (2009). Studying the behaviour and sensory ecology of marine mammals using acoustic recording tags: A review. *Marine Ecology Progress Series*, 395, 55–73.

Jones, B. A., Stanton, T. K., Lavery, A. C., Johnson, M. P., Madsen, P. T., & Tyack, P. L. (2008). Classification of broadband echoes from prey of a foraging Blainville's beaked whale. *Journal of the Acoustical Society of America*, 123, 1753–1762.

Jones, G., & Siemers, B. (2011). The communication potential of bat echolocation pulses. *Journal of Comparative Physiology A: Neuroethology, Sensory, Neural, and Behavioral Physiology*, 197, 447–457.

Jung, K., Kalko, E. K. V., & von Helversen, O. (2007). Echolocation calls in Central American emballonurid bats: Signal design and call frequency alternation. *Journal of Zoology*, 272(2), 125–137.

Kalko, E. K. V. (1995). Insect pursuit, prey capture and echolocation in Pipistrelle bats (Microchiroptera). *Animal Behaviour*, 50, 861–880.

Kalko, E. K. V., & Schnitzler, H.-U. (1993). Plasticity in echolocation signals of European pipistrelle bats in search flight: Implications for habitat use and prey detection. *Behavioral Ecology and Sociobiology*, 33, 415–428.

Kalko, E. K. V., & Condon, M. (1998). Echolocation, olfaction, and fruit display: How bats find fruit of flagellichorous cucurbits. *Functional Ecology*, 12, 364–372.

Kalko, E. K. V., Schnitzler, H.-U., Kaipf, I., & Grinnell, A. D. (1998). Echolocation and foraging behavior of the lesser bulldog bat, *Noctilio albiventris*: Preadaptations for piscivory? *Behavioral Ecology and Sociobiology*, 42, 305–319.

Ketten, D. R. (2000). Cetacean ears. In W. W. L. Au, A. N. Popper, & R. R. Fay (Eds.), *Hearing by whales and dolphins* (pp. 43–108), New York: Springer-Verlag.

Kick, S. A. (1982). Target-detection by the echolocating bat, *Eptesicus fuscus*. *Journal of comparative Physiology A: Neuroethology, Sensory, Neural, and Behavioral Physiology*, 145, 431–435.

Kick, S. A., & Simmons, J. A. (1984). Automatic gain control in the bat's sonar receiver and the neuroethology of echolocation. *Journal of Neuro*science, 4, 2725–2737.

Kingston, T., & Rossiter, S. J. (2004). Harmonic hopping in Wallacea's bats. *Nature*, 429, 654–657.

Kingston, T., Jones, G., Akbar, Z., & Kunz, T. H. (2003). Alternation of echolocation calls in five species of aerial-feeding insectivorous bats from Malaysia. *Journal of Mammalogy*, 84, 205–215.

Kobler, J. B., Wilson, B. S., Henson, O. W., & Bishop, A. L. (1985). Echo intensity compensation by echolocating bats. *Hearing Research*, 20, 99–108.

Konishi, M., & Knudsen, E. I. (1979). The oilbird: hearing and echolocation. *Science*, 204, 425–427.

Krützen, M., Mann, J., Heithaus, M. R., Connor, R. C., Bejder, L., & Sherwin, W. B. (2005). Cultural transmission of tool use in bottlenose dolphins. *Proceedings of the National Academy of Sciences of the USA*, 102, 8939–8943.

Kyhn, L. A., Tougaard, J., Jensen, F., Wahlberg, M., Stone, G., Yoshinaga, A., Beedholm, K., & Madsen, P. T. (2009). Feeding at a high pitch: Source parameters of narrow band, high frequency clicks from echolocating off-shore hourglass dolphins and coastal Hector's dolphins. *Journal of the Acoustical Society of America*, 125, 1783–1791.

Kyhn, L. A., Jensen, F. H., Beedholm, K., Tougaard, J., Hansen M., & Madsen, P. T. (2010). Echolocation in sympatric Peale's dolphins (*Lagenorhynchus australis*) and Commerson's dolphins (*Cephalorhynchus commersonii*) producing narrow-band high-frequency clicks. *Journal of Experimental Biology*, 213, 1940–1949.

Lawrence, B. D., & Simmons, J. A. (1982). Measurements of atmospheric attenuation at ultrasonic frequencies and the significance for echolocation by bats. *Journal of the Acoustical Society of America*, 71, 585–590.

Lazure, L., & Fenton, M. B. (2011). High duty cycle echolocation and prey detection by bats. *Journal of Experimental Biology*, 214, 1131–1137.

Lemonds, D. W., Kloepper, L. N., Nachtigall, P. E., Au, W. W. L, Vlachos, S. A., & Branstetter, B. K. (2011). A re-evaluation of auditory filter shape in delphinid odontocetes: Evidence of constant-bandwidth filters. *Journal of the Acoustical Society of America*, 130, 3107–3114

Linnenschmidt, M., Beedholm, K., Wahlberg, M., Højer-Kristensen, J., & Nachtigall, P. E. (2012). Keeping returns optimal: Gain control exerted through sensitivity adjustments in the harbour porpoise auditory system. *Proceedings of the Royal Society of London B: Biological Sciences*, 279, 2237–2245.

Liu, Y., Rossiter, S. J., Han, X., Cotton, J. A., & Zhang, S. (2010). Cetaceans on a molecular fast track to ultrasonic hearing. *Current Biology*, 20, 1834–1839.

Ma, J., & Műller, R. (2011). A method for characterizing the biodiversity in bat pinnae as a basis for engineering analysis. *Bioinspiration and Biomimetics*, 6, 1–12.

Madsen, P. T., Payne, R., Kristiansen, N. U., Wahlberg, M., Kerr, I., & Møhl, B. (2002). Sperm whale sound production studied with ultrasound time/depth-recording tags. *Journal of Experimental Biology*, 205, 1899–1906.

Madsen, P. T., Carder, D. A., Au, W. W. L., Nachtigall, P. E., Møhl, B., & Ridgway, S. H. (2003). Sound production in neonate sperm whales. *Journal of the Acoustical Society of America*, 113, 2988–2991.

Madsen, P. T., Kerr, I., & Payne, R. (2004). Echolocation clicks of two free-ranging, oceanic delphinids with different food preferences: False killer whales *Pseudorca crassidens* and Risso's dolphins *Grampus griseus. Journal of Experimental Biology*, 207, 1811–1823.

Madsen, P. T., Johnson, M., Aguilar de Soto, N., Zimmer, W. M. X., & Tyack, P. (2005a). Biosonar performance of foraging beaked whales (*Mesoplodon densirostris*). *Journal of Experimental Biology*, 280, 181–194.

Madsen, P. T., Carder, D. A., Beedholm, K., & Ridgway, S. (2005b). Porpoise clicks from a sperm whale nose: Convergent evolution of toothed whale echolocation clicks? *Bioacoustics*, 15, 195–206.

Madsen, P. T., Wilson, M., Johnson, M., Hanlon, R. T., Bocconcelli, A., Aguilar de Soto, N., & Tyack, P. L. (2007). Clicking for calamari: Toothed whales can echolocate squid *Loligo pealeii. Aquatic Biology*, 1, 141–150.

Madsen, P. T., Wisniewska, D. M., & Beedholm, K. (2010). Single source sound production and dynamic beam formation in echolocating harbour porpoises (*Phocoena phocoena*). *Journal of Experimental Biology*, 213, 3105–3110.

Madsen, P. T., Jensen, F. H., Carder, D., & Ridgway, S. (2011). Dolphin whistles: A functional misnomer revealed by heliox breathing. *Biology Letters*, 8 (2), 211–213.

Marques, T. A., Thomas, L., Ward, J., DiMarzio, N., & Tyack, P. L. (2009). Estimating cetacean population density using fixed passive acoustic sensors: An example with Blainville's beaked whales. *Journal of the Acoustical Society of America*, 125, 1982–1994.

McGowen, M. R., Spaulding, M., & Gatesy, J. (2009). Divergence date estimation and a comprehensive molecular tree of extant cetaceans. *Molecular Phylogenetics and Evolution*, 53, 891–906.

Melcon, M. L., Failla, M., & Iniguez, M. A. (2012). Echolocation behavior of franciscana dolphins (*Pontoporia blainvillei*) in the wild. *Journal of the Acoustical Society of America*, 131, EL448–EL453.

Meredith, R. W., Janecka, J. E., Gatesy, J., Ryder, O. A., Fisher, C. A., Teeling, E. C., Goodbla, A., Eizirik, E., Siamo, T. L. L., Stadler, T., Rabosky, D. L., Honeycutt, R. L., Flynn, J. J., Ingram, C. M., Steiner, C., Williams, T. L., Robinson, T. J., Burk-Herrick, A., Westerman, M., Ayoub, N. A., Springer, M. S., & Murphy, W. J. (2011). Impacts of the Cretaceous terrestrial revolution and KPg extinction on mammal diversification. *Science*, 334, 521–524.

Miller, P. J. O., Johnson, M. P., & Tyack, P. L. (2004). Sperm whale behaviour indicates the use of rapid echolocation click buzzes 'creaks' in prey capture. *Proceedings of the Royal Society of London B: Biological Sciences*, 271, 2239–2247.

Møhl, B., Wahlberg, M., Madsen, P. T., Heerfordt, A., & Lund, A. (2000). Sperm whale clicks: Directionality and source level revisited. *Journal of the Acoustical Society of America*, 107, 638–648.

Møhl, B., Wahlberg, M., & Heerfordt, A. (2001). A large-aperture array of nonlinked receivers for acoustic positioning of biological sound sources. *Journal of the Acoustical Society of America*, 109, 434–437.

Møhl, B., Wahlberg, M, Madsen, P. T., Heerfordt, A., & Lund, A. (2003). The monopulsed nature of sperm whale clicks. *Journal of the Acoustical Society of America*, 114, 1143–1154.

Möhres, F. P. (1966). Communicative characters of sonar signals in bats. In R.-G. Busnel (Ed.), *Animal sonar systems: Biology and bionics*, Vol. 2. Jouy-en-Josas, France: Laboratoire de Physiologie Acoustique.

Mora, E. C., & Macias, S. (2007). Echolocation calls of Poey's flower bat (*Phyllonycteris poeyi*) unlike those of other phyllostomids. *Naturwissenschaften*, 94, 380–383.

Mora, E. C., Ibanez, C., Macias, S., Juste, J., Lopez, I., & Torres, L. (2011). Plasticity in the echolocation inventory of *Mormopterus minutus* (Chiroptera, Molossidae). *Acta Chiropterologica.*, 13, 179–187.

Morisaka, T., & Connor, R. C. (2007). Predation by killer whales (*Orcinus orca*) and the evolution of whistle loss and narrow-band high frequency clicks in odontocetes. *Journal of Evolutionary Biology*, 20, 1439–1458.

Moss, C. F., Bohn, K., Gilkenson, H., & Surlykke, A. (2006). Active listening for spatial orientation in a complex auditory scene. *PLOS Biology*, 4(4), 615–626.

Moss, C. F., Chiu, C., & Surlykke, A. (2011). Adaptive vocal behavior drives perception by echolocation in bats. *Current Opinion in Neurobiology*, 21, 1–8.

Mũller, R. (2004). A numerical study of the role of the tragus in the big brown bat. *Journal of the Acoustical Society of America*, 116, 3701–3712.

Mũller, R., Lu, H., Zhang, S., & Peremans, H. (2006). A helical biosonar scanning pattern in the Chinese noctule, *Nyctalus plancyi. Journal of the Acoustical Society of America*, 119, 3083–3092.

Nachtigall, P. E., & Supin, A. Y. (2008). A false killer whale adjusts its hearing when it echolocates. *Journal of Experimental Biology*, 211, 1714–1718.

Nakamura, K., Akamatsu, T., & Shimazaki, K. (1998). Threat clicks of captive harbor porpoises, *Phocoena phocoena. Bulletin of the Faculty of Fisheries Hokkaido University*, 49, 91–105.

Neuweiler, G., Bruns, V., & Schuller, G. (1980). Ears adapted for the detection of motion, or how echolocating bats have exploited the capacities of the mammalian auditory system. *Journal of the Acoustical Society of America*, 68, 741–753.

Norris, K. S. (1968). The evolution of acoustic mechanisms in odontocete cetaceans. In E. T. Drake (Ed.), *Evolution and environment* (pp. 297–324). New Haven, CT: Yale University Press.

Norris, K. S., Prescott, J. H., Asa-Dorian, P. V., & Perkins, P. (1961). An experimental demonstration of echo-location behavior in the porpoise, *Tursiops truncatus* (Montagu). *Biological Bulletin*, 120, 163–176.

Nummela, S., Thewissen, J. G. M., Bajpal, S., Hussain, S. T., & Kumar, K. (2004). Eocene evolution of whale hearing. *Nature*, 430, 776–778.

Obrist, M. K., Fenton, M. B., Eger, J. L., & Schlegel, P. A. (1993). What ears do for bats: A comparative study of pinna sound pressure transformation in Chiroptera. *Journal of Experimental Biology*, 180, 119–152.

Page, R. A., & Ryan, M. J. (2006). Social transmission of novel foraging behavior in bats: Frog calls and their referents. *Current Biology*, 16, 1201–1205.

Pedersen, S. C. (1998). Morphometric analysis of the chiropteran skull with regard to mode of echolocation. *Journal of Mammalogy*, 79, 91–103.

Podlutsky, A. J., Khritankov, A. M., Ovodov, N. D., & Austad, S. N. (2005). A new field record for bat longevity. *Journal of Gerontology*, 60, 1366–1368.

Popov, V. V., Supin, & Ya, A. (1990). Localization of the acoustic window at the dolphin's head. In J. A. Thomas & R. A. Kastelein (Eds.), *Sensory abilities of cetaceans: Laboratory and field evidence* (pp. 417–426). New York: Plenum Press.

Popov, V. V., Supin, & Ya, A., Wang, D., & Wang, K. (2006). Nonconstant quality of auditory filters in the porpoises, *Phocoena phocoena* and *Neophocaena phocaenoides* (Cetacea, Phocoenidae). *Journal of the Acoustical Society of America*, 119, 3173–3180.

Popper, A. N., Plachta, D. T. T., Mann, D. A., & Higgs, D. (2004). Response of clupeid fish to ultrasound: A review. *Journal of Marine Science*, 61, 1057–1061.

Ratcliffe, J. M., & Nydam, M. L. (2008). Multimodal warning signals for a multiple predator world. *Nature*, 455, 96–99.

Ratcliffe, J. M., Raghuram, H., Marimuthu, G., Fullard, J. H., & Fenton, M. B. (2005). Hunting in unfamiliar space: Echolocation in the Indian false vampire bat, *Megaderma lyra*, when gleaning prey. *Behavioral Ecology and Sociobiology*, 58, 157–164.

Ratcliffe, J. M., Jakobssen, L., Kalko, E. K. V., & Surlykke, A. (2011). Frequency alternation and an offbeat rhythm indicate foraging behavior in the echolocating bat, *Saccopteryx bilineata. Journal of Comparative Physiology A: Neuroethology, Sensory, Neural, and Behavioral Physiology*, 197(5), 413–423.

Rhinelander, M. Q., & Dawson, S. M. (2004). Measuring sperm whales from their clicks: Stability of interpulse intervals and validation that they indicate whale length. *Journal of the Acoustical Society of America*, 115, 1826–1831.

Ridgway, S. H., & Carder, D. A. (1988). Nasal sound production and pressure in an echolocating white whale. In P. Nachtigall & P. Moore (Eds.), Animal sonar: Processes and performance (pp. 53–60). New York: Plenum Press.

Ridgway, S. H., & Carder, D. A. (2001). Assessing hearing and sound production in cetaceans not available for behavioral audiograms: Experiences with sperm, pygmy sperm, and gray whales. *Aquatic Mammals*, 27, 267–276.

Ridgway, S. H., Elsberry, W. R., Blackwood, D. J., Kamolnick, T., Todd, M., Carder, D. A., Chaplin, M., & Cranford, T. W. (2012). Vocal reporting of echolocation targets: Dolphins often report before click trains end. *Journal of the Acoustical Society of America*, 131(1), 593–598.

Roeder, K. D. (1967). *Nerve cells and insect behavior*, revised edition. Cambridge, MA: Harvard University Press.

Rossbach, K. A., & Herzing, D. L. (1997). Underwater observations of benthic-feeding bottlenose dolphins (*Tursiops truncatus*) near Grand Bahama Island, Bahamas. *Marine Mammal Science*, 13, 498–504.

Roverud, R. C., & Grinnell, A. D. (1985). Frequency tracking and Doppler shift compensation in response to an artificial CF/FM echolocation sound in the CF/FM bat, *Noctilio albiventris. Journal of Comparative Physiology A: Neuroethology, Sensory, Neural, and Behavioral Physiology*, 156, 471–475.

Ruczynski, I., Kalko, E. K. V., & Siemers, B. (2007). The sensory basis of roost finding in a bat, *Nyctalus noctula. Journal of Experimental Biology*, 210, 3607–3615.

Schmieder, D. A., Kingston, T., Hashim, R., & Siemers, B. (2010). Breaking the trade-off: rainforest bats maximize bandwidth and repetition rate of echolocation calls as they approach prey. *Biology Letters*, 23, 604–609.

Schnitzler, H.-U. (1973). Control of Doppler shift compensation in the greater horseshoe bat, *Rhinolophus ferrumequinum. Journal of Comparative Physiology A: Neuroethology, Sensory, Neural, and Behavioral Physiology*, 82, 79–92.

Schnitzler, H.-U., & Flieger, E. (1983). Detection of oscillating target movements by echolocation in the greater horseshoe bat. *Journal of Comparative Physiology A: Neuroethology, Sensory, Neural, and Behavioral Physiology*, 153, 385–391.

Schnitzler, H.-U., & Kalko, E. K. V. (2001). Echolocation by insect-eating bats. *BioScience*, 51, 557–569.

Siemers, B. M., & Schnitzler, H.-U. (2000). Natterer's bat (*Myotis nattereri* Kuhl 1818) hawks for prey close to vegetation using echolocation signals of very broad bandwidth. *Behavioral Ecology and Sociobiology*, 47, 400–412.

Siemers, B. M., & Schnitzler, H.-U. (2004). Echolocation signals reflect niche differentiation in five sympatric congeneric bat species. *Nature*, 429, 657–661.

Siemers, B. M., Kalko, E. K. V., & Schnitzler, H.-U. (2001). Echolocation behavior and signal plasticity in the Neotropical bat *Myotis nigricans* (Schinz, 1821) (Vespertilionidae): A convergent case with European species of *Pipistrellus*? *Behavioral Ecology and Sociobiology*, 50(4), 317–328.

Similä, T., & Ugarte, F. (1993). Surface and underwater observations of cooperatively feeding killer whales in northern Norway. *Canadian Journal of Zoo*logy, 71, 1494–1499.

Simmons, J. A., & Stein, R. A. (1980). Acoustic imaging in bat sonar: Echolocation signals and the evolution of echolocation. *Journal of Comparative Physiology A: Neuroethology, Sensory, Neural, and Behavioral Physiology*, 135, 61–84.

Simmons, N. B., Seymour, K. L., Habersetzer, J., & Gunnell, G. F. (2008). Primitive early Eocene bat from Wyoming and the evolution of flight and echolocation. *Nature*, 451, 818–821.

Simon, R., Holderied, M. W., Koch, C. U., & von Helversen, O. (2011). Floral acoustics: Conspicuous echoes of a dish-shaped leaf attract bat pollinators. *Science*, 333, 631–633.

Smolker, R. A., Richards, A., Connor, R., Mann, J., & Berggren, P. (1997). Sponge carrying by dolphins (Delphinidae, Tursiops sp.): A foraging specialization involving tool use? *Ethology* 103, 454–465.

Suga, N., & Jen, P. H. S. (1975). Peripheral control of acoustic-signals in auditory system of echolocating bats. *Journal of Experimental Biology*, 62, 277–311.

Surlykke, A., & Moss, C. F. (2000). Echolocation behavior of big brown bats, *Eptesicus fuscus*, in the field and the laboratory. *Journal of the Acoustical Society of America*, 108, 2419–2429.

Surlykke, A., & Kalko, E. K. V. (2008). Echolocating bats cry out loud to detect their prey. PLoS ONE, 3(4), e2036.

Surlykke, A., Ghose, K., & Moss, C. F. (2009). Acoustic scanning of natural scenes by echolocation in the big brown bat, *Eptesicus fuscus. Journal of Experimental Biology*, 212, 1011–1020.

Suthers, R. A. (1967). Comparative echolocation by fishing bats. *Journal of Mammalogy*, 48, 79–87.

Suthers, R. A. (1988). The production of echolocation signals by bats and birds. In P. E. Nachtigall, & P. W. B. Moore (Eds.), *Animal sonar processes and performance* (pp. 23–45). New York: Plenum Press.

Suthers, R. A., Thomas, S. P., & Suthers, B. J. (1972). Respiration, wing-beat and ultrasonic pulse emission in an echo-locating bat. *Journal of Experimental Biology*, 56, 37–48.

Swartz, C., Tressler, J., Keller, H., Vanzant, M., Ezell, S., & Smotherman, M. (2007). The tiny difference between foraging and communication buzzes uttered by the Mexican free-tailed bat, *Tadarida brasiliensis. Journal of Comparative Physiology A: Neuroethology, Sensory, Neural, and Behavioral Physiology*, 193, 853–863.

Szymanski, M. D., Bain, D. E., Kiehl, K., Pennington, S., Wong, S., & Henry, K. R. (1999). Killer whale (*Orcinus orca*) hearing: Auditory brainstem response and behavioral audiograms. *Journal of the Acoustical Society of America*, 106, 1134–1141.

Teeling, E. (2009). Hear, hear: The convergent evolution of echolocation in bats? *Trends in Ecology and Evolution*, 24, 251–254.

Teeling, E. C., Springer, M. S., Madsen, O., Bates, P., O'Brien, S. J., & Murphy, W. (2005). A molecular phylogeny for bats illuminates biogeography and the fossil record. *Science*, 307, 580–584.

Thies, W., Kalko, E. K. V., & Schnitzler, H.-U. (1998). The roles of echolocation and olfaction in two Neotropical fruit-eating bats, *Carollia perspicillata* and *C. castanea*, feeding on *Piper. Behavioral Ecology and Sociobiology*, 42, 397–409.

Thomassen, H. A., & Povel, G. D. E. (2006). Comparative and phylogenetic analysis of the echolocation and social vocalizations of swiftlets (Aves: Apodidae). *Biological Journal of the Linnean Society*, 88, 631–643.

Thorpe, C. W., Bates, R. H., & Dawson, S. M. (1991). Intrinsic echolocation capability of Hector's dolphin, *Cephalorhynchus hectorii. Journal of the Acoustical Society of America*, 90, 2931–2934.

Turner, D. C. (1975). *The vampire bat: A field study in behavior and ecology*. Baltimore: Johns Hopkins University Press.

Tyack, P. L. (1997). Studying how cetaceans use sound to explore their environment. *Perspectives in Ethology*, 12, 251–297.

Ulanovsky, N., Fenton, M. B., Tsoar, A., & Korine, C. (2004). Dynamics of jamming avoidance in echolocating bats. *Proceedings of the Royal Society of London B: Biological Sciences*, 271, 1467–1475.
Urick, R. J. (1995). *Principles of underwater sound*, 3rd ed. Los Altos Hills: Peninsula Publishing.
Vanderelst, D., de Mey, F., Peremans, H., Geipel, I., Kalko, E. K. V., & Firzlaf, U. (2010). What noseleaves do for FM bats depends on their degree of sensorial specialization. *PLoS One* 5(8), e11893.
Varanasi, U., Feldman, H. R., & Malins, D. C. (1975). Molecular basis for formation of lipid sound lens in echolocating cetaceans. *Nature*, 255, 340–343.
Verfuss, U. K., Miller, L. A., Pilz, P. K. D., & Schnitzler, H. U. (2009). Echolocation by two foraging harbor porpoises. *Journal of Experimental Biology*, 212, 823–834.
Veselka, N., McErlain, D. D., Holdsworth, D. W., Eger, J. L., Chhem, R. K., Mason, M. J., Brain, K. L., Faure, P. A., & Fenton, M. B. (2010). A bony connection signals laryngeal echolocation in bats. *Nature*, 463, 939–942.
Villadsgaard, A., Wahlberg, M., & Tougaard, J. (2007). Echolocation signals of wild harbour porpoises, *Phocoena phocoena. Journal of Experimental Biology*, 210, 56–64.
Voight-Heucke, S., Taborsky, M., & Dechmann, D. K. N. (2010). A dual function of echolocation: Bats use echolocation calls to identify familiar and unfamiliar individuals. *Animal Behaviour*, 80, 59–67.
von Helversen, D., & von Helversen, O. (1999). Acoustic guide in bat-pollinated flower. *Nature*, 398, 759–760.
Wahlberg, M., Jensen, F. H., Aguilar de Soto, N., Beedholm, K., Bejder, L., Oliveira, C., Rasmussen, M., Simon, M., Villadsgaard, A., & Madsen, P. T. (2011a). Source parameters of echolocation clicks from wild bottlenose dolphins (*Tursiops aduncus* and *Tursiops truncatus*). *Journal of the Acoustical Society of America*, 130(4), 2263–2274.
Wahlberg, M., Beedholm, K., Heerfordt, A., & Møhl, B. (2011b). Characteristics of biosonar signals from the northern bottlenose whale, *Hyperoodon ampullatus. Journal of the Acoustical Society of America*, 130, 3077–3084.
Watkins, W.A. and Schevill, W.E. (1972). Sound source location by arrival times on a non-rigid three-dimensional hydrophone array. Deep-Sea Res. 19, 691–706.
Watkins, W. A., & Schevill, W. E. (1977). Sperm whale codas. *Journal of the Acoustical Society of America*, 62, 1486–1490.
Weilgart, L., & Whitehead, H. (1997). Group-specific dialects and geographical variation in coda repertoire in South Pacific sperm whales. *Behavioral Ecology and Sociobiology*, 40, 277–285.
Werth, A. J. (2006a). Mandibular and dental variation and the evolution of suction feeding in odontoceti. *Journal of Mammalogy*, 87, 579–588.
Werth, A. J. (2006b). Odontocete suction feeding: Experimental analysis of water flow and head shape. *Journal of Morphology*, 267, 1415–1428.
Whitehead, H. (2002). Estimates of the current global population size and historical trajectory for sperm whales. *Marine Ecology Progress Series*, 242, 295–304.
Wilson, M., Acolas, M.-L., Bégout, M.-L., Madsen, P. T., & Wahlberg, M. (2008). Allis shad (*Alosa alosa*) exhibit an intensity-graded behavioral response when exposed to ultrasound. *Journal of the Acoustical Society of America*, 124(4), EL243–EL247.
Wilson, M., Schack, H. B., Madsen, P. T., Surlykke, A., & Wahlberg, M. (2011). Directional escape behavior in allis shad (*Alosa alosa*) exposed to ultrasonic clicks mimicking an approaching toothed whale. *Journal of Experimental Biology*, 214, 22–29.
Xitco, M. J., & Roitblat, H. L. (1996). Object recognition through eavesdropping: Passive echolocation in bottlenose dolphins. *Animal Learning & Behavior*, 24, 355–365.
Yovel, Y., Moss, C. F., & Ulanovsky, N. (2010). Optimal localization by pointing off axis. *Science*, 327, 701–704.
Yovel, Y., Falk, B., Moss, C. F., & Ulanovsky, N. (2011a). Active control of acoustic field-of-view in a biosonar system. *PLoS ONE*, 9, e1001150.

Yovel, Y., Sagiv, M. G., & Ulanovsky, N. (2011b). Click-based echolocation in bats: not so primitive after all. *Journal of Comparative Physiology A: Neuroethology, Sensory, Neural, and Behavioral Physiology*, 197, 515–510.

Zhuang, Q., & Mueller, R. (2006). Noseleaf furrows in a horseshoe bat act as resonance cavities shaping the biosonar beam. *Physics Review Letters*, 97, 218701–1to 4.

Zimmer, W. M. X., Tyack, P. L., Johnson, M., & Madsen, P. (2005a). 3-Dimensional beam pattern of regular sperm whale clicks confirms bent-horn hypothesis. *Journal of the Acoustical Society of America*, 117, 1473–1485.

Zimmer, W. M. X., Johnson, M., Madsen, P. T., & Tyack, P. L. (2005b). Echolocation clicks of Cuvier's beaked whales (*Ziphius cavirostris*). *Journal of the Acoustical Society of America*, 117, 3919–3927.

Chapter 3
Production of Biosonar Signals: Structure and Form

Whitlow W.L. Au and Roderick A. Suthers

Keywords Sound production • Echolocation • Sonar pulse • Clicks • Phonic lips • Melon • Larynx • Frequency spectrum • Waveform • Laryngeally echolocating bats • Vocal membrane • Acoustic energy • Acoustic filter • FM signal • CF-FM signal • Search phase • Approach phase • Terminal buzz phase

3.1 Introduction

The biosonar capabilities of any animal will depend on many characteristics of their biosonar system such as the acuity of the auditory system, the information carrying capacity of the projected signals, the spatial resolution of the biosonar process, the amount of auditory memory, the speed of auditory recall, the degree of coupling between the biomechanics and signal processing systems, and so on. The first of all these factors is the ability to generate or produce signals that will enable the animal to perform the necessary sonar tasks for its survival. This includes the generation of sufficiently intense signals to receive echoes from prey that are above the background ambient noise environment and the production of signals with the proper information carrying capacity for the specific task at hand. Needless to say, the environments that dolphins and whales live in cannot be more different than that of bats, and each environment imposes much different constraints on the biosonar system. At 20 °C the density (ρ) of air at sea level is 1.21 kg m^{-3} and the speed of sound

W.W.L. Au (✉)
Hawaii Institute of Marine Biology, University of Hawaii,
46-007 Lilipuna Road, Kaneohe, HI 96744, USA
e-mail: wau@hawaii.edu

R.A. Suthers
Department of Biology, School of Medicine, Indiana University,
1001 E. Third Street, Bloomington, IN 47405, USA
e-mail: suthers@indiana.edu

A. Surlykke et al. (eds.), *Biosonar*, Springer Handbook of Auditory Research 51,
DOI 10.1007/978-1-4614-9146-0_3, © Springer-Verlag New York 2014

(c) is 344 m s^{-1} compared with 1,026 kg m^{-3} for the density of seawater and a speed of sound of 1,500 m s^{-1}. Therefore, the characteristic impedance (ρc) is 416 Pa·s m^{-1} for air and 1.5×10^6 Pa·s m^{-1} for seawater, which means that for a given acoustic pressure, the intensity in seawater will be 3,606 times, or 36 dB higher than in air.

The foraging process is also very different for dolphins and bats. Acoustic energy is absorbed about 100 times or 40 dB more per meter in air than in seawater (Au 1993). Therefore, dolphins can forage for fish prey that may be several tens of meters away whereas bats are restricted to seeking insect prey within a few meters. Finally, the prey of bats tend to dart about much more rapidly to avoid capture and to in turn capture fast moving prey than do the prey of dolphins. The environmental, behavioral, anatomical, and size differences between dolphins and bats have led to the evolution of very different types of biosonar for these groups.

Research on sound production in dolphins and bats has proceeded in different directions, as will be obvious by the content of this chapter. A considerable amount of physiological and functional anatomical research has been directed toward understanding sound production in bats, whereas only a few studies have been devoted to the physiology of sound production in dolphins. There are many reasons for these differences in approach and the consequent level of understanding. The most obvious is that dolphins are protected by the Marine Mammal Protection Act of 1972 in the United States and by similar laws in many other countries that limit the number of dolphins available for research. Captive dolphins are expensive to maintain and require expensive facilities in which to live for research. Therefore, research in biosonar signal production by dolphins has been focused mainly on understanding the characteristics of biosonar signals after they have been transmitted into the water.

3.2 Signal Production in Dolphins

This section concentrates on signal production and the role this plays in the biosonar process. Signal production includes the production of the initial acoustic energy within the head of an animal, the shape of the initial energy, how that energy changes with time as it propagates through the head, the manner in which the energy emerges from the head into the water, and, finally, the spatial distribution of the energy propagating into the water. Knowledge of the process exists on a gross scale but is lacking much of the finer details, especially on the propagation process within the head of dolphins.

3.2.1 Site of the Sound Source

The general practice for those studying social sounds of dolphins is to refer to these sounds as vocalizations. This terminology probably began with describing human voice sounds and has carried over to acoustic research with toothed whales and dolphins (odontocetes). The term vocalization means to give voice and refers to the fact that voiced sounds are produced by the vocal cords in the larynx of humans and other animals, or in the syrinx of birds. This is an unfortunate choice of terminology because research on odontocetes, using widely different techniques and instrumentation, has produced data that clearly rule out the larynx and instead implicate the nasal system as the site of sound production. The proponents of a laryngeal sound production mechanism based their arguments mainly on anatomical considerations and experiments with dead animals (Purves 1967; Schenkkan 1973; Purves and Pilleri 1983).

The research on sound production in dolphins has been focused on determining the location or site of sound production. Diercks et al. (1971) used an array of contact hydrophones to measure biosonar signals on the head of the Atlantic bottlenose dolphin (*Tursiops truncatus*).Norris et al. (1971) and Dormer (1979) used high-speed X-ray motion pictures to observe movements of the laryngeal and nasal region associated with sound production in live phonating *T. truncatus*, Pacific bottlenose dolphin (*Tursiops gilli*), and the spinner dolphin (*Stenella longirostris*). Hollien et al. (1976) used soft tissue spot X-ray images to observe the larynx and melon of two phonating *T. truncatus*. All of these investigators observed movements in the nasal system that correlated with sound production and did not see any correlated movements in the larynx. Ridgway et al. (1980) measured the muscle activities (using electromyographic techniques) and air pressures associated with sound production in five *T. truncatus*. Ridgway and Carder (1988) did similar measurements with a white whale (*Delphinapterus leucas*). They found that the anterior internus, the posterior internus, diagonal membrane muscle, and the nasal plug muscles (all associated with the nasal system) fired just before and during the production of sounds whereas the hyoepiglottal muscle and the intercostal muscles (associated with the larynx) did not. They also found that pressure increased in the nares and premaxillary sacs before sound production and dissipated after the sound. There were no pressure changes in the trachea during sound production Amundin and Andersen (1983) measured the muscle activity of the nasal plug muscle and air pressure in the bony nares of a harbor porpoise (*Phocoena phocoena*) and a *T. truncatus* and obtained results similar to those of Ridgway et al. (1980). Elsberry (2003) used two pressure catheters inserted into each naris to measure the intranarial pressure while three bottlenose dolphins performed a biosonar task. He also simultaneously measured emitted clicks with a broadband hydrophone and whistles with a low-frequency hydrophone. The animals were required to echolocate on various targets and respond by whistling if a certain target was present or remain silent if there were no targets. His results are also included in Cranford et al. (2011) and an example of the results Elsberry obtained is shown in Fig. 3.1. The intranarial pressure had to increase to a particular value above the basal pressure level before clicks were generated. The pressure had to increase even further before whistles were produced. The data shown in Fig. 3.1 demonstrate the role of pneumatics in sound generation by dolphins. All of the measurements discussed in this section clearly indicate that the larynx is not involved in the production of either click or tonal sounds. Elsberry's results provided a general understanding of the role of intranarial pressure on sound production. The temporal resolution of pressure catheters is not

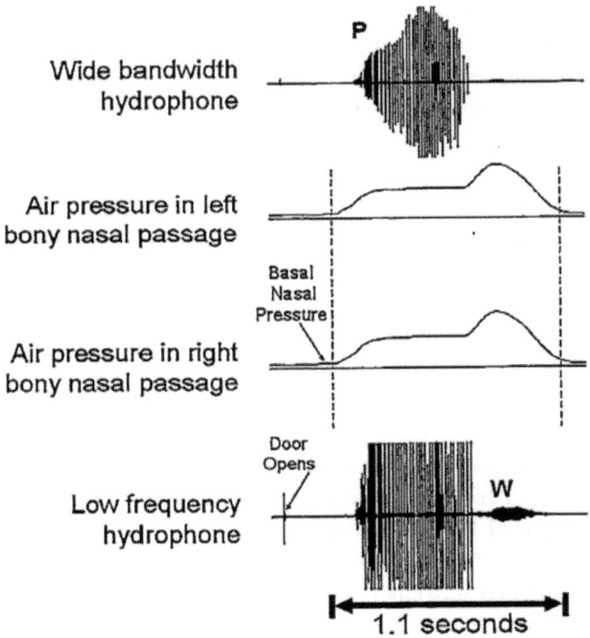

Fig. 3.1 Simultaneous display of acoustic and internarial pressure events for a *Tursiops truncates* emitting sounds. (From Cranford et al. 2011)

nearly fine enough to capture the short pressure transients that must be associated with the production of clicks that are extremely short, on the order of 70–100 μs.

Cranford et al. (1996) theorized that sounds were generated by pushing air across two sets of internal "lips" also referred to as the monkey lips. A schematic showing the location of the phonic lips (renamed by Cranford 2000) is shown in Fig. 3.2a. A photograph of the phonic lips is also shown in Fig. 3.2b. On each wall of the phonic lips are ridges that interlock with the ridges on the opposite side to form a tight fitting set of lips. Using an endoscope, they observed synchronized vibration of phonic lips, one located on either side and just above the membranous nasal septum, with the production of acoustic signals (Cranford 2000; Cranford et al. 1997). When air pressure in the nares increases to a point that the pneumatic force is momentarily greater than the muscular force keeping the phonic lips closed, a pulse of air is shot through the lips, causing the structure connecting the lips to vibrate. Acoustic pulses occurred in coincidence with one oscillatory cycle of the lips, whereas no other structures were found to vibrate in synchrony with each acoustic pulse event. Simultaneous recordings of sonar pulses with endoscopic images of lip movements showed good correspondence between these events (Cranford et al. 1997). Although Cranford et al. (1997) were able to localize the site of biosonar sound production, there are still many unknowns concerning the actual mechanisms involved. The high-speed video camera with a frame rate up to 400 frames per second (2.5 ms per frame) was not fast enough to capture events associated with producing a 70–100 μs (0.07–0.1 ms) duration click. How the structures around the phonic lips are involved in producing each click is still unknown. How click trains exhibiting different interclick intervals are produced is also a mystery. There are many other processes in the biosonar sound production that are unknown.

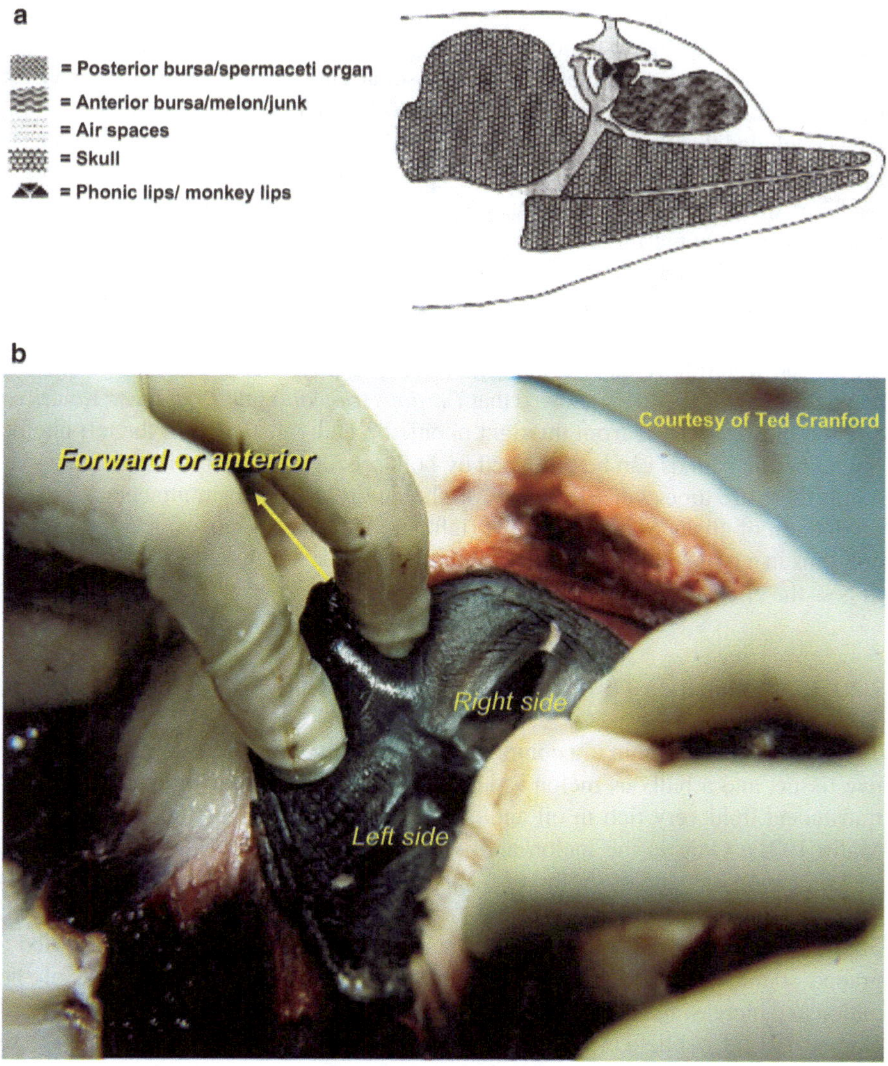

Fig. 3.2 (**a**) Schematic diagram of structures in the forehead of the dolphin. (Adapted from Cranford 2000.) (**b**) Phonic lips exposed from above for a bottlenose dolphin. (Courtesy of Ted Cranford.) On either side of each lip are ridges that interlock with the ridges on the opposite side to form a tight interface when the lips are in the closed position

Cranford et al. (1997, 2011) found that biosonar signals could be produced by each set of phonic lips, and although they had only one endoscope and could observe only one set of phonic lips at a time, they speculated that both phonic lips could also be operated together. Lammers and Castellote (2009) provided data that suggested the operation of two sound sources by an echolocating beluga whale. They used two hydrophones, one directly in front of the animal stationed in a hoop and one at variable angles around the animal. They clearly detected two clicks for angles greater than ±45° while the hydrophone directly in front of the animal detected only a single click. They argued that both phonic lips produced the double clicks. Madsen et al. (2010) placed three suction cup hydrophones on the head of a *Phocoena phocoena* and simultaneously measured the biosonar clicks of the animal. By measuring the time differences between the arrival of the clicks to the three hydrophones, they concluded that the signals were being produced by the right phonic lips and not by two sets of phonic lips operating in synchrony. Au et al. (2012) examined double clicks at wide off-axis angle for a *T. truncatus* and found that they were not the results of two phonic lips functioning in synchrony but were probably caused by internal reflections from the different air sacs within the head of the animal. Nakamura et al. (1998) reconstructed the air sac geometry of a common dolphin using silicon foam and showed that the phonic lips are surrounded by air sacs and the air sac would have an effect as signals propagate from the phonic lips into the water. Madsen et al. (2013) showed that *Tursiops truncatus* and *Pseudoca crassidens* emitted biosonar clicks from the right phonic lip and whistles from the left phonic lip. Although sounds can be produced by both sets of phonic lips, it is still not certain if there are times when both sets of phonic lips are used simultaneously in the production of clicks or how often the right and left or both sets of lips are used by dolphins in a typical biosonar task or if there might be a difference between species in how the clicks are produced.

3.2.1.1 Propagation Through the Head of Dolphins

The forehead of a dolphin is a complex structure with air passages, air sacs, connective tissue, and a bulbous melon as shown in Fig. 3.3. The melon is composed of translucent lipid very rich in oil and is referred to as "acoustic fat" (Varanasi and Malin 1971). Only in the melon and lower jaw can this lipid material be found. Wood (1964) was one of the first to suggest that the fatty melon of the dolphin's forehead may be used to couple sounds from inside the animal's head into the water. Sound velocity measurements of tissue samples from a dolphin melon indicated a graded sound velocity profile, with a low-velocity core near the midline of the melon and increasing velocity outwards toward the surface of the melon (Norris and Harvey 1972; Litchfield et al. 1973). A low-density core that likely corresponds to the low velocity pathway through the melon has been shown in X-ray computer tomography (CT) scans (Cranford 1988; Cranford et al. 1996). Sounds propagating in an inhomogeneous sound velocity structure will be governed by Snell's law so that as sound propagates through a medium of changing sound velocity, the sound will refract or bend toward the lower velocity region (Urick 1983). In the dolphin's melon, the refraction should be toward the low-velocity core. Because of this structure of lipid fractions, Norris (1968; Norris and Harvey 1972) believed that the melon is primarily responsible for refraction of sound to form the sonar beam (Norris 1968). However, recent measurements by Au et al. (2010) with suction cup hydrophones on the head of two echolocating *T. truncatus* produced results that suggest that echolocation signals are already focused before propagating any distance into the melon. The results of Au et al. (2010) showing the relative peak-to-peak

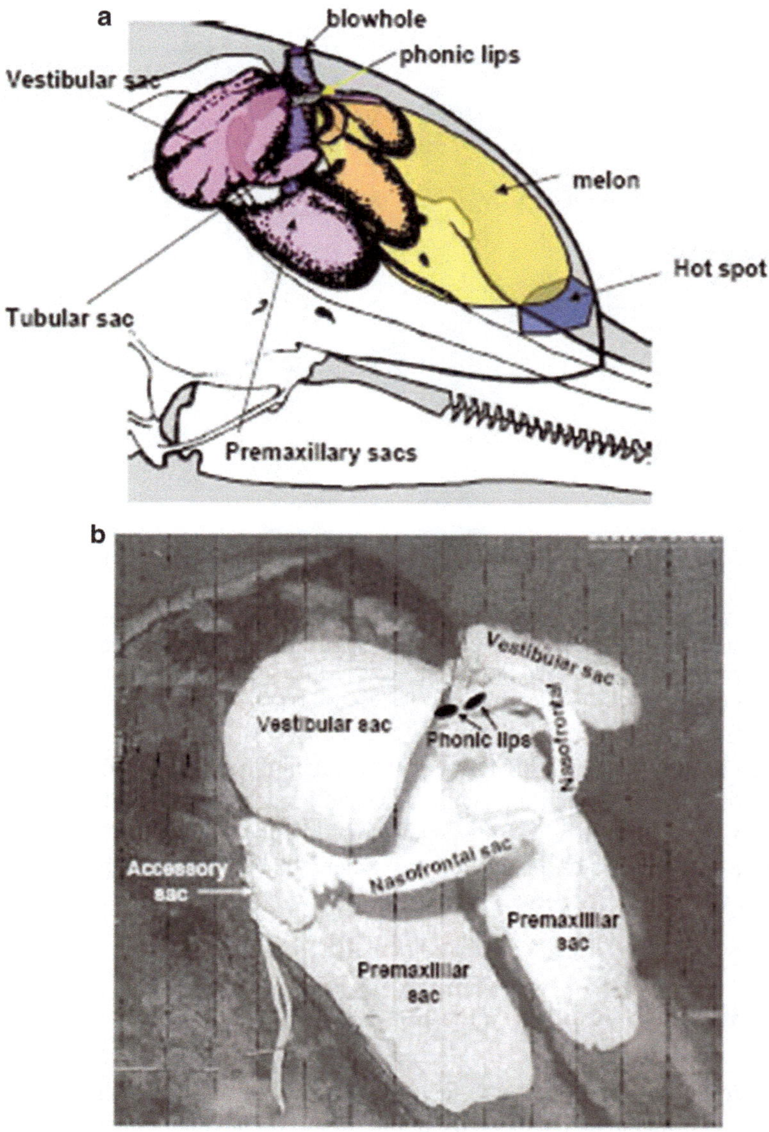

Fig. 3.3 (**a**) Schematic showing the relative position of the air sacs and other internal structure of the forehead. (Adapted from Purves and Pilleri 1983, by Alexis Rudd.) (**b**) Silicon cast of Nakamura et al. (1998) for a common dolphin. The left nasofrontal sac and phonic lips were drawn in by one of the authors

amplitude of the sound field on the surface of two dolphins head are shown in Fig. 3.4. These results are consistent with the numerical simulation of Aroyan et al. (1992) which indicated that "the skull and the air sacs appear to be acting in concert as an acoustical mirror which reflects sounds originating from the region of the

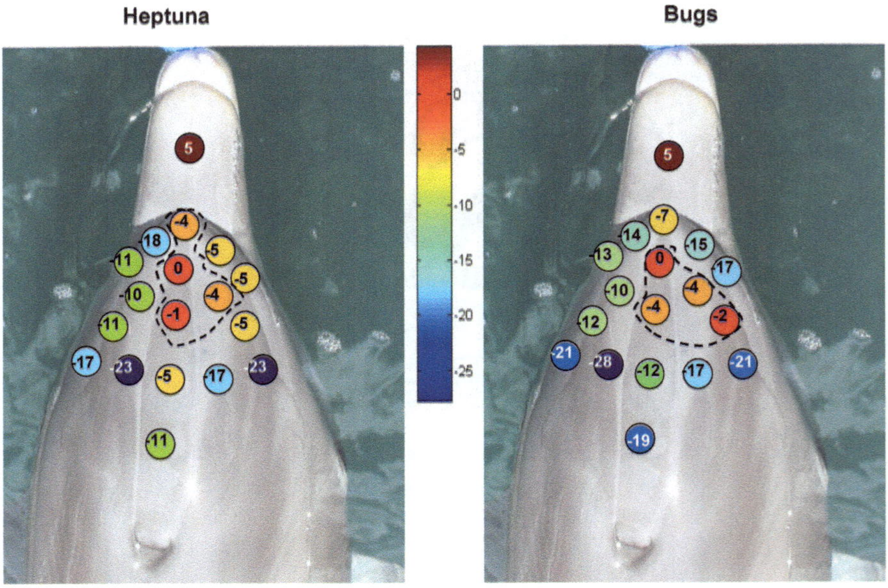

Fig. 3.4 The average relative amplitude of the echolocation signals on the head of two dolphins showing that echolocation amplitude is already reduced toward the rear of the melon location suggesting that the signal is already focus before reaching the melon (from Au et al. 2010) forward beams. A melon velocity profile of the magnitude tested is capable of mild focusing

monkey lips dorsa bursae (MLDB) complex into highly directed forward beams." A melon velocity profile of the magnitude tested is capable of a mild focusing effect and may serve to 'fine focus' the beam formed by the skull and air sacs (p. 2543). Figure 3.4 also indicates the general area where the biosonar signals exit the melon into the water.

3.3 Signal Production in Bats

Sophisticated biosonar systems, such as those found in laryngeally echolocating bats, place stringent demands on the ability of the vocal system to generate acoustically appropriate sonar pulses and must often push the limits of vocal performance. Unlike toothed whales, laryngeally echolocating bats do rely on their vocal system to produce echolocation signals. Bat sonar signals are thought to have evolved from longer, more complex, less stereotyped vocalizations used for social communication (Fenton 1984; Clement et al. 2006; Ma et al. 2006). Recent molecular evidence of bat phylogeny (Teeling et al. 2005; Jones and Teeling 2006) indicates that sonar pulse design has undergone extensive convergent adaptive radiation to maximize sonar performance for different ecological conditions (e.g., Neuweiler 2003; Moss and Surlykke 2010). It is not surprising therefore that the vocal anatomy and dynamics of sonar pulse production vary between, and even within, taxa.

The physiological and acoustic mechanisms of vocal production have been investigated in only a few laryngeally echolocating bats. For the purpose of this review, it is convenient to categorize echolocating bats according to the salient characteristics of their sonar pulses. Some species, such as the big brown bat (*Eptesicus fuscus*: Vespertilionidae), produce relatively short, frequency modulated (FM) sonar pulses. These pulses are well designed to measure time intervals between the emitted pulse and its returning echo, which are proportional to the distance between the bat and its target prey. Other species, such as the greater horseshoe bat (*Rhinolophus ferrumequinum*: Rhinolophidae), and the mustache bat (*Pteronotus parnellii*:Mormoopidae; formerly *Chilonycteris rubiginosa*), produce sonar calls that include a relatively long constant frequency (CF) portion followed by a downward FM sweep, which are referred to here as "long CF-FM" pulses. The CF component may also be preceded by a brief upward FM sweep. The auditory system of these bats is sensitive to small Doppler shifts in the frequency of the CF component, such as those generated by fluttering wings of insect prey, thus facilitating the detection of flying insects in cluttered environments (Neuweiler 2003).

In both FM and long CF-FM bats, the production of sonar pulses requires the close coordination between at least three major vocal subsystems: (1) the respiratory system, which powers vocalization; (2) the vocal organ, which generates the sound; and (3) the vocal tract, which filters and modifies the spectral properties of the laryngeal source. Understanding sonar pulse production requires an understanding of how each of these vocal subsystems contributes to sonar pulse production (see also Schuller and Moss 2004; Metzner and Schuller 2010).

3.3.1 Respiratory Dynamics of Sonar Pulse Production

Respiratory pressure provides the driving force for vocalization. Because sonar pulses are produced during expiratory airflow (Schnitzler 1968; Roberts 1972; Suthers 1988), the respiratory motor pattern must be modified and coordinated with the other vocal subsystems during the different phases of insect pursuit in order to meet the competing requirements for adequate pulmonary gas exchange in addition to those for acoustic orientation.

As insectivorous bats, such as rhinolophids, mormoopids, and vespertilionids, forage for prey, they adjust the tempo, duration, and bandwidth of their sonar pulses according to the changing need for detailed sonar information. The typical sequence of these changes involves search, approach, and terminal buzz phases (Griffin 1958). In *E. fuscus*, the search phase pulse repetition rate is low (5–10 Hz), pulse duration is relatively long (15–20 ms), and the bandwidth of the FM sweep is only a couple of kHz (from about 26 kHz down to 24 kHz). This narrow bandwidth improves the detection of distant targets. The approach phase begins with the detection and pursuit of an insect. During this phase pulse repetition rate increases and pulse duration decreases. The shallow FM pulses of the search phase are replaced by a fundamental that sweeps steeply from about 65 kHz down to 25 kHz. Prey capture is accompanied by a terminal buzz phase, in which the duration of the steep FM pulses is reduced to 1 ms or less and pulse repetition rate rises to about 170 pulses per second (Surlykke and Moss 2000; Moss and Surlykke 2010).

Long CF-FM bats make qualitatively similar changes in pulse emission during insect pursuit, but retain a CF component in their sonar pulse. The pulse duration and pulse repetition rate during the terminal phase of the mustache bat (*Pteronotus parnellii*) are about 6–8 ms and 80–100 Hz, respectively. During the terminal buzz of the greater horseshoe bat (*Rhinolophus ferrumequinum*), the pulse duration is reduced to about 10 ms and the repetition rate increases to >80 Hz, causing the sonar pulse duty cycle to increase from about 60 to 90 % (Novick and Vaisnys 1964; Tian and Schnitzler 1997). The execution of this vocal tour de force depends on the ability of the respiratory muscles to provide precisely timed expiratory pressure pulses for phonation in coordination with the motor pattern of the laryngeal muscles that control the glottal opening and tension on the sound-producing vibratory membranes. Respiratory muscles have key roles in determining the repetition rate and intensity of sonar pulses.

3.3.1.1 Timing of Sonar Pulses in Respiratory Cycle

Most sonar pulses are produced during the expiratory phase of the respiratory cycle. Sonar pulses which occasionally occur during the inspiratory phase are probably accompanied by a brief puff of expiratory air, which momentarily interrupts the inspiration (Suthers et al. 1972). At low sonar pulse repetition rates during the search phase a single pulse is typically produced during each expiration (Fig. 3.5a) and some expirations may be silent. As pulse repetition rate increases, each expiration contains multiple sonar pulses. Pulse duration is reduced and the duration of the expiratory phase may be increased, enabling the bat to achieve a wide range of pulse repetition rates (Fig. 3.5b). Schnitzler (1968) recorded up to 30 short FM pulses in a single expiration, from a stationary *Myotis myotis*, vocalizing spontaneously.

3.3.1.2 Respiration and Sonar Pulse Intensity

Sonar pulses are accompanied by a marked increase in subglottal pressure due to increased expiratory effort and constriction of the glottal opening by laryngeal adductor muscles. In FM and long CF-FM bats, peak subglottic pressure is positively correlated with the maximum sound pressure level (SPL) of sonar pulses (Fattu and Suthers 1981; Suthers 1988). Subglottic pressure is thus one of the important factors that determine the intensity of a sonar pulse and therefore the range at which a target can be detected. The first sonar pulse in each expiration is preceded by an increase in subglottic pressure as the expiratory effort increases and the glottal opening is reduced. If multiple sonar pulses are emitted during one expiration, subglottic pressure remains high until the last pulse is finished (Fig. 3.5).

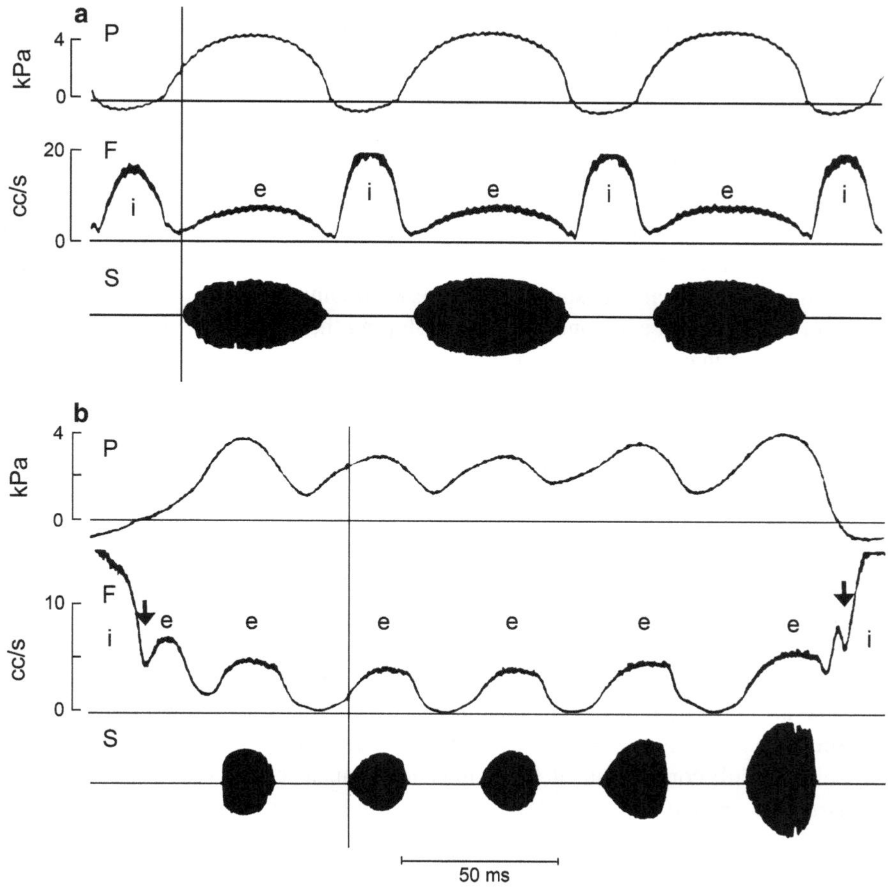

Fig. 3.5 Dynamics of glottal airflow and subglottic pressure during the production of sonar signals. (**a**) *Rhinolophus hildebrandti* emitting one sonar pulse per breath. (**b**) *Pteronotus parnellii* emitting a train of five sonar pulses during one exhalation. *P* subglottic pressure, *F* rate of laryngeal airflow, *i* inspiration, *e* expiration. *Arrows* in (**b**) mark points of flow reversal between inspiration and expiration. Response time of microbead thermistor, 90 % full scale in 6 ms or less, was not fast enough to return to zero flow at reversal point. (Adapted from Suthers 1988, with permission)

Subglottic pressure typically rises from a few hundred Pascals during quiet respiration to about 3 or 4 kPa during pulse emission, and can occasionally exceed 6 kPa. Subglottic pressure during human speech at normal conversation levels is about 0.8 kPa (Ladefoged 1968) and reaches 2–3 kPa only when shouting at maximum intensity (Isshiki 1964). Peak subglottic pressure during echolocation may sometimes approach that of the pulmonary circulation, raising the possibility that the maximum sonar pulse intensity may ultimately be limited by the ability to maintain pulmonary circulation during the production of intense sonar pulses with high duty cycles (Suthers 1988).

3.3.2 Respiratory Muscle Specializations for Echolocation

To investigate how respiratory muscles adjust their activation patterns to the special respiratory demands of echolocation, Lancaster et al. (1995) telemetered the electromyograms (EMG) of these muscles in *Pteronotus* and in certain FM bats (Lancaster and Speakman 2001). They found that contraction of expiratory muscles in the lateral abdominal wall of *Pteronotus* (Lancaster et al. 1995) and some FM bats (Lancaster and Speakman 2001) is highly correlated with pulse production at rest and in flight. Their data suggest that lateral abdominal wall muscles are specialized for echolocation and contribute to developing the high subglottal pressure that accompanies sonar pulse emission.

3.3.3 Respiration, Wingbeat Cycle, and Sonar Pulse Emission

Flying bats emit most sonar calls during the upstroke of the wings, which coincides with expiration (Schnitzler 1971; Suthers et al. 1972). Koblitz et al. (2010) found that the source sound level of sonar pulses emitted by *Eptesicus* as they approach a landing varies up to 12 dB (average 4 dB) within the group of sonar pulses emitted during a single wingbeat cycle. Furthermore, the source level of each sonar pulse depends on its position in the wingbeat cycle. This variation in source sound level is likely due to wingbeat-related changes in subglottic pressure associated with contraction of flight muscles (Lancaster et al. 1995). The wingbeat-related change in source level is superimposed on a previously discovered approximately 6 dB decrease in source level per halving of the target range, which occurs as bats approach a target, and is thought to maintain a relatively constant echo level at the bat's ears by compensating for the reduction in transmission losses as the distance to the target decreases (Kobler et al. 1985; Hartley et al. 1989). Koblitz et al. (2010) suggest that the dependence of source level on wing position might allow bats to change the level of single calls by emitting them at different phases of the wing's upstroke but its implications for the intensity compensation hypothesis need further study.

3.4 The Larynx

3.4.1 Anatomy

The laryngeal vocal subsystem contains the vibratory structures that generate sound when positioned in the expiratory air stream and the muscles that control their configuration and tension. The laryngeal anatomy of echolocating bats has been described for a number of species (e.g., Robin 1881; Elias 1907; Sprague 1943;

Fischer and Gerken 1961; Fischer and Vomel 1961; Denny 1976; Griffiths 1983; Durrant 1988), but the mechanism of sound production and its control has been studied in only a few of these.

The larynx of laryngeally echolocating bats is formed on a framework of four main cartilages (Fig. 3.6). A ring-shaped cricoid cartilage lies at the caudal end of the larynx where it is attached to the cranial end of the trachea. The cricoid cartilage articulates with the thyroid cartilage, the laminae of which support the lateral walls of the larynx and fuse anteriorly to form a bridge. The cricoid also articulates with a pair of arytenoid cartilages. Portions of these cartilages become ossified early in development compared to other mammals, presumably as an adaptation to withstand the mechanical stresses associated with the sonar pulse production (Denny 1976; Schuller and Moss 2004; Metzner and Schuller 2010).

3.4.2 *Innervation*

This cartilaginous and partially ossified skeleton of the larynx supports several intrinsic and extrinsic muscles (Fig. 3.6a, b) that are ipsilaterally innervated via the superior and recurrent laryngeal branches of the vagus nerve. The only exception to this ipsilateral innervation is the bilateral innervation of the interarytenoid muscle, a vocal fold abductor, by the recurrent nerve. The superior laryngeal nerve divides into an internal branch, which carries sensory fibers from the larynx, and an external branch, which provides the motor innervation of the cricothyroid muscle. This superficial intrinsic laryngeal muscle originates on the cricoid cartilage and inserts on the thyroid cartilage. The cricothyroid is a complex muscle with multiple bellies that vary across species. The recurrent laryngeal nerve provides the motor innervation of all the other intrinsic laryngeal muscles, except the cricothyroid (Griffin 1958; Novick and Griffin 1961; Quay 1970). Data from other mammals indicate some sensory fibers also travel in the recurrent laryngeal nerve (e.g., Wyke and Kirchner 1976).

3.4.3 *Sensory Feedback*

Sensory feedback plays an important role in the coordination of different muscle groups and vocal subsystems, but information on sensory receptors in the vocal system of echolocating bats is limited. Bass et al. (2008) have shown that fish and terrestrial vertebrates share an ancestral origin of their vocal motor networks that undergo similar development in the caudal hindbrain and rostral spinal cord. It is reasonable to assume that the sensory innervation of the bat's larynx and vocal system is broadly similar to that found in other mammals (Wyke and Kirchner 1976; Smotherman 2007). Most of this sensory information appears to go to the solitary tract nucleus in the brain stem, which projects back onto laryngeal motoneurons in nucleus ambiguus. Sensory information also goes to brain nuclei involved in controlling the respiratory rhythm (Smotherman 2007).

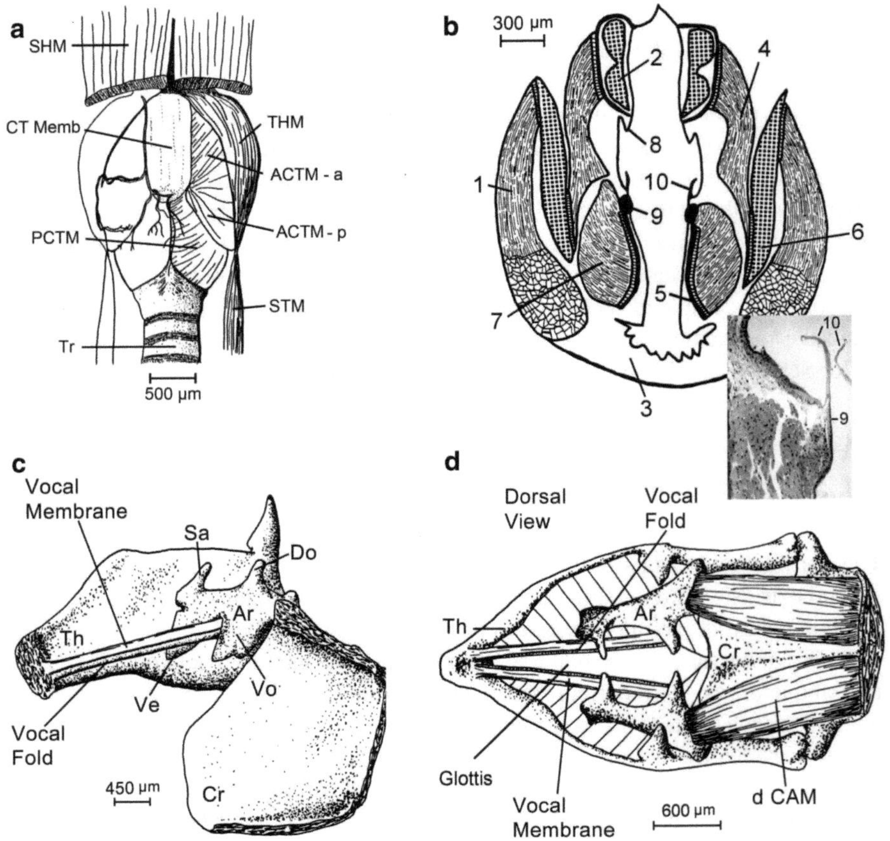

Fig. 3.6 Laryngeal anatomy of the big brown bat, *Eptesicus fuscus*. (**a**) Ventral view of larynx showing superficial muscles. The sternohyoid muscles have been cut. (**b**) Cross section of larynx in transverse plane of the bat. *Double lines with cross hatching* are origins and insertions of laryngeal muscles. *Heavy black lines* are tendons or ligaments. *Inset* is section through *E. fuscus* vocal fold showing thin vocal membranes. (**c**) Interior view of larynx cut in half at mid-sagittal plane with muscles removed, showing attachment points of vocal fold and vocal membranes on anterior bridge of thyroid cartilage and arytenoid cartilages. (**d**) Dorsal view showing glottal opening between vocal membranes and attachments of the dorsal cricoarytenoid muscles. *ACTM-a* anterior portion of anterior belly of cricothyroid muscle, *CMemb* cricothyroid membrane, *ACTM-p* posterior portion of anterior belly of cricothyroid muscle, *Ar* arytenoid cartilage, *Cr* cricoid cartilage, *dCAM* cricoarytenoid muscle, *Do* dorsal process, *PCTM* posterior belly of cricothyroid muscle, *Sa* process of Santorini, *SHM* sternohyoid muscle, *STM* sternothyroid muscle, *Th* thyroid cartilage, *Tr* trachea, *Ve* ventricular process, *Vo* vocal process. (1) anterior portion of anterior cricothyroid muscle; (2) arytenoid cartilage; (3) cricothyroid membrane; (4) dorsal thyroarytenoid muscle; (5) elastic tendon; (6) thyroid cartilage; (7) major branch of ventral thyroarytenoid muscle; (8) ventricular membrane; (9) vocal fold; (10) vocal membrane. (From Durrant 1988; *inset* in (**b**) from Suthers and Fattu 1973, with permission)

3.4.4 *Vocal Membranes: The Laryngeal Sound Source*

The discovery that bats emit ultrasonic vocalizations (Pierce 1938) and its use for acoustic orientation (Griffin and Galambos 1941) raised the question of how this ultrasound is produced. The vocal folds, which form the glottis and sound source in most other mammals, are also present in laryngeally echolocating bats, but seem too massive to vibrate at ultrasonic frequencies. The issue was resolved when Griffin (1946, 1958) discovered a very thin, almost transparent membrane of connective tissue only 6-μm thick, extending from the edge of each vocal fold (Fig. 3.6b–d). These vocal membranes seem well suited to function as low mass oscillators capable of vibrating at ultrasonic frequencies. Tearing or cutting the vocal membranes in various FM or long CF-FM bats either abolishes sonar pulses or significantly reduces sonar pulse frequency, whereas damaging the ventricular membranes has little effect on the sonar pulses (e.g., Griffin 1958; Novick and Griffin 1961; Durrant 1988). Similar vocal folds, or "lips" (not to be confused with the phonic lips of dolphins), have been reported in a number of mammals, including various primates whose calls contain very high, sometimes ultrasonic, frequencies (Mergell et al. 1999). It is interesting that the ultrasonic vocalizations of small rodents, which lack thin membranes on their vocal folds, are not affected by vocal tract cavity resonances, suggesting rodents produce ultrasound by a mechanism that is fundamentally different from that underlying both their audible vocalizations and the ultrasonic vocalizations of laryngeally echolocating bats (Roberts 1975a).

Griffin (1958) estimated that the resonant frequency of the vocal membranes in *Eptesicus* should be about 45 kHz, which is well within the fundamental frequency range of most of this bat's sonar pulses. His calculations, however, considered the vocal membranes as isolated structures unaffected by any dynamic interaction with the vocal folds to which they are attached. To investigate the dynamic behavior of the vocal membrane–vocal fold complex, Mergell et al. (1999) used a two-mass computational model of the "standard mammalian" vocal folds, to which the vocal membranes were added as an upward extension of the folds. Depending, in part, on the angle and amount by which the membranes extend beyond the edge of the vocal fold, the model predicts that the vocal membranes may not only increase the range of frequencies that the larynx can produce but may also increase vocal efficiency by lowering the minimum subglottal pressure required to sustain vibration.

3.4.5 *Laryngeal Control of Sonar Pulse Timing: The Laryngeal Gate*

In laryngeally echolocating bats, the timing of phonation, including pulse duration and repetition rate, is determined by the activation of laryngeal adductor and abductor muscles that control the glottal aperture. The adductors move the vocal folds and membranes medially into the laryngeal lumen where they approximate each other in the midline and expiratory airflow induces a self-sustained oscillation. The abductors terminate phonation by moving the vocal folds and membranes out of the expiratory air stream. The glottis thus behaves like a pneumatic valve that gates phonation. This gating function of the larynx is typical of terrestrial mammals, but differs from that in dolphins where phonic lips in the nasal passages appear to function as valves gating sound production (Au and Hastings 2008).

A number of experiments have taken advantage of the fact that sonar-like pulses can be elicited from restrained bats by delivering trains of electrical stimuli to the midbrain in the region of the periaqueductal grey (Suga et al. 1973). However, Fenzl and Schuller (2005, 2007) have shown that the periaquaductal gray and paralemniscal area of *Phyllostomus discolor* contain separate central pathways for the production of communication calls and sonar pulses. Because brain stimulation in this part of the midbrain may activate either or both of these circuits, it is necessary to use caution when interpreting EMGs elicited by electrical stimulation in this region of the brain stem (Schuller and Moss 2004).

The principal glottal adductor muscle in most bats studied is the cricothyroid. This muscle is hypertrophied in *E. fuscus* and covers most of the ventral surface of the larynx (Fig. 3.6a). It becomes electrically active just before each sonar pulse but this high-amplitude EMG ends a few milliseconds before phonation starts (Suthers and Fattu 1973; Durrant 1988) (Figs. 3.7 and 3.8). Contraction of the anterior portion of the cricothyroid muscles rotates the ventral portion of the cricoid cartilage toward the thyroid cartilage (Fig. 3.7). This action both stretches and adducts the vocal and ventricular folds and vocal membranes, which extend from the anterior bridge of the thyroid cartilage to the arytenoid cartilage (Fig. 3.6c, d) and flexes the thyroid laminae (Griffiths 1983; Durrant 1988). These actions increase laryngeal resistance to airflow and contribute to the rise in subglottic pressure.

Durrant (1988) cut a "window" in the elastic cricothyroid membrane, which forms the ventral wall of the laryngeal cavity below the glottis of *E. fuscus*. This window enabled him to directly observe adduction of the vocal folds during fictive phonation elicited by brain stimulation. Although no sound was produced in this experiment, because subglottal air escaped through the window, the vocal folds were adducted into the midline each time the bat attempted to vocalize, as indicated by EMG bursts in both cricothyroid muscles. A similar adduction of the vocal folds accompanied bilateral electrical stimulation of the superior laryngeal nerves. The vocal membrane and ventricular folds were obscured from view by the vocal folds.

Some deep intrinsic laryngeal muscles may also assist in glottal adduction, but their biomechanical function is controversial. Contraction of the cricothyroid muscles in the *E. fuscus* is accompanied by contraction of the ventral thyroarytenoid muscles (Durrant 1988) (Fig. 3.8). The biomechanical action of the ventral thyroarytenoid muscles is uncertain. Griffiths (1983) concluded that the ventral thyroarytenoid muscles in *E. fuscus* is modified to dilate the glottis, a role it may have evolved to oppose the powerful glottal constriction by the cricothyroid muscles. Durrant (1988), however, believes that the ventral thyroarytenoid is a glottal adductor that partially adducts the vocal folds by a mechanism separate from that powered by the cricothyroid muscles and may be involved in controlling pulse duration.

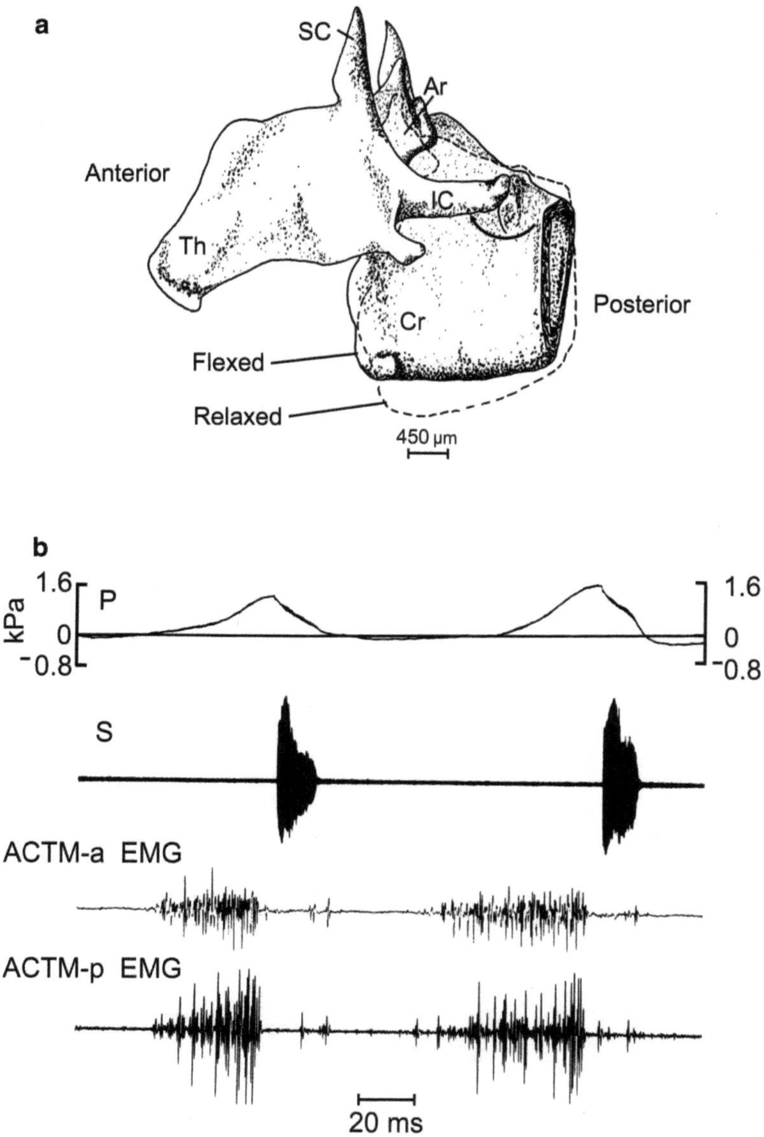

Fig. 3.7 (**a**) Lateral view of *Eptesicus fuscus* larynx showing the position of the cricoid and arytenoid cartilages in a relaxed position and after flexion of the cricothyroid joint. (**b**) Subglottic pressure (P) and timing of EMG in anterior (ACTM-a) and posterior (ACTM-p) portion of the anterior cricothyroid muscle during emission of two downward sweeping FM sonar pulses. Sonar pulses are produced as subglottic pressure declines after ACTM EMG ends. *Ar* arytenoid cartilage, *Cr* cricoid cartilage, *IC* inferior cornu, *Sc* superior cornu, *Th* thyroid cartilage. (From Durrant 1988)

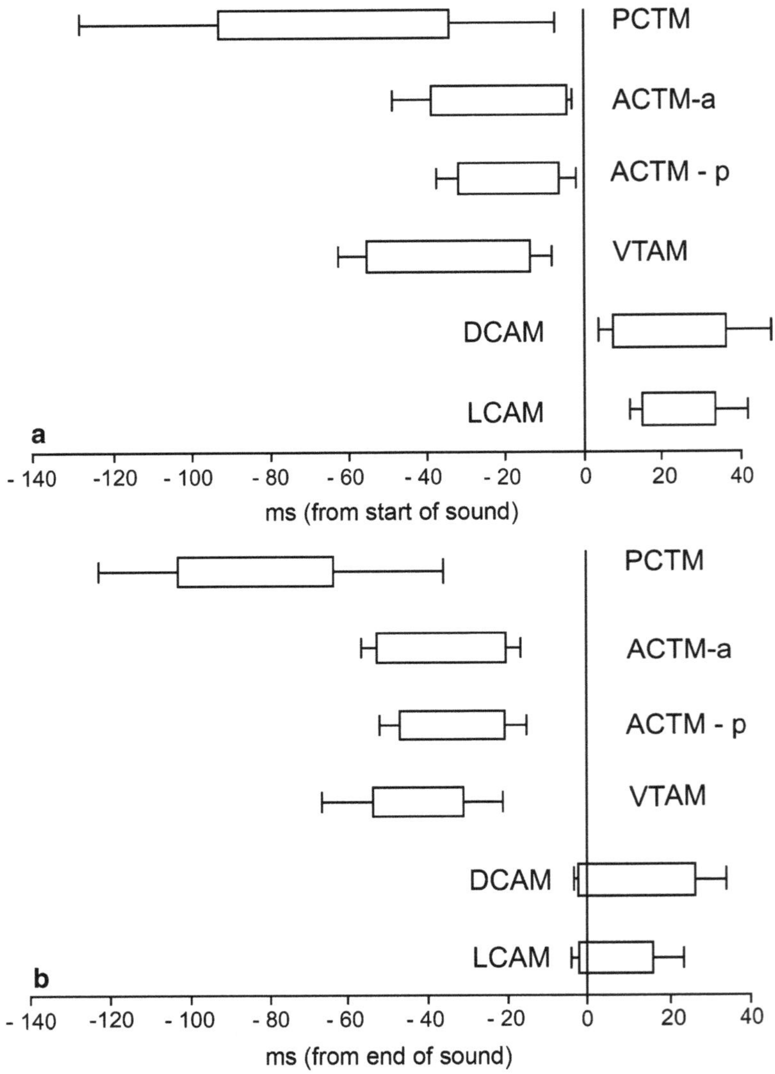

Fig. 3.8 Summary of timing of EMG activity in *Epesticus fuscus* laryngeal muscles relative to the beginning (**a**) or ending (**b**) of FM sonar pulses. Composite collection from many different bats. All data are from single sonar pulses preceded and followed by an inspiration. Mean EMG activity ± 1 SD. (From Durrant 1988)

The cricothyroid muscle may not be an important glottal adductor in all CF-FM bats. Based on laryngeal anatomy, Griffiths (1983) predicted that the cricothyroid muscle of *Pteronotus parnellii* has little effect on the glottal aperture, but may instead participate in fine tuning the fundamental frequency. Glottal constriction in the *P. parnellii* appears to be mediated, at least in part, by the transverse arytenoid muscle, innervated by the recurrent laryngeal nerve (Griffiths 1983).

The cricoarytenoid muscles (Fig. 3.6d) play a role in abducting the glottis. The dorsal and lateral cricoarytenoid muscles begin firing in *E. fuscus* a few milliseconds before the end of the sonar pulse and continue firing until about 20–30 ms after the end of phonation (Figs. 3.6d and 3.8). The dorsal cricoarytenoid muscle is a strong glottal abductor and may determine the end of the pulse. The lateral cricoarytenoid muscle may also act as a glottal abductor, but Durrant (1988) believed it mainly functions to stabilize the cricoarytenoid joint when the dorsal cricoarytenoid muscle is contracting. Recordings from the intact recurrent laryngeal nerve of *Rhinolophus ferrumequinum* show peaks of neural activity immediately preceding the onset of vocalization and just before its termination, which may represent, respectively, the pre-phonatory activation of glottal adductor muscles initiating phonation followed by activation of glottal abductors to terminate sound production (Rübsamen and Schuller 1981).

Bilateral neurotomy of the recurrent laryngeal nerve, which paralyzes all intrinsic laryngeal muscles except the cricothyroid, has surprisingly little effect on sonar pulses in most species tested (e.g., Griffin 1958; Novick and Griffin 1961; Suthers and Fattu 1982). The ability of most bats to produce seemingly normal sonar pulses after section of the recurrent laryngeal nerve is in contrast to the human larynx in which damage to this nerve severely disrupts speech. The limited effect of recurrent laryngeal nerve section in most bats may reflect the ability of the cricothyroid muscles and extrinsic laryngeal muscles to make compensatory motor adjustments or it may produce subtle changes in the quality of the sonar pulse or its acoustic coupling to the trachea, which are hard to detect.

This compensatory ability is absent in the greater horseshoe bat, which suffocates after bilateral recurrent laryngeal nerve section, presumably because the paralyzed deep intrinsic glottal abductor muscles can no longer oppose the closure of the glottis by the powerful cricothyroid muscles. The bilateral innervation of the interarytenoid muscle, a vocal fold abductor, may explain why unilateral recurrent laryngeal nerve section does not cause suffocation and has little or no detectable effect on the temporal pattern, fundamental frequency, or Doppler shift compensation (Schuller and Suga 1976; Schuller and Moss 2004). Other laryngeal muscles in addition to the cricothyroid, considered by Elias (1907) to be glottal adductors in *Rhinolophus*, include the cricoarytenoid and lateral thyroarytenoid. Rübsamen and Schuller (1981) also counted the lateral cricoarytenoid as an adductor, but believed the dorsal (=posterior) cricoarytenoid was a glottal abductor.

3.5 The Biosonar Signal in Dolphins

3.5.1 Wave Shapes and Frequency Spectra

There are basically two general types of biosonar signals produced by odonocetes: those produced by dolphins that can also emit whistle signals and those that do not emit any whistles. Dolphins that produce whistles emit broadband clicks, with only

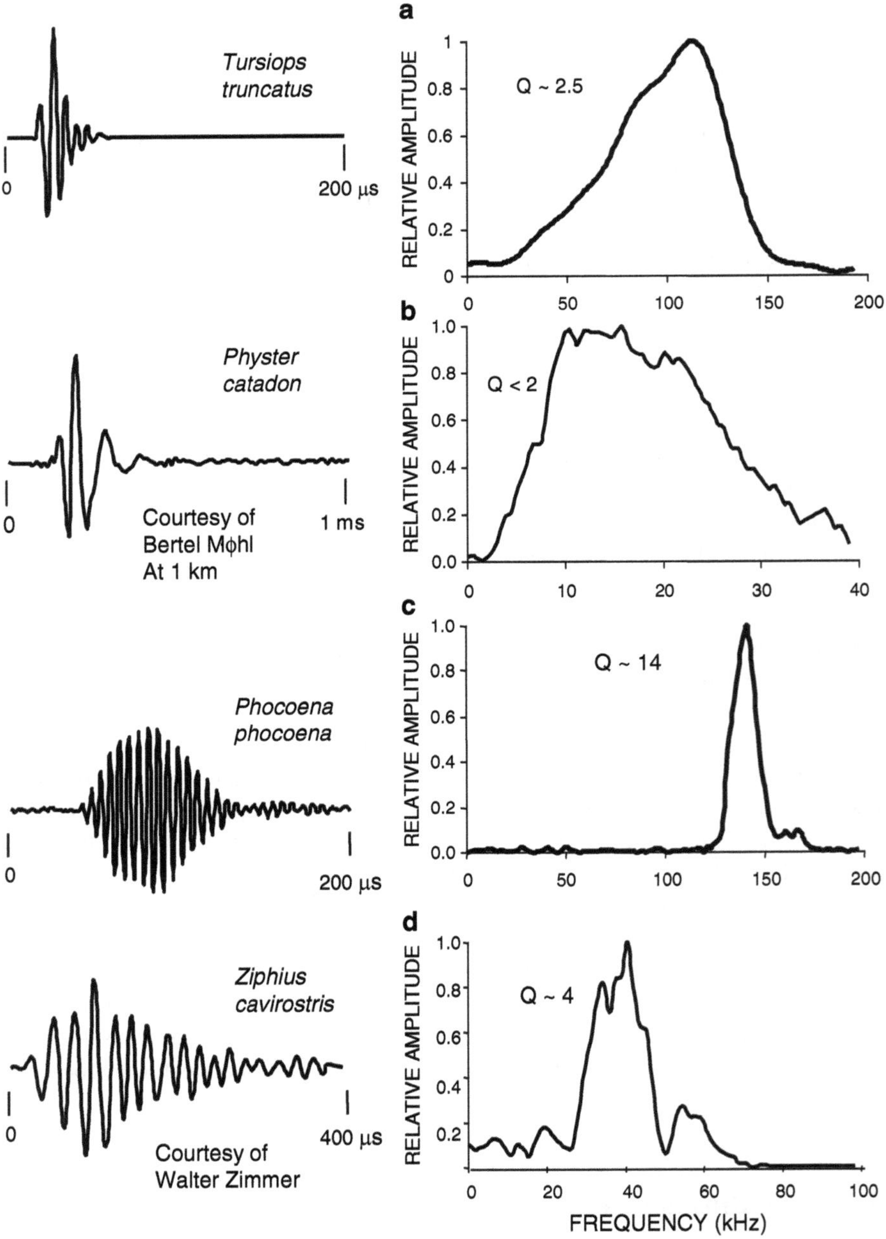

Fig. 3.9 Examples of broadband biosonar clicks emitted by (**a**) a whistling dolphin and (**b**) by a sperm whale, the only nonwhistling odontoceres to emit broadband clicks. (**c**) Narrowband polycyclic click produced by a non-whistling harbor porpoise. (**d**) Borderline broadband polycyclic click produced by a non-whistling Cuvier beaked whale. (From Zimmer et al. 2005)

sperm whales being an exception. Sperm whales do not emit whistles but only emit broadband clicks. Examples of representative broadband clicks produced by a bottlenose dolphin and a sperm whale are shown in Fig. 3.9a, b, where the parameter Q is defined as either the ratio of the peak frequency over 3-dB bandwidth or the ratio of the center frequency over the root mean square (RMS) bandwidth. Whether a signal is broadband or not is somewhat arbitrary; the larger the Q value the narrower the signal. Broadband signals generally have Q values less than about 4 and narrow band signals have Q greater than about 7.

Odontocetes that do not produce whistles emit biosonar signals that can be described as polycyclic, a term introduced by Cranford (2000) to describe the biosonar signals of porpoises and other odontocetes that emit only narrowband click signals. Others have recently used the term narrow band high frequency (NBHF) to describe biosonar signals produced by porpoises and other non-whistling odontocetes. An example of a representative polycyclic biosonar signal produced by a harbor porpoise is shown in Fig. 3.9c. Its polycyclic character is apparent in the waveform shown on the left and its narrowband property is apparent in the spectrum shown on the right. Species of odontocetes that produce polycyclic or NBHF clicks include at least three different groups: Kogiidae, Phocoenidae, and genus *Cephalorhynchus*.

Beaked whales are the exception in this category of non-whistling odontocetes. They emit clicks that can be considered marginally broadband. An example of the biosonar signal of a Cuvier beaked whale is shown in Fig. 3.9d. This example is assumed to be on-axis although it is very difficult to confirm this since beaked whales are deep divers and do not start to emit biosonar signals until they pass the depth of about 400 m (Madsen et al. 2005). Unfortunately, this example may be the sole example of an on-axis beaked whale signal. The polycyclic nature of the Cuvier beaked whale signal can be seen in the waveform of Fig. 3.9d. The Q of the signal is approximately 4, right on the upper edge of our criterion for a broadband signal. The spectrum of the beaked whale signal is certainly wider than that of the harbor porpoise of Fig. 3.9c, but certainly narrower than the spectra shown in Fig. 3.9a, b. Beaked whales are also the only adontocete species that emit frequency modulated biosonar signals.

3.5.2 *Transmission Beam Pattern*

Most of energy contained in the biosonar signal exits through the forehead of the bottlenose dolphin in the area marked off by the dashed lines superimposed on the heads of the two dolphins in Fig. 3.4. The signal is emitted in a beam with its major axis approximately 5° above the horizon in the vertical plane and is directly forward of the animal in the horizontal plane of the dolphin at the location marked as 0 dB in Fig. 3.4. The 3-dB beamwidth in both planes is approximately 10°. The signals become progressively distorted as the angle moves away from the beam axis in both planes. This is not surprising because broadband signals emitted by linear transducer also become progressively more distorted as the angle moves away from the beam axis in both planes. A transducer can be viewed as a low-pass filter with the cutoff frequency decreasing with azimuth as the angle moves away from the beam axis. Therefore, the bandwidth of the transducer becomes progressively narrower with the angle away from the beam axis and cannot faithfully project a broadband signal.

The biosonar signals of harbor porpoises are also emitted in a directional beam that is wider than that of the *T. trunctatus* in both planes. The 3-dB beamwidth of the harbor porpoise, *Phoecena phoecena*, in both planes is approximately 16° compared to the 10° for *T. truncatus*. The signal shown in the vertical plane is longer than the signal shown for the horizontal plane. This difference is just a reminder of natural variations that exist in sound production, emphasizing the fact that biosonar system has natural variations that do not exist with technological sonars. The signals do not show any discernable distortion even as the emission angle moves away from the beam axis. This is because the signal has a narrow bandwidth. However, we can expect distortion to creep in as the angle increases beyond a particular point.

3.5.3 Relationship Between Source Level and Center Frequency

The first controlled measurements of the biosonar signals of bottlenose dolphins in an open water environment produced some very startling results. The averaged peak-to-peak source levels (SL) were in the order of 220 dB re 1 μPa for two bottlenose dolphins and the peak frequencies of the clicks were between 110 and 130 kHz (Au et al. 1974). Before that study, SL measured with dolphins in tanks produced results of about 170–180 dB and peak frequencies between 30 and 50 kHz (Evans 1973). Measurements of a beluga whale biosonar signal in San Diego and in Hawaii indicated that in San Diego, the whale's largest SL for a single trial was 203 dB while in Hawaii it was 214 dB (Au et al. 1985). However, the peak frequencies of the biosonar signals measured in San Diego were between 40 and 60 kHz versus 110–120 kHz in Hawaii. The data of Thomas et al. (1988) and Thomas and Turl (1990) also indicated a similar relationship between intensity and spectrum. The biosonar signals of a false killer whale measured in a tank had peak frequencies between 20 and 60 kHz and SL of approximately 180 dB. Most of the biosonar signals used by another false killer whale in Kaneohe Bay had peak frequencies between 100 and 110 kHz and source levels between 220 and 225 dB (Thomas and Turl 1990). These three studies in Hawaii suggested that there might be a connection between source levels and peak frequencies: the higher source levels the higher the peak frequency (see also Wahlberg & Surlykke, Chap. 3).

Results with a false killer whale showed a clear relationship between the frequency content of echolocation signals and source level (Au et al. 1995). The *Pseudorca crassidens* emitted four basic type of signals, shown in Fig. 3.10a. The four signal types have spectra that are bimodal (having two peaks in the spectrum). The type I signals were defined as those with a low frequency peak (below 70 kHz)

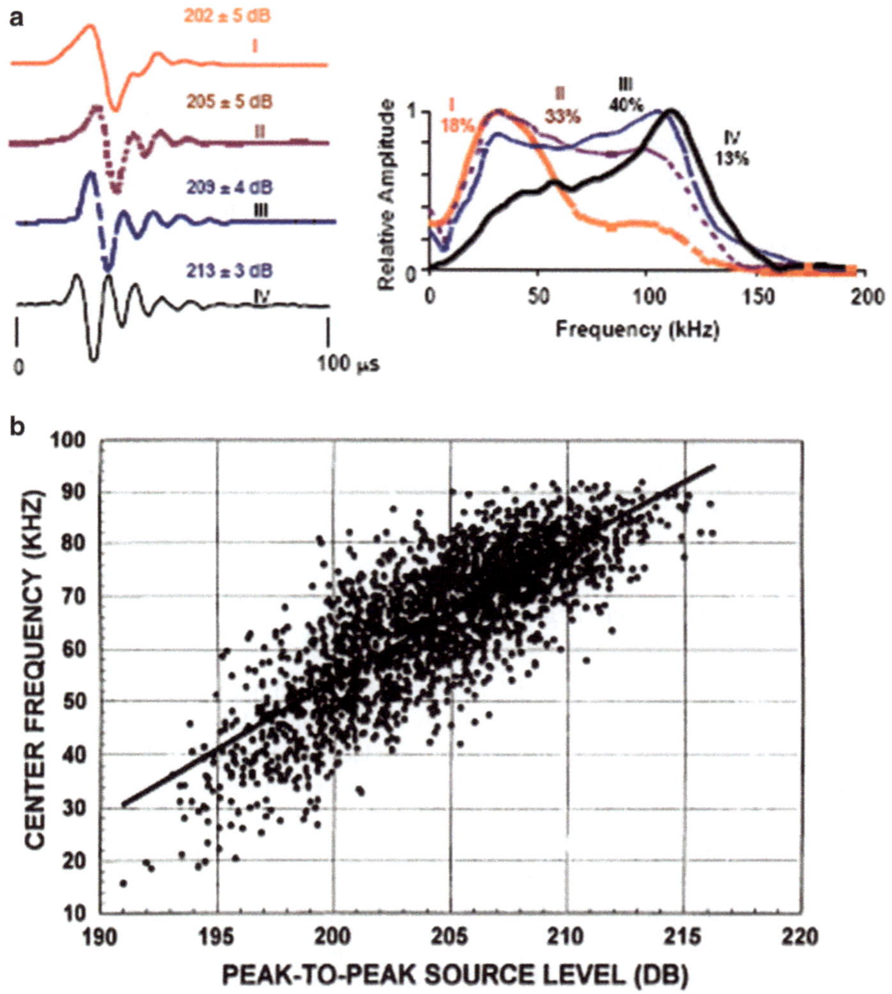

Fig. 3.10 (**a**) Examples of *Pseudorca crassidens* biosonar signal type. SL is the averaged peak-to-peak source level in dB re 1 μPa. (After Au et al. 1995.) (**b**) Center frequency of the biosonar signal emitted by a false killer whale as a function of the peak-to-peak source level in dB re 1 μPa. (From Au et al. 1995)

being the primary peak and a secondary peak at higher frequencies. Type II signals were defined as those with a low-frequency primary peak and a high frequency secondary peak that was within 3 dB of the primary peak. Type III signals were those with a high-frequency primary click (above 70 kHz) and a low-frequency secondary peak with an amplitude within 3 dB of the primary peak. Finally, type IV signals were those with a high-frequency primary peak that was at least 3 dB higher than that of any secondary low-frequency peak. A bimodal spectrum is best described by its center frequency, which is defined as the centroid of the spectrum, and is the frequency that divides the spectrum into two parts with equal energy. The relationship between source level and center frequency that was found by Au et al. (1995) is shown in Fig. 3.10b. The solid line in the figure is a linear regression fit of the data and has an r^2 value of 0.89 indicating a strong linear relationship between source level and center frequency.

The bimodal property of the biosonar signal in Fig. 3.12 suggests that the response of the sound generator may be determined by the intensity of the driving force that eventually produces the biosonar signal. When the intensity of the driving force is low, only signals with low amplitude and low peak frequency are produced. As the driving force increases to a moderate level, the low-frequency peak also increases in amplitude, and the high-frequency portion of the signal begins to kick in. As the driving force increases further, the amplitude of the high-frequency peak becomes larger than that of the low-frequency peak. As the driving force continues to increase to a high level, the amplitude of the high-frequency peak becomes much greater than the amplitude of the low-frequency peak and completely dominates the low-frequency peak, causing the bimodal feature to be somewhat suppressed.

3.5.4 *Effects of Hearing Loss*

Hearing loss among a number of individuals within a mammalian population should not be a surprising occurrence. This is likely true for any dolphin species. The issue is how a dolphin adjusts its biosonar signal production after suffering hearing loss. A bottlenose dolphin was used in an echolocation experiment in 1998 in Kaneohe Bay and was emitting biosonar signals with peak frequencies in the vicinity of 120–135 kHz (Aubauer et al. 2000). The animal's audiogram was measured using an evoked potential technique in 2001 and 2005 and her upper frequency of hearing was about 45 kHz and the animal was now projecting much lower frequency biosonar signals. Although an audiogram was not obtained before 2001, the shift in the frequency of the biosonar signal certainly suggests that it suffered high-frequency hearing loss between about 1998 and 2001. It would make very little sense to emit biosonar clicks with peak frequencies greater than 120 kHz if its upper limit of hearing was only 45 kHz. The peak frequency of its clicks dropped from 135 to 40 kHz, a 3.5 octave drop (Ibsen et al. 2007). It seems obvious that BJ shifted to a lower frequency in order to compensate for a high-frequency hearing loss. However, by shifting to a lower frequency, the SL also shifted downwards. In 1998/1999 BJ had an average peak-to-peak source level of 206 dB compared to 187 dB in 2003/2004, a drop of 19 dB in source level.

Another example of how an odontocete adjusted its echolocation signals after suffering high-frequency hearing loss was reported by Kloepper et al. (2010). The subject was a false killer whale participating in a masked study experiment (Thomas et al. 1990) demonstrating that it could hear frequencies as high as 110 kHz. A hearing test in 2004 showed that the animal's high-frequency hearing threshold was reduced to approximately 50 kHz (Yuen et al. 2005). The false killer whale shifted its signals to lower frequencies and most of the signals had a peaked frequency of approximately 40 kHz. Before hearing loss, the animal emitted 81 % of its signals with a high-frequency peak at approximately 100–110 kHz. After hearing loss, 58 % of its signals had a peak frequency of 40 kHz compared to 18 % and the high-frequency peak dropped to about 80–90 kHz. Performance on an echolocation discrimination performance task correspondingly dropped dramatically.

3.6 Biosonar Signal of Bats

3.6.1 Achieving High Pulse Repetition Rates

Echolocating bats may be unique among terrestrial vertebrates in the speed and temporal precision, which high sonar pulse repetition rates require of the glottal valve. FM bats are capable of pulse repetition rates in excess of 150 Hz with pulse durations less than 1 ms during the terminal buzz phase. *Craseonycteris thonlongyai*, the world's smallest mammal, holds the record pulse repetition rate of 215 Hz (Surlykke et al. 1993). Long CF-FM bats also increase their pulse repetition rate during prey capture but, owing to the importance of Doppler shift information in their sonar system (see later in this section), their sonar pulses are not as short and their maximum pulse repetition rates are not nearly as high as those achieved by FM bats.

High pulse repetition rates imply fast contraction times for the cricothyroid muscle in its role as the primary adductor of the glottal valve. Revel (1962) reported that sarcoplasmic reticulum of the cricothyroid muscle in *Myotis lucifugus* and *E. fuscus* is exceptionally well developed. Suthers and Fattu (1973) measured a tetanus fusion frequency between 220 and 230 Hz for the cricothyroid muscle of *E. fuscus*. Schuller and Suga (1976) reported that electrical stimulation of this muscle in *Rhinolophus ferrumequinum* was accompanied by corresponding rates of frequency modulation in the CF sonar pulse up to 400 Hz when bursts of electrical stimuli were delivered during orientation sounds.

Elemans et al. (2011) showed free-flying FM bats (*Myotis daubentonii*: Vestpertilionidae), achieve pulse repetition rates of 180+ pulses per second during the terminal portion of their buzz—much faster than the contraction rate of typical synchronous skeletal muscles. Elemans et al. analyzed the contractile properties of isolated muscle fibers from the anterior portion of the anterior cricothyroid muscle of *M. daubentonii*. This is the part of the cricothyroid muscle that is primarily responsible for tensing the vocal membranes prior to each sonar pulse by flexing the cricothyroid joint (Fig. 3.7; Durrant 1988). Elemans et al. measured a contraction half-time of 4.72 ms at 39 °C for isometric twitches of isolated muscle fibers. Using work-loop experiments in which isolated muscle fibers were subjected to different strain cycles and patterns of stimulation, Elemans et al. demonstrated that this laryngeal muscle produces power and work at contraction rates up to 180 Hz (200 Hz in one bat). Muscles that produce work at contraction rates greater than 100 per second are considered superfast muscles. The bat cricothyroid appears to be the first superfast muscle identified in mammals. The close agreement between the maximum contraction speed of the circothyroid muscle and the maximum call repetition rate of the buzz supports the hypothesis that the buzz repetition rate is limited by the performance of this laryngeal muscle. Superfast muscles have also been found in the songbird syrinx (Elemans et al. 2008) and sound-producing organs of toadfish (*Opsanus tau*) (Rome 2006) and rattlesnakes (Rome et al. 1996).

Much still remains to be learned about how bats achieve such high pulse repetition rates. If superfast cricothyroid muscles adduct the labia to initiate each pulse, what is the source of the abducting force that terminates the pulse? Neither the activity of laryngeal muscles nor the dynamics of glottal airflow and pressure have been recorded during the terminal buzz and the contractile properties of deep laryngeal glottal abductor muscles are not known. The buzz appears to be produced during a single expiration, there being no inspiratory pause in the pulse repetition rate. The sustained positive subglottal air pressure produced by expiratory muscles during the buzz will tend to force the vocal folds apart. The rapid cycling of the glottal valve between open and closed states, in which each cycle of airflow contributes a sonar pulse to the terminal buzz, may depend on a delicate balance in the coactivation of antagonistic muscles such that a slight change in one of these opposing forces crosses a phonation threshold and switches the larynx between phonatory and nonphonatory configurations.

3.6.2 *Control of Fundamental Frequency*

Echolocating bats can precisely control the frequency of their sonar pulses to optimize them for particular perceptual tasks (Wadsworth and Moss 2000). The fundamental frequency of the human voice is determined in part by contraction of the cricothyroid muscle and the vocalis branch of the thyroarytenoid muscle. By altering the length and/or stiffness of the vocal folds, these muscles contribute to the complex control of fundamental frequency (Titze 1994). In echolocating bats the thyroarytenoid muscle does not have a vocalis branch so the fundamental frequency is controlled mainly by the cricothyroid muscles.

3.6.2.1 Long CF-FM Bats

Doppler shift compensation is a striking example of frequency control by the cricothyroid muscles. It has evolved in long CF-FM Old World families Rhinolophidae (Schnitzler 1968, 1972; Metzner et al. 2002) and Hipposideridae (Habersetzer et al. 1984) and in two members of the New World Mormoopidae: *P. parnellii* (Schnitzler

1970) and *P. personatus* (Smotherman and Guillen-Servent 2008) as a means of stabilizing the frequency of the returning echo so it falls within a sharply tuned, low-threshold "acoustic fovea" in the cochlea. In *Rhinolophus ferrumequinum* the firing rate of motor axons in the superior laryngeal nerve is linearly proportional to the CF (Schuller and Suga 1976; Schuller and Rübsamen 1981; Schuller and Moss 2004). This bat can compensate with an accuracy of ±30 to 50 Hz, for upward frequency shifts of the returning echo by decreasing the firing rate of cricothyroid motoneurons just enough to keep the CF of the Doppler-shifted echo close to the 83 kHz reference frequency (Schnitzler 1968) (see Surlykke and Nachtigall, Chap. 1). Horseshoe bats rely on their sensitivity to small Doppler shifts to detect spectral changes in echoes produced by wingbeat patterns of insect prey.

Rhinolophus can also experience negative Doppler shifts as it reduces its flight speed when approaching prey or during somersault landings, but these negative shifts are only partially compensated (Metzner et al. 2002).

3.6.2.2 FM Bats

The cricothyroid muscle is also involved in controlling the fundamental frequency of sonar pulses emitted by FM bats. In *E. fuscus* bilateral denervation of the cricothyroid results in a large drop in the frequency of the sonar pulse, eliminates most frequency modulation, and abolishes the correlation between frequency range and the concurrent change in subglottic pressure (Fattu and Suthers 1981; Suthers and Fattu 1982; Durrant 1988).

Because EMG activity in the cricothyroid muscle of intact *E. fuscus* precedes the sonar pulse and ends a few milliseconds before the start of phonation, the relationship between the firing rate of this muscle and fundamental frequency has not been quantified. Both subglottic pressure and vocal membrane tension are presumably near their maximum value when airflow through the glottis initiates phonation (Suthers and Fattu 1973; Durrant 1988; Suthers 1988). Relaxation of the cricothyroid muscle in *Eptesicus* allows the glottis to open and produces the downward FM sweep as tension on the vocal membranes decreases (Fig. 3.6). If the timing of this laryngeal gating process is disrupted, causing the glottis to open as tension in the cricothyroid muscle is rising, sonar pulses with abnormal upward frequency sweeps may be produced (Suthers and Fattu 1982).

The means by which *Eptesicus* actively controls the slope or other acoustic parameters of its FM sonar pulse needs further study, however. Although Suthers and Fattu (1973) recorded EMG's from the posterior and lateral portion of the cricothyroid that continued during pulse emission, this recording needs to be repeated since none of the laryngeal muscles recorded by Durrant were consistently active during the downward sweeping sonar pulse. The neuromuscular basis for fine motor control of the acoustic properties of ongoing sonar pulses by *Eptesicus* remains to be determined.

3.6.3 Acoustic Filters of Laryngeal Sound

3.6.3.1 Vocal Tract Filters

The valve-like action of the glottis during phonation produces a broadband, periodic sound containing a fundamental and its higher harmonics. As sound from the larynx travels through the supralaryngeal vocal tract its frequency spectrum can be significantly altered by the tract's resonance filters. The suppression of the fundamental frequency in the sonar pulses of long CF-FM bats in the Rhinolophidae, Hipposideridae, and some members of the Mormoopidae, clearly demonstrates the potential of the vocal tract resonances, by attenuating some frequencies and supporting others, to sculpt the sound emanating from the glottal source, into a vocalization with very different spectral properties. Vocal tract resonance filters are also important in FM bats. The vespertilionid, *Myotis myotis*, for example, can recognize individual conspecifics based on the formant structure of their sonar pulses (Yovel et al. 2009).

In the echolocating bats studied, as in humans, the acoustic coupling between the laryngeal sound source and the vocal tract appears to be weak so that the column of air oscillating at its resonant frequency in the vocal tract does not significantly affect the frequency at which the vocal folds or membranes vibrate. This independence between the laryngeal source, which determines the fundamental frequency, and the vocal tract filter, which determines the formant frequencies, also applies to human speech where it is known as the source-filter model (Fant 1960). To understand how sonar pulses are produced in these sonar systems we need to know the acoustic properties of the vocal tract, including the anatomical basis of the filter and the extent to which the tuning of its resonant frequency is subject to active neuromuscular control.

The possible importance of vocal tract resonance can be assessed by comparing the amplitude spectrum of vocalizations produced in air with that produced while the animal is breathing a light gas mixture such as heliox (20 % oxygen; 80 % helium). The speed of sound is about 74 % higher in heliox than in air and because vocal tract resonance depends on the relationship between the dimensions of the vocal tract and the wavelength of the sound, light gas will shift formant frequencies and harmonic emphasis (but not the fundamental frequency) upward almost an octave if vocal tract resonance is important. Light gas experiments by Pye (1968), Schnitzler (1973), and Roberts (1973) were among the first to demonstrate the importance of vocal tract resonance in the long CF-FM bats.

Adult rhinolophids and hipposiderids emit their sonar pulses through their nostrils with the mouth closed. The epiglottis fits snugly into the nasopharynx to form a tight laryngo-nasal junction during phonation, excluding the oral cavity from the vocal tract (Möhres 1953; Matsumura 1979). The nasal passages of these bats are equipped with prominent rigid bony chambers not present in orally emitting bats (Fig. 3.11a, b). In *Rhinolophus hildebrandti* the nasal passage of each nostril is connected to a dorsal nasal chamber and a pair of lateral nasal chambers in addition to

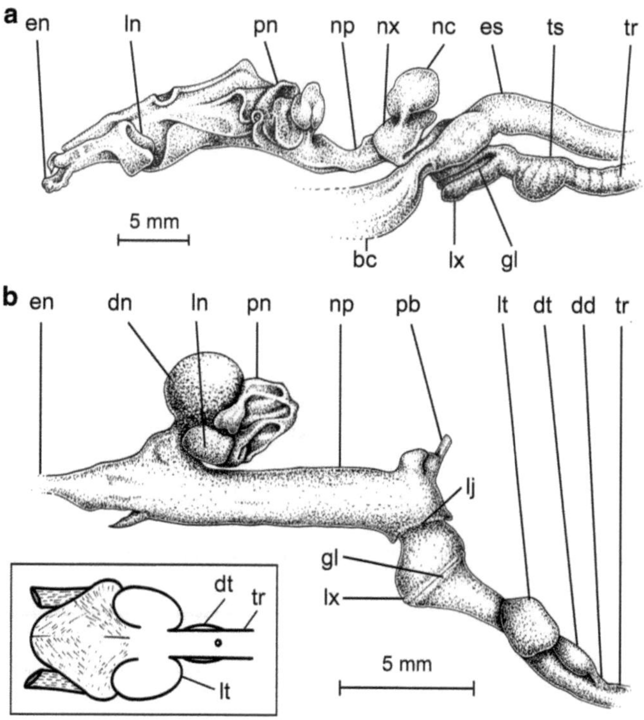

Fig. 3.11 Lateral view of casts of the vocal tract of orally emitting *Pteronotus parnellii* (**a**) and nasally emitting *Rhinolophus hildebrandti* (**b**) showing nasal and tracheal chambers. *Inset* is cutaway ventral view of rigid tracheal pouches in *R. hildebrandti* showing their connection to trachea. *bc* buccal cavity, *dd* duct of dorsal chamber, *dn* dorsal nasal chamber, *dt* dorsal tracheal chamber, *en* external nares, *es* esophagus, *gl* glottis, *lj* laryngonasal junction, *ln* lateral nasal chamber, *lt* lateral tracheal chamber, *lx* larynx, *nc* nasopharyngeal chambers, *np* nasal passage, *nx* nasopharynx, *pb* duct of bursa pharyngealis, *pn* posterior nasal chamber, *tr* trachea, *ts* tracheal sac. ((**a**) From Hartley and Suthers 1990, with permission; (**b**) from Hartley and Suthers 1988, with permission; *inset* from Suthers 1988, with permission)

a posterior olfactory chamber. Forcing *R. hildebrandti* to vocalize through its mouth, by plugging both nostrils, produces a 10- to 21-dB increase in the emitted fundamental frequency and fourth harmonic with no significant change in the second harmonic, suggesting the nasal chambers are an important part of the vocal tract filter. However, the effect on the sonar pulse spectrum of filling the dorsal nasal chambers is variable and difficult to interpret (Suthers et al. 1988).

Hartley and Suthers (1988, 1990) used incremental changes in the helium content of inspired gas to obtain a profile of the transfer function for the vocal tract of *R. hildebrandti* and *P. parnellii*. The transfer function of the intact vocal tract of *R. hildebrandti* has a broad transmission maximum at the 48-kHz second harmonic and sharply tuned transmission minima at the fundamental and third harmonic (Fig. 3.12). Hartley and Suthers (1988) hypothesized that the nasal chambers might have a role in impedance matching between the nostrils and outside environment for

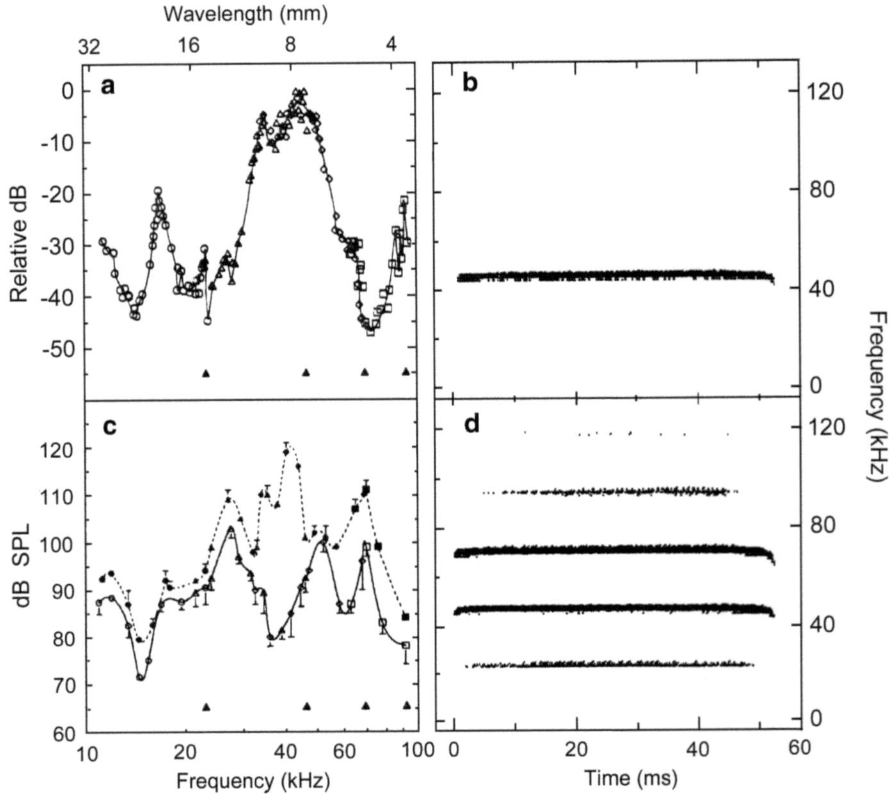

Fig. 3.12 Vocal tract and tracheal resonance filters in *Rhinolophus hildebrandti*. (**a**) Transfer function of intact supralaryngeal vocal tract attenuates the fundamental and third harmonic and accentuates the second harmonic. (**b**) Spectrogram of emitted pulse. (**c**) Subgottal filter function with intact vocal tract (*solid line*, *open symbols*) and with sublaryngeal tracheal chambers partially filled (*broken line*, *solid symbols*). (**d**) Spectrogram of subglottal sound for same sonar pulse as (**b**). *Triangles* just above abscissa in (**a**) and (**c**) mark frequency of CF harmonic components. Ordinate of (**c**) gives SPL for fundamental and second harmonic sections of transfer function. See source reference for third and fourth harmonic correction. (From Hartley and Suthers 1988, with permission)

the second harmonic. They proposed that the suppression of the fundamental and third harmonic may be the result of reflections between acoustic impedance boundaries elsewhere in the vocal tract, such as at the nostrils, the openings to the nasal cavities, the bifurcation of the nasal passages, the laryngonasal junction, and/or the glottis. Further experiments are needed to determine the precise anatomical basis of the rhinolophid vocal tract filter.

Sonar pulse emission by rhinolophids is synchronized with back and forth antiphasic movements of the pinnae. Vanderelst et al. (2011) modeled the sound field of the sonar system around the head and noseleaf of *Rhinolophus rouxi*. This bat emits long CF sonar calls having a fundamental at about 40 kHz accompanied by a second harmonic, which is usually at a higher amplitude. Pinna movements produce frequency-dependent amplitude modulation of sound reaching the ears. Their calculations show that each azimuth-elevation position of a target with respect to the bat produces a different frequency-dependent pattern of amplitude modulation Vanderelst et al. (2011) concluded that the head morphology and the harmonic structure of its sonar system provide *R. rouxi* with two simultaneous views of its environment: the fundamental frequency provides a wide view conveying "a general impression of the environment including parameters such as density of the clutter, level of confinement and nearness of large reflectors" while the 80-kHz second harmonic provides a more focused view containing detailed information about the precise location of a specific target in the bat's region of interest with minimal interference from clutter. The head and noseleaf morphology thus appears to be tuned to the second harmonic in a way that enables information from pinna movements to reject clutter and optimize target localization.

Little is known about how the dimensions, and therefore the tuning, of vocal tract resonance filters adjusts to differences in body size or fundamental frequency. *R. philippinensis*, the large-eared horseshoe bat of Southeast Asia, has evolved three different size morphs. The emitted CF of each morph corresponds to a different CF harmonic of the large morph's 13.5-kHz fundamental frequency (i.e., the second, third, or fourth harmonic) for the large, intermediate, and small morph, respectively. The emitted CF frequency of the large, intermediate, and small morphs is therefore about 27.2, 41.7, and 53.6 kHz, respectively. In each morph, the emitted frequency is the second harmonic of its own suppressed fundamental. This remarkable evolutionary "harmonic hopping" must have been accompanied by parallel changes in the frequency tuning of the cochlear acoustic fovea and the acoustic filters in the vocal tract. It is thought to have contributed to the rapid radiation of rhinolophid bats in Southeast Asia (Kingston and Rossiter 2004).

Like *Rhinolophus*, *Pteronotus* emit long CF-FM sonar pulses in which the fundamental and third harmonic are attenuated and most of the energy is in the second harmonic. Unlike rhinolophids and hipposiderids, *Pteronotus* emits its sonar pulses orally and lacks the enlarged nasal chambers present in the rhinolophids (Fig. 3.11). The vocal tract transfer function of *P. parnellii* has only one transmission minimum and it is tuned to the 30-kHz fundamental, which is 34–50 dB below a broad transmission maximum at the second and third harmonics (Hartley and Suthers 1990). Hartley and Suthers (1990) speculate that suppression of the fundamental frequency is most likely caused by a combination of antiphasic reflections from the nasal passages and an area-function resonance due to a change in the diameter of the vocal tract in the pharynx.

3.6.3.2 Subglottal Filters

Some long CF-FM bats have prominent tracheal chambers or sacs below the glottis. Rhinolophids and hipposiderids, for example, have rigid subglottal chambers that open into the trachea a short distance below the larynx. The high acoustic impedance of the glottis during phonation acoustically isolates the subglottal vocal tract from the supraglottal vocal tract. Filling the tracheal chambers of *R. hildebrandti* with dental impression medium has negligible effect on the transfer function of the supralaryngeal vocal tract, but increases the fundamental in the trachea 15–19 dB (Fig. 3.12). Conversely, filling the nasal chambers has little effect on sound in the trachea (Hartley and Suthers 1988; Suthers et al. 1988).

In *R. hildebrandti*, the distance between the glottis and the lateral tracheal chambers is one half the wavelength of the second harmonic and that between the glottis and the dorsal chamber is equal to the wavelength of the fundamental (Fig. 3.11b inset). The relationship of these distances to the wavelength of the CF component suggests that tracheal chambers in long CF bats evolved to reduce the backward propagated CF component by reflecting it back toward the glottis with a phase shift that provides a positive feedback to the vibration of the vocal membranes and increases vocal efficiency (Hartley and Suthers 1988).

The *P. parnellii* lacks rigid tracheal chambers but has instead an elastic expandable section of trachea, the tracheal sac, just behind the larynx (Fig. 3.11b). Griffiths (1978) hypothesized that the expanded tracheal sac might form part of a Helmholtz resonator tuned to amplify the second harmonic. However, when the tracheal sac is prevented from expanding, by the application of tissue adhesive, there is no significant effect on the transfer function of the supraglottal vocal tract (Hartley and Suthers 1990). By reducing the reflection of backward propagated sound from the lungs, tracheal sacs and chambers may also reduce the temporal smearing of the sonar signal.

3.6.3.3 Beamforming of the Sonar Signal

A final step in the emission of a sonar pulse involves focusing its sound energy in front of the bat toward the area of interest. The optimal beam width varies with the bat's feeding ecology and the perceptual task at hand. A broad emission beam provides a wider "field of view" but focusing the energy into a narrow beam provides a longer detection range and reduces clutter from objects off the beam's axis.

The pattern of sound radiation from orally emitting FM vespertilionid bats can be modeled, to a first approximation, as a piston source having a radius about equal to that of the open mouth and an infinite baffle (Strother and Mogus 1970; Hartley and Suthers 1989; Surlykke et al. 2009). This is in contrast to most nasally emitting bats such as those in the Rhinolophidae and Phyllostomidae in which each nostril acts as a point source of sound. Many of these nasally emitting bats have evolved elaborate structures around their nostrils and on their face that have an important role in determining the shape of their sonar beam.

Hartley and Suthers (1987) measured the sound emission pattern of nasally emitted FM sonar pulses from the leafnosed bat *Carollia perspicillata* (Phyllostomidae). Sonar pulses were elicited by midbrain stimulation while the head was held in a fixed position in front of recording microphones. The forward directed sonar beam had half-amplitude beam widths of ±25 to ±30° at 90 kHz, the highest frequency emitted. Bending the tip of the lancet, the blade-shaped dorsal portion of the noseleaf, backward and cementing it onto the forehead caused the sonar beam to become much broader vertically with little change in the horizontal width, indicating that the lancet focuses the vertical dimension of the sonar beam. Following a model proposed by Strother and Mogus (1970), Hartley and Suthers postulated that the beam width in the horizontal plane is limited by destructive interference between sound emanating from the two nostrils, which are about one-half wavelength apart.

Recent experiments by Brinkløv et al. (2011), who recorded the echolocation calls of *C. perspicillata* as they flew toward a microphone array, have provided important new insights into this bat's ability to control the shape of their sonar beam. Brinkløv et al. found that when *Carollia* are actively echolocating during flight they can reduce the half-amplitude angle of the sonar beam to about 16° horizontally and 14° vertically at the peak FM frequency of 90 kHz. This is about half the width recorded by Hartley and Suthers (1987) and is the narrowest beam yet measured for an echolocating bat. This ability of *Carollia* to focus the acoustic energy of its sonar pulse into a narrow beam may also explain the finding of Brinkløv et al. that the on-axis sound pressure level of sonar pulses emitted during flight was considerably higher than the relatively low intensity calls previously recorded from non-flying *Carollia*. The difference between the directivity of the sonar beam measured by Hartley and Suthers (1987) and that measured by Brinkløv et al. (2011) is doubtless due to differences in experimental design and emphasizes the importance of keeping experimental conditions as natural as possible when testing sonar performance.

It is not clear how flying bats achieve such a highly focused sonar beam. Brinkløv et al. hypothesize that flying *Carollia* may use facial muscles to actively focus their sonar beam by modifying the shape of the noseleaf. Hartley and Suthers (1987) observed that *Carollia* have at least a limited ability to rotate the whole noseleaf several degrees around the vertical axis and to move the tip of the lancet backward and forward. These movements may enable *Carollia* to steer the direction of its sonar beam independently of head orientation and give them the best of both worlds—a narrow, highly focused sonar beam capable of sweeping over a wide angle of view.

The acoustic function of the noseleaf can differ depending on the sensory ecology of the owner. Vanderelst et al. (2010) used a computational method on 3D models of the head to simulate the spatial sensitivity of hearing, the head-related transfer function, and sonar beam emission in two other species of phyllostomids that have morphologically similar noseleaves, but occupy very different ecological niches. *Micronycteris microtis* has a small hunting area where it specializes in gleaning insects from forest vegetation and must detect small stationary insects in the presence of a lot of clutter from surrounding foliage. *Phyllostomus discolor* is omnivorous and hunts over large areas where it supplements echolocation with olfaction and vision. Vanderelst et al. wanted to know if the noseleaf of these species has evolved different acoustic functions in response to their very different ecological niches. Both of these species produce FM pulses through their nostrils but *M. microtis* is much smaller than *P. discolor* and its sonar pulses extend to much higher frequencies—about 50–140 kHz for *M. microtis* compared to 40–90 kHz for *P. discolor*. As in the case of *Carollia*, removal of the lancet widened the sonar beam in the vertical direction in both species, but the lancet has a bigger role in focusing the sound energy in the frontal direction in *M. microtis* than it does in *P. discolor*. This added focusing power is due to the fact that although the lancet is the same size in both of these species, it is more effective in focusing the short wavelength, high frequencies of *M. microtis*. Vanderelst et al. conclude that the added energy of this highly focused beam increases the ability to detect faint echoes from stationary insects on foliage at the center of the beam while reducing the clutter from echoes of objects in the beam's periphery.

Müller and his colleagues (Zhuang and Müller 2006, 2007; Müller 2010) applied numerical methods to investigate how the facial morphology of bats might affect the properties of the near-field sound, which in turn could influence beamforming of the sonar pulse in the far-field. Most rhinolophids, for example, have prominent half-open cavities (called "cells") in the lancet of their noseleaf. Zhuang and Müller's numerical analysis of these cavities in the lancet of the preserved long CF-FM, nasally emitting bat, *R. rouxi*, suggests they, behave acoustically as half-open acoustic resonance cavities having a resonant frequency near that at the low frequency end of the downward FM sweep at the end of their sonar pulse. Further analysis suggests that excitation of the cavities at their near-field resonant frequency causes the far-field beam pattern to widen. It thus appears that the lancet cavities may provide a way to quickly change sonar beam width as a function of frequency. If so, this might enable the bat to combine a narrow beam during the CF portion of the call with a wide beam during the FM sweep.

3.7 Echolocation in Air with Clicks

Sonar systems based on brief click-like signals have evolved independently in Old World fruit bats belonging to the genus *Rousettus* (family Pteropodidae) (Jones and Teeling 2006) and in two groups of birds—neotropical oilbirds (*Steatornis caripensis*: Steatornithidae) and certain swiftlets (Apodidae).

3.7.1 Lingual Sonar Clicks

Like the other non-echolocating Pteropodidae, *Rousettus* has well developed eyes specialized for nocturnal vision. Unlike other Pteropodidae, *Rousettus* forms diurnal roosts in caves where it navigates acoustically with clicks. Möhres and Kulzer (Kulzer 1956, 1960; Möhres and Kulzer 1956) showed that these clicks are produced by the tongue, first on one side of the mouth, then on the other, without involvement of the larynx. Each click consists of a few high-amplitude cycles followed by several more cycles at much lower amplitude. The mean click duration is only about 75–534 μs, depending on the noise level. Clicks appear in spectrograms as broadband transients with a peak frequency between about 34 and 39 kHz (Holland et al. 2004). Like dolphins, *Rousettus* clicks are produced in pairs with an intra pair interval of about 20–40 ms. In light gas, the energy peaks are shifted upward as expected if the click excites buccal cavity resonances (Roberts 1975b).

Yovel et al. (2010) showed that when *Rousettus* needs precise sonar information about the spatial location of a target, they aim their sonar beam so that the target is slightly off its axis. This causes the target to intercept the edge of the beam where the slope of its envelope is greatest, that is, where the sound intensity changes most rapidly with small differences in target position. The sonar beam of clicks generated on the left side of the mouth have their central axis aimed slightly to the left of the target and clicks generated on the right side are aimed slightly to the right of the target. Off-axis pointing is the optimal strategy for target localization, but it comes at the cost of a decrease in the signal-to-noise ratio because the axis of the beam, which coincides with its maximum intensity, is not aimed at the target. Yovel et al. report that in the presence of masking noise *Rousettus* can switch to an optimal strategy for target detection in noise by aiming the sonar beam from clicks on both sides of the mouth directly toward the target so they intercept the target near the peak amplitude of the beam.

Compared to marine mammals that use similar clicks, *Rousettus* is at a disadvantage owing to the impedance mismatch between the sound source and air. Holland et al. (2004) calculated that the energy flux of a *Rousettus* sonar signal is at least an order of magnitude less than that of dolphins and laryngeally echolocating bats. In the case of dolpins this difference is due mainly to the impedance mismatch with the terrestrial environment, whereas in the case of the laryngeally echolocating bats it is due to the difference in the duration of the signal. The ability of *Rousettus* to detect echoes from small-diameter wire obstacles is surprisingly good despite their low energy flux (Griffin et al. 1958; Waters and Vollrath 2003). Part of this unexpected performance may be due to head-related changes in sensitivity of the pinnae as they move back and forth together in synchrony with click production (Herbert 1985; Holland and Waters 2005).

3.7.2 Syringeal Sonar Clicks

Two avian taxa are known to use echolocation. These include some of the cave-dwelling swiftlets of Southeast Asia and Australia (Medway and Pye 1977) and the cave-dwelling oilbird of the New World tropics (Griffin 1953). Both produce impulsive click-like sounds, which are generated in the syrinx, the avian vocal organ, to navigate in the dark caves where they form breeding colonies. Swiftlets are diurnal insectivores; oilbirds are nocturnal frugivores and though they have large eyes adapted for nocturnal vision they sometimes also echolocate on dark nights when flying to fruiting trees outside the cave.

The mechanism of click production has been studied in the Australian (or gray) swiftlet, *Aerodramus terraereginae* (formerly *Collocalia spodiopygia*) (Suthers and Hector 1982). Clicks have a broad bandwidth with most of their energy below 6 kHz (Coles et al. 1987). Click duration is typically a few milliseconds and they are usually produced in pairs. Contrary to statements by Fullard et al. (1993) and Thomassen et al. (2004), intact *A. terraereginae* also occasionally produce single clicks. The interval between the members of a pair varies, but is often about 25 ms. The swiftlet syrinx is at the junction of the bronchi and trachea. Unlike songbirds, there are no intrinsic syringeal muscles. Clicks are produced by the sequential action of two pairs of extrinsic muscles. The first click of the click pair is produced as the sternotrachealis muscles contract, pulling the trachea toward the syrinx. This reduces the longitudinal tension on the bronchi and causes the first bronchial cartilage at the cranial end of each bronchus to rotate into the bronchial lumen. Smyth (1979) refers to this cartilage as BC2 and considers BC1 to be a cartilage that is fused to the drum and cannot rotate into the lumen or close the bronchus. The first click is produced just before the bronchial lumen closes, by vibration of one or more of the following: the medial (internal) tympaniform membranes, the lateral (external) labia and/or the medial labia. Both bronchi remain closed, preventing airflow, despite a positive subsyringeal pressure, during the silent interval between members of a click pair. The second click is produced as the sternotrachealis muscles relax, and the tracheolateralis muscles contract, stretching the bronchi, causing the bronchial valve to open abruptly and produce the second click as air starts to flow through the bronchus.

Oilbirds differ from swiftlets in having a bronchial syrinx. Each primary bronchus contains a semi-syrinx consisting of a medial and lateral tympaniform membrane that divides the bronchus into a cranial and caudal portion. The semi-syrinx acts as a pneumatic valve controlled by a muscle (M. syringealis of King 1989 = M. broncholateralis of Suthers and Hector 1985) that extends from the base of the trachea along the cranial portion of the bronchus to a bronchial cartilage on the cranial edge of the lateral tympaniform membrane. As in swiftlets, contraction of the sternotrachealis muscle reduces the longitudinal tension on the bronchus and causes the lateral tympaniform membrane to fold into the bronchial lumen, restricting or blocking airflow to produce a click. The click is terminated by contraction of the syringealis muscle that opens the bronchial valve formed by the infolded tympaniform membrane. The duration of the click and whether it is single or double depend on the dynamics of the valve action. Clicks are produced synchronously in both semi-syringes; the members of a double click are not produced on different sides of the syrinx (Suthers and Hector 1985). The length of the cranial portion of the bronchus differs on each side and between birds. Formants corresponding to the predicted resonance of each cranial bronchus are present in some clicks and calls and may be useful for individual identification (Suthers and Hector 1988; Suthers 1994).

3.8 Conclusions

Now that the site of biosonar sound production has been identified in dolphins, research can proceed, but in a limited way, into the physiology and functional anatomy of biosonar signal production. There are many questions to be answered and unknowns to be discovered about the forces and mechanisms involved with sound generation in the head of dolphins. One intriguing topic relates to the physical characteristics of the tissues connected to the phonic lips, which are set into vibration by pulses of air moving rapidly through these lips. These vibrations can be damped out in about 70–100 μs. How are signals with large peak-to-peak source levels, on the order of 210–225 dB re 1 μPa, generated by pressurized air that forces the lips open long enough for a pulse of air to shoot by them? What muscular forces are needed to keep the phonic lips closed while the pressure below them is building up to force the lips open and let a puff generate a sonar pulse as it passes between the lips?

Another area in need of study is concerned with how signals, generated by air moving past the phonic lips, are coupled into the melon. The head of a dolphin is very complex with connective tissues, air sacs, bony structures all in the near vicinity of the phonic lips. It is not clear how the vibrations of the structures abutting the phonic lips are coupled into the melon and how sounds propagate through the melon into the water. More mathematical modeling is needed using the technique of Aroyan (2001), who numerically solved the inhomogeneous three-dimensional wave equation using a finite difference approximation or by Krysl et al. (2005), who used a finite element solution in attempting to understand hearing in the Cuvier beaked whale. Improved imaging techniques are needed, including CT scans on live animals, instead of dead frozen carcasses. Scans of living dolphins can provide more accurate information on the geometry and volume of the air sacs and the position of various tissues within the head. Such information is needed to understand how sonar signals generated at the sonic lips propagate through the head.

Important differences in the biosonar signals of dolphins and bats include the energy content, frequency spectra and duration and shape of the signals and these differences require different types of generation mechanisms. Dolphin biosonar signals are generated within the nasal system and the mechanisms that determine how the signal is transmitted through the head and eventually enters the water are not well understood. Most bat biosonar signals are produced in the larynx and how the signals are projected into the air is fairly well understood. The duration of the biosonar pulses are also very different in dolphins and bats. Dolphin biosonar signals can best be described as clicks that vary about 75–400 μs whereas the duration of bat biosonar pulses from as little as about 200 μs in the terminal buzz to as long as 20–100 ms. The energy content of the biosignals in dolphins is much higher than that of bats. For example, the energy of a *Tursiops truncatus* click can be as much as 40,000 times larger than the energy of an *Eptesicus fuscus* (Au 2004) pulse. These differences between the dolphin and bat biosonar signals are associated with living in seawater versus air and how the respective animals' biosonar systems are utilized, especially in foraging (see also Madsen and Surlykke, Chap. 8).

The vocal system of laryngeally echolocating bats provides a valuable experimental model in which to study the coordination of diverse muscle groups in the execution of complex motor patterns, operating at the extremes of speed and sensitivity, to generate a well-defined acoustic behavior. Although the basic features of sonar pulse production have been worked out in a few species, very little is known about the activity or vocal function of deep intrinsic laryngeal muscles, extrinsic laryngeal muscles or muscles of the supralaryngeal vocal tract that may influence its filter properties and, in orally emitting species, the directionality of the sonar beam. Quantitative data on somatosensory feedback and on the contractile properties including contraction times and the temporal relationship between the EMG and force development would be an important step toward developing a quantitative dynamic biomechanical model of sonar pulse production that integrates the respiratory, laryngeal, and vocal tract subsystems.

References

Amundin, M., & Andersen, S. H. (1983). Bony nares air pressure and nasal plug muscle activity during click production in the harbour porpoise, *Phocoena phocoena*, and the bottlenosed dolphin, *Tursiops truncates*. *Journal of Experimental Biology*, 105, 275–282.

Aroyan, J. L. (2001). Three-dimensional modeling of hearing in *Delphinus delphis*. *Journal of the Acoustical Society of America*, 110, 3305–3318.

Aroyan, J. L., Cranford, T. W., Kent, J., & Norris, K. S. (1992). Computer modeling of acoustic beam formation in *Delphinus delphis*. *Journal of the Acoustical Society of America*, 95, 2539–2545.

Au, W. W. L. (1993). *The sonar of dolphins*. New York: Springer Verlag.

Au, W. W. L. (2004). A comparison of the sonar capabilities of bats and dolphins. In J. Thomas, C. Moss, & M. Vater (Eds.), *Echolocation in bats and dolphins* (pp. xiii–xxvii). Chicago: University of Chicago Press.

Au, W. W. L., & Hastings, M. C. (2008). *Principles of marine bioacoustics*. New York: Springer.

Au, W. W. L., Floyd, R. W., Penner, R. H., & Murchison, A. E. (1974). Measurement of echolocation signals of the Atlantic bottlenose dolphin, *Tursiops truncatus* Montagu, in open waters. *Journal of the Acoustical Society of America*, 56, 1280–1290.

Au, W. W. L., Carder, D. A, Penner, R. H., & Scronce, B. L. (1985). Demonstration of adaptation in beluga whale echolocation signals. *Journal of the Acoustical Society of America*, 77, 726–730.

Au, W. W. L., Pawloski, J. L., Nachtigall, P. E., Blonze, M., & Gisiner, R. C. (1995). Echolocation signals and transmission beam pattern of a false killer whale (*Pseudorca crassidens*). *Journal of the Acoustical Society of America*, 98, 51–59.

Au, W. W. L., Houser, D. S., Finneran, J., Lee, W-J, Talmadge, L. A., & Moore, P. W. (2010). The acoustic field on the forehead of echolocating Atlantic bottlenose dolphins (*Tursiops truncatus*). *Journal of the Acoustical Society of America*, 128, 1426–1434.

Au, W. W. L., Branstetter, B., Moore, P. W., & Finneran, J. J. (2012). Dolphin biosonar signals measured at extreme off-axis angles: Insights to sound propagation in the head. *Journal of the Acoustical Society of America*, 132, 1119–1206.

Aubauer, R., Au, W. W. L., Nachtigall, P. E., Pawloski, J. A., Pawloski, D. A., & DeLong, C. (2000). Classification of electronically generated phantom targets by an Atlantic bottle nose dolphin (*Tursiops truncatus*). *Journal of the Acoustical Society of America*, 107, 2750–2754.

Bass, A. H., Gilland, E. H., & Baker, R. (2008). Evolutionary origins for social vocalization in a vertebrate hindbrain-spinal compartment. *Science*, 321, 417–421.

Brinkløv, S., Jakobsen, L., Ratcliffe, J. M., Kalko, E. K. V., & Surlykke, A. (2011). Echolocation call intensity and directionality in flying short-tailed fruit bats, *Carollia perspicillata* (Phyllostomidae). *Journal of the Acoustical Society of America*, 129, 427–435.

Clement, M. J., Gupta, P., Dietz, N., & Kanwal, J. S. (2006). Audiovocal communication and social behavior in mustached bats. In J. S. Kanwal & G. Ehret (Ed.), *Behavior and neurodynamics for auditory communication* (pp. 57–84). New York: Cambridge University Press.

Coles, R. B., Konish, M., & Pettigrew, J. D. (1987). Hearing and echolocation in the Australian Grey Swiftlet, *Collocalia Spodiopygia. Journal Experimental Biology*, 129, 365–371.

Cranford, T. W. (1988). The anatomy of acoustic structures in the spinner dolphin forehead as shown by X-ray computed tomography and computer graphics. In P. E. Nachtigall, & P. W. B. Moore (Eds.), *Animal sonar: Processes and performance* (pp. 67–77). New York: Plenum Press.

Cranford, T. W. (2000). In search of impulse sound sources in Odontocetes. In W. W. L. Au, A. N. Popper, & Fay, R. R. (Eds.) *Hearing by whales and dolphins* (pp. 109–156). New York: Springer-Verlag.

Cranford, T. W., Amundin, M., & Norris, K. S. (1996). Functional morphology and homology in the odontocete nasal complex: Implications for sound generation. *Journal of Morphology*, 228, 223–285.

Cranford, T. W., Van Bonn, W. G., Chaplin, M. S., Carr, J. A., & Kamolnick, T. A. (1997). Visualizing dolphin sonar signal generation using high-speed video endoscopy (A). *Journal of the Acoustical. Society of America* 102, 3123–3123.

Cranford, T. W., Elsberry, W. R., Van Bonn, W. G., Jeffress, J. A., Chaplin, M. S., Blackwood, D. J., Carder, D. A., Kamolnick, T., Todd, M. A., & Ridgway, S. H. (2011). Observation and analysis of sonar signal generation in the bottlenose dolphin (*Tursiops truncates*): Evidence of two sonar sources. *Journal of Experimental Marine Biology and Ecology*, 407, 81–96.

Denny, S. (1976). Comparative morphology of the larynx. In R. Hinchcliffe & D. Harrison (Eds.), *Scientific foundations of otolaryngology* (pp. 536–545). Chicago: Year Book.

Diercks, K. J., Trochta, R. T., Greenlaw, C. F., & Evans, W. E. (1971). Recording and analysis of dolphin echolocation signals. *Journal of the Acoustical Society of America*, 49, 1729–1733.

Dormer, K. J. (1979). Mechanism of sound production and air recycling in delphinids: Cineradiographic evidence. *Journal of the Acoustical Society of America*, 65, 229–239.

Durrant, G. E. (1988). *Laryngeal control of the duration and frequency of emitted sonar pulses in the echolocating bat,* Eptesicus fuscus. Ph.D. dissertation, Indiana University, Bloomington. Available at http://proquest.umi.com/pqdweb?did=745547241&sid=1&Fmt=6&clientId=12010&RQT=309&VName=PQD

Elemans, C. P. H., Mead, A. F., Rome, L. C., & Goller, F. (2008). Superfast vocal muscles control song production in songbirds. *PLoS ONE* 3, e2581, 2581–2586.

Elemans, C. P. H., Mead, A. F., Jakobsen, L., & Ratcliffe, J. M. (2011). Superfast muscles set maximum call rate in echolocating bats. *Science*, 333, 1885–1888.

Elias, H. (1907). Zur Anatomie des kehlkopfes der Mikrochiropteren. *Morphologischen Jahrbuch*, 37, 70–119.

Elsberry, W. R. (2003). *Interrelationships between interanarial pressure and biosonar clicks in bottlenose dolphins* (Tursiops truncatus). Ph.D. dissertation, Texas A&M University, Galveston.

Evans, W E. (1973). Echolocation by marine delphinids and one species of fresh-water dolphin. *Journal of the Acoustical Society of America*, 54, 191–199.

Fant, G. (1960). *Acoustic theory of speech production.* The Hague: Mouton.

Fattu, J. M., & Suthers, R. A. (1981). Subglottic pressure and the control of phonation by the echolocating bat, *Eptesicus. Journal of Comparative Physiology*, 143, 465–475.

Fenton, M. B. (1984). Echolocation: Implications for ecology and evolution of bats. *Quarterly Review of Biology*, 59, 33–52.

Fenzl, T., & Schuller, G. (2005). Echolocation calls and communication calls are controlled differentially in the brainstem of the bat *Phyllostomus discolor. BMC Biology*, 3, 1–12.

Fenzl, T., & Schuller, G. (2007). Dissimilarities in the vocal control over communication and echolocation calls in bats. *Behavioural Brain Research*, 182,173–179.

Fischer, H., & Gerken, H. (1961). Le larynx de la chauve-souris (*Myotis myotis*) et le larynx humain. *Annales d'Oto-Laryngologie*, 78, 577–585.

Fischer, H., & Vomel, H. J. (1961). Der Ultraschallapparat des Larynx von *Myotis myotis*. Eine morphologische Studie über einen primitiven Saugerkehlkopf. *Morphologische Jahrbuch*, 102, 201–226.

Fullard, J. H., Barclay, R. M. R., & Thomas, D.W. (1993). Echolocation in free-flying Atiu swiftlets (*Aerodramus sawtrelli*). *Biotropica*, 25, 334–339.
Griffin, D. R. (1946). The mechanism by which bats produce supersonic sounds. *Anatomical Record*, 96, 519.
Griffin, D. R. (1953). Acoustic orientation in the oilbird, *Steatornis*. *Proceedings of the National Academy of Sciences of the USA*, 39, 884–893.
Griffin, D. R. (1958). *Listening in the dark*. New Haven, CT: Yale University Press.
Griffin, D. R., & Galambos, R. (1941). The sensory basis of obstacle avoidance by flying bats. *Journal of Experimental Zoology*, 86, 481–506.
Griffin, D. R., Novick, A., & Kornfield, M. (1958). The sensitivity of echolocation in the fruit bat *Rousettus*. *Biological Bulletin*, 155, 107–113.
Griffiths, T. A. (1978). Modification of m. cricothyroideus and the larynx in the Mormoopidae, with reference to amplification of high-frequency pulses. *Journal of Mammalogy*, 59, 724–730.
Griffiths, T. A. (1983). Comparative laryngeal anatomy of the big brown bat, *Eptesicus fuscus*, and the mustached bat, *Pteronotus parnellii*. *Mammalia*, 47, 377–394.
Habersetzer, J., Schuller, G., & Neuweiler, G. (1984). Foraging behavior and Doppler shift compensation in echolocating hipposiderid bats, *H. bicolor* and *H. speoris*. *Journal of Comparative Physiology A*, 155, 559–567.
Hartley, D. J., & Suthers, R. A. (1987). The sound emission pattern and the acoustical role of the noseleaf in the echolocating bat, *Carollia perspicillata*. *Journal of the Acoustical Society of America*, 82, 1892–1900.
Hartley, D. J., & Suthers, R. A. (1988). The acoustics of the vocal tract in the horseshoe bat, *Rhinolophus hildebrandti*. *Journal of the Acoustical Society of America*, 84, 1201–1213.
Hartley, D. J., & Suthers, R. A. (1989). The sound emission pattern of the echolocating bat, *Eptesicus fuscus*. *Journal of the Acoustical Society of America*, 85, 1348–1351.
Hartley, D. J., & Suthers, R. A. (1990). Sonar pulse radiation and filtering in the mustached bat, *Pteronotus parnellii rubiginosus*. *Journal of the Acoustical Society of America*, 87, 2756–2772.
Hartley, D. J., Campbell, K. A., & Suthers, R. A. (1989). The acoustic behavior of the fish-catching bat, *Noctilio leporinus*, during prey capture. *Journal of the Acoustical Society of America*, 86, 8–27.
Herbert, H. (1985). Echolocation behavior in the megachiropteran bat, *Rousettus aegyptiacus*. *Zeitschrift für Saugetierkunde—International Journal of Mammalian Biology*, 50,141–152.
Holland, R. A., & Waters, D. A. (2005). Echolocation signals and pinnae movement in the fruitbat *Rousettus aegyptiacus*. *Acta Chiropterologica*, 7, 83–90.
Holland, R. A., Waters, D. A., & Rayner, J. M. V. (2004). Echolocation signal structure in the Megachiropteran bat *Rousettus aegyptiacus* Geoffroy 1810. *Journal of Experimental Biology*, 207, 4361–4369.
Hollien, H., Hollien, P., Caldwell, D. K., & Caldwell, M. C. (1976). Sound production by the Atlantic bottlenose dolphin, *Tursiops truncates*. *Cetology*, 26, 1–7.
Ibsen, S. D., Au, W. W. L., Nachtigall, P. E., Delong C. M., & Breese, M. (2007). Changes in signal parameters over time for an echolocating Atlantic bottlenose dolphin performing the same target discrimination task. *Journal of the Acoustical Society of America*, 112, 2446–2450.
Isshiki, N. (1964). Regulatory mechanism of voice intensity variation. *Journal of Speech and Hearing Research*, 7, 17–29.
Jones, G., & Teeling, E. C. (2006). The evolution of echolocation in bats. *Trends in Ecology and Evolution*, 21, 149–156.
King, A. S. (1989). Functional anatomy of the syrinx. In A. S. King & J. McLelland (Eds.), *Form and function in birds*, Vol. 4 (pp. 105–220). New York: Academic Press.
Kingston, T., & Rossiter, S. J. (2004). Harmonic-hopping in Wallacea's bats. *Nature*, 429, 654–657.
Kloepper, L. N., Nachtigall, P. E., & Breese, M. (2010). Change in echolocation signals with hearing loss in a false killer whale (*Pseudorca crassidens*). *Journal of the Acoustical Society of America*, 129, 2233–2237.

Kobler, J. B., Wilson, B. S., Henson, O. W., Jr., & Bishop, A. L. (1985). Echo intensity compensation by echolocating bats. *Hearing Research*, 20, 99–108.

Koblitz, J. C., Stilz, P., & Schnitzler, H-U. (2010). Source levels of echolocation signals vary in correlation with wingbeat cycle in landing big brown bats (*Eptesicus fuscus*). *Journal of Experimental Biology*, 213, 3263–3268.

Krysl, P., Cranford, T. W., Wiggins, S. M., & Hildebrand, J. (2005). Three-dimensional modeling of hearing in *Delphinus delphis*. *Journal of the Acoustical Society of America*, 120, 2328–2339.

Kulzer, E. (1956). Flughunde erzeugen Orientierungslaute durch Zungenschlag. *Naturwissenschaften*, 43, 117–118.

Kulzer, E. (1960). Physiologische und morphologische Untersuchungen über die Erzeugung der Orientierungslaute von Flughunden der Gattung *Rousettus*. *Zeitschrift für vergleichende Physiologie*, 43, 231–268.

Ladefoged, P. (1968). Linguistic aspects of respiratory phenomena. *Annals of the New York Academy of Sciences*, 155, 141–151.

Lammers, M. O., & Castellote, M. (2009). The beluga whale produces two pulses to form its sonar signal. *Biological Letters*, 5, 297–301.

Lancaster, W. C., & Speakman, J. R. (2001). Variations in respiratory muscle activity during echolocation when stationary in three species of bat (Microchiroptera: Vespertilionidae). *Journal of Experimental Biology*, 204, 4185–4197.

Lancaster, W. C., Henson, O. W., & Keating, A. W. (1995). Respiratory muscle activity in relation to vocalization in flying bats. *Journal of Experimental Biology*, 198, 175–191.

Litchfield, C., Karol, R., & Greenberg, A. J. (1973). Compositional topography of melon lipids in the Atlantic bottlenose dolphin (*Tursiops truncatus*): Implications for echolocation. *Marine Biology*, 52, 285–290.

Ma, J., Kobayasi, K., Zhang, S. Y., & Metzner, W. (2006). Vocal communication in adult greater horseshoe bats, *Rhinolophus ferrumequinum*. *Journal of Comparative Physiology A*, 192, 535–550.

Madsen, P. T., Johnson, M., Aguilar de Soto, N., Ximmer, W. M., &Tyack, P. (2005). Biosonar performance of foraging beaked whales (*Mesoplodon densirostris*). *Journal of Experimental Biology*, 208, 181–194.

Madsen, P. T., Wisniewska, D., & Breedholm, K. (2010). Single source sound production and dynamic beam formation in echolocating harbour porpoises (*Phocoena phocoena*). *The Journal of Experimental Biology*, 213, 3105–3110.

Madsen, P. T., Lammers, M., Wisniewska, D., Beedholm, K. (2013). "Nasal sound production in echolocating delphinids (*Tursiops truncatus* and *Pseudorca crassidens*) is dynamic, but unilateral: clicking on the right side and whistling on the left side," *The Journal Experimental Biology*, 216, 4091–4102.

Matsumura, S. (1979). Mother-infant communication in a horseshoe bat (*Rhinolophus ferrumequinum nippon*): Development of vocalizations. *Journal of Mammalogy*, 60, 76–84.

Medway, L., & Pye, D. (1977). Echolocation and the systematics of swiftlets, In B. Stonehouse & C. Perrins (Eds.), *Evolutionary biology* (pp. 225–238). London: Macmillan.

Mergell, P., Fitch, W. T., & Herzel, H. (1999). Modeling the role of nonhuman vocal membranes in phonation. *Journal of the Acoustical Society of America*, 105, 2020–2028.

Metzner, W., & Schuller, G. (2010). Vocal control in echolocating bats. In S. M. Brudzynski (Ed.), *Handbook of mammalian vocalization* (pp. 403–415). New York: Elsevier.

Metzner, W., Zhang, S. Y., & Smotherman, M. (2002). Doppler-shift compensation behavior in horseshoe bats revisited: Auditory feedback controls both a decrease and an increase in call frequency. *Journal of Experimental Biology*, 205, 1607–1616.

Möhres, F. P. (1953). Über die Ultraschallorientierung der Hufeisen-nasen (Chiroptera-Rhinolophinae). *Zeitschrift für vergleichende Physiologie*, 34,547–588.

Möhres, F. P., & Kulzer, E. (1956). Über die Orientierung der Flughunde (Chiroptera-Pteropodidae). *Zeitschrift für vergleichende Physiologie*, 38, 1–29.

Moss, C., & Surlykke, A. (2010). Probing the natural scene by echolocation in bats. *Frontiers in Behavioral Neuroscience*, 4, Article 33.
Müller, R. (2010). Numerical analysis of biosonar beamforming mechanisms and strategies in bats. *Journal of the Acoustical Society of America*, 128, 1414–1425.
Nakamura, K., Yamada, T. K., & Shimazaki, K. (1998). Measurements of the nasal sacs of individual common dolphin, *Delphinus delphis*, and Dall's porpoise, *Phocoenoides dalli*, by means of silicon reconstruction. *Mammal Study*, 23, 119–122.
Neuweiler, G. (2003). Evolutionary aspects of bat echolocation. *Journal of Comparative Physiology A*, 189, 245–256.
Norris, K. S. (1968). The evolution of acoustic mechanisms in odontocete cetaceans. In E. T. Drake (Ed.), *Evolution and environment* (pp. 297–324). New Haven, CT: Yale University Press.
Norris, K. S., & Harvey, G. W. (1972). A theory of the function of the spermaceti organ of the sperm whale (*Physeter catodon* L.), NASA SP-26.
Norris, K. S., Dormer K. J., Pegg, J., & Liese, G. T. (1971). The mechanism of sound production and air recycling in porpoises: A preliminary report. In *Proceeding of the VIIIth Conference on Biology of Sonar Diving Mammals*, Menlo Park, CA.
Novick, A., & Griffin, D. R. (1961). Laryngeal mechanisms in bats for the production of orientation sounds. *Journal of Experimental Zoology*, 148, 125–145.
Novick, A., & Vaisnys, J. R. (1964). Echolocation of flying insects by the bat, *Chilonycteris parnellii*. *Biological Bulletin*, 127, 478–488.
Pierce, G. W. (1938). Experimental determination of supersonic notes emitted by bats. *Journal of Mammalogy*, 19, 454–455.
Purves, P. E. (1967). Anatomical and experimental observations on the Cetacean sonar system. In R. G. Busnel (Ed.), *Animal sonar systems: Biology and bionics* (pp. 197–270). Jouy-en-Jouy, France : Laboratoire de Physiologie Acoustique.),
Purves, P. E., & Pilleri, G. (1983). *Echolocation in whales and dolphins*. London: Academic Press.
Pye, J. D. (1968). Animal sonar in air. *Ultrasonics*, 6, 32–38.
Quay, W. B. (1970). Peripheral nervous system. In W. A. Wimsatt. (Ed.), *Biology of bats*, Vol. 3 (pp. 153–179). New York: Academic Press.
Revel, J. P. (1962). The sarcoplasmic reticulum of the bat cricothyroid muscle. *Journal of Cell Biology*, 12, 571–588.
Ridgway, S. H., & Carder, D. A. (1988). Nasal pressure and sound production in an echolocating white whale, *Delphinapterus leucas*. In P. E. Nachtigall & P. W. B. Moore (Eds.), *Animal sonar: Processes and performances* (pp. 53–60). New York: Plenum Press.
Ridgway, S. H., Carder, D. A., Green, R. F., Gaunt, A. S., Gaunt, S. L. L., & Evans, W. E. (1980). Electromyographic and pressure events in the nasolaryngeal system of dolphins during sound production. In R. G. Busnel, & J. F. Fish (Eds.), *Animal sonar systems* (pp. 239–249). New York: Plenum Press.
Roberts, L. H. (1972). Correlation of respiration and ultrasound production in rodents and bats. *Journal of Zoology (London)*, 168, 439–449.
Roberts, L. H. (1973). Cavity resonances in the production of orientation cries, *Periodicum Biologorum*, 75, 27–32.
Roberts, L. H. (1975a). The functional anatomy of the rodent larynx in relation to audible and ultrasonic cry production. *Zoological Journal of the Linnean Society*, 56, 255–264.
Roberts, L. H. (1975b). Confirmation of the echolocation pulse production mechanism of *Rousettus*. *Journal of Mammalogy*, 56, 218–220.
Robin, M. H. A. (1881). Recherches anatomiques sur les mammiferes de l'ordre des Chiropteres. *Ann Sci Nat (Zool)*, 12, 1–180.
Rome, L. C. (2006). Design and function of superfast muscles: New insights into the physiology of skeletal muscle. *Annual Review of Physiology*, 68, 193–221.
Rome, L. C., Syme, D. A., Hollingworth, S., Lindstedt, S. L., & Baylor, S. M. (1996). The whistle and the rattle: The design of sound producing muscles. *Proceedings of the National Academy of Sciences of the U SA*, 93, 8095–8100.

Rübsamen, R., & Schuller, G. (1981). Laryngeal nerve activity during pulse emission in the CF-FM bat, *Rhinolophus ferrumequinum.* I. The recurrent laryngeal nerve. *Journal of Comparative Physiology*, 143, 323–327.
Schenkkan, E. J. (1973). On the comparative anatomy and function of the nasal tract in odontocetes (Mammalia, Cetaea). *Bijdragen tot de Dierkunde*, 43, 127–159.
Schnitzler, H. U. (1968). Die Ultraschall-Ortungslaute der Hufeisen-Fledermäuse (Chiroptera-Rhinolophidae) in verschiedenen Orientierungssituationen. *Zeitschrift für vergleichende Physiologie*, 57, 376–408.
Schnitzler, H. U. (1970). Echoortung bei der fledermäus *Chilonycteris rubiginosa. Zeitschrift für vergleichende Physiologie*, 68, 25–38.
Schnitzler, H. U. (1971). Fledermäuse im Windkanal. *Journal of Comparative Physiology A*, 73, 209–221.
Schnitzler, H. U. (1972). Control of Doppler shift compensation in the greater horseshoe bat, *Rhinolophus ferrumequinum. Journal of Comparative Physiology A*, 82, 79–92.
Schnitzler, H. U. (1973). Die echoortung der fledermäuse und ihre horphysiologischen grundlagen. *Fortschritte der Zoologie*, 21, 136–189.
Schuller, G., & Suga, N. (1976). Laryngeal mechanisms for the emission of CF-FM sounds in the Doppler-shift compensating bat, *Rhinolophus ferrumequinum. Journal of Comparative Physiology*, 107, 253–262.
Schuller, G., & Rübsamen, R. (1981). Laryngeal nerve activity during pulse emission in the CF-FM bat, *Rhinolophus ferrumequinam* I. Superior laryngeal nerve (external motor branch). *Journal of Comparative Physiology*, 143, 317–321.
Schuller, G., & Moss, C. (2004). Vocal control and acoustically guided behavior in bats. In J. A. Thomas, C. F. Moss, & M. Vater (Eds.), *Echolocation in bats and dolphins* (pp. 3–16). Chicago: University of Chicago Press.
Smotherman, M. (2007). Sensory feedback control of mammalian vocalizations. *Behavioural Brain Research*, 182(2), 315–326.
Smotherman, M., & Guillen-Servent, A. (2008). Doppler-shift compensation behavior by Wagner's mustached bat, *Pteronotus personatus. Journal of the Acoustical Society of America*, 123, 4331–4339.
Smyth, D. M. (1979). *Studies on echolocation by the grey swiftlet* Aerodramus spodiopygius. PhD thesis. Department of Zoology, James Cook University of North Queensland. Townsville, Queensland.
Sprague, J. M. (1943). The hyoid region of placental mammals with special reference to the bats. *American Journal of Anatomy*, 72, 385–472.
Strother, G. K., & Mogus, M. (1970). Acoustical beam patterns for bats: Some theoretical considerations. *Journal of the Acoustical Society of America*, 48, 1430–1432.
Suga, N., Schlegel, P., Shimozawa, T., & Simmons, J. A. (1973). Orientation sounds evoked from echolocating bats by electrical stimulation of the brain. *Journal of the Acoustical Society of America*, 54, 793–797.
Surlykke, A., & Moss, C. F. (2000). Echolocation behavior of big brown bats, *Eptesicus fuscus*, in the field and the laboratory. *Journal of the Acoustical Society of America*, 108, 2419–2429.
Surlykke, A., Miller, L., Møhl, B., Andersen, B., Christensen-Dalsgaard, J., & Jørgensen, M. (1993). Echolocation in two very small bats from Thailand: *Craseonycteris thonglongyai* and *Myotis siligorensis. Behavioral Ecology and Sociobiology*, 33, 1–12.
Surlykke, A., Pedersen, S. B., & Jakobsen, L. (2009). Echolocating bats emit a highly directional sonar sound beam in the field. *Proceedings of the Royal Society B: Biological Sciences*, 276, 853–860.
Suthers, R. A. (1988). The production of echolocation signals by bats and birds. In P. E. Nachtigall & P. W. B. Moore (Eds.), *Animal sonar processes and performance* (pp. 23–45). New York: Plenum Press.
Suthers, R. A. (1994). Variable asymmetry and resonance in the avian vocal tract: A structural basis for individually distinct vocalizations. *Journal of Comparative Physiology A*, 175, 457–466.

Suthers, R. A., & Fattu, J. M. (1973). Mechanisms of sound production by echolocating bats. *American Zoology*, 13, 1215–1226.
Suthers, R. A., & Fattu, J. M. (1982). Selective laryngeal neurotomy and the control of phonation by the echolocating bat, *Eptesicus*. *Journal of Comparative Physiology A*, 145, 529–537.
Suthers, R. A., & Hector, D. H. (1982). Mechanism for the production of echolocating clicks by the grey swiftlet, *Collocalia spodiopygia*. *Journal of Comparative Physiology A*, 148, 457–470.
Suthers, R. A., & Hector, D. H. (1985). The physiology of vocalization by the echolocating oilbird, *Steatornis caripensis*. *Journal of Comparative Physiology A*, 156, 243–266.
Suthers, R. A., & Hector, D. H. (1988). Individual variation in vocal tract resonance may assist oilbirds in recognizing echoes of their own sonar clicks. In P. E. Nachtigall & P. W. B. Moore (eds.), *Animal sonar: Processes and performance* (pp. 87–91). New York: Plenum Press.
Suthers, R. A., Thomas, S. P., & Suthers, B. J. (1972). Respiration, wing-beat and ultrasonic pulse emission in an echolocating bat. *Journal of Experimental Biology*, 56, 37–48.
Suthers, R. A., & Hartley, D. J., & Wenstrup, J. J. (1988). The acoustic role of tracheal chambers and nasal cavities in the production of sonar pulses by the horseshoe bat, *Rhinolophus hildebrandti*. *Journal of Comparative Physiology A*, 162, 799–813.
Teeling, E. C., Springer, M. S., Madsen, O., Bates, P., O'Brian, J., & Murphy, W. J. (2005). A molecular phylogeny for bats illuminates biogeography and the fossil record. *Science*, 307, 580–584.
Thomas, J., Stoermer, M., Bowers, C., Anderson, L., & Garver, A. (1988). Detection abilities and signal characteristics of echolocating false killer whales (*Pseudorca crassidens*). In P. E. Nachtigall & P. W. B. Moore (Eeds.), *Animal sonar processing and performance* (pp. 323–328). New York: Plenum Press.
Thomas, J., Pawloski, J., & Au, W. W. L. (1990). Masked hearing abilities of a false killer whale (*Pseudorca crassidens*). In J. Thomas & R. Kastelien, (Eds.) *Sensory abilities of ceta-ceans* (pp. 395–404). New York: Plenum Press.
Thomas, J. A., & Turl, C. W. (1990). Echolocation characteristics and range detection by a false killer whale (*Pseudorca crassidens*). In J. A. Thomas & R. A. Kastelein (Eds.), *Sensory abilities of cetaceans* (pp. 321–334). New York: Plenum Press.
Thomassen, H. A., Djasim, U. M., & Povel, D. E. (2004). Echoclick design in swiftlets: Single as well as double clicks. *Ibis*, 146, 173–174.
Tian, B., & Schnitzler, H. U. (1997). Echolocation signals of the greater horseshoe bat (*Rhinolophus ferrumequinum*) in transfer flight and during landing. *Journal of the Acoustical Society of America*, 101, 2347–2364.
Titze, I. R. (1994). *Principles of voice production*. Englewood Cliffs, NJ: Prentice-Hall.
Urick, R. J. (1983). *Principles of underwater sound*. New York: McGraw–Hill.
Vanderelst, D., De May, F., Peremans, H., Geipel, I., Kalko, E., & Firzlaff, U. (2010). What noseleaves do for FM bats depends on their degree of sensorial specialization. *PLoS ONE*, 5(8), e11893.
Vanderelst, D., Reijniers, J., Steckel, J., & Peremans, H. (2011). Information generated by the moving pinnae of *Rhinolophus rouxi*: Tuning of the morphology at different harmonics. *PLoS ONE*, 6(6), e20627.
Varanasi, U., & Malin, D. C. (1971). Unique lipids of the porpoise (*Tursiops gilli*): Differences in triacylglycerols and wax esters of acoustic (mandibular canal and melon) and bubble tissues. *Biochemica Biophysica Acta*, 231, 415–418.
Wadsworth, J., & Moss, C. F. (2000). Vocal control of acoustic information for sonar discriminations by the echolocating bat, *Eptesicus fuscus*. *Journal of the Acoustical Society of America*, 107, 2265–2271.
Waters, D. A., & Vollrath, C. (2003). Echolocation performance and call structure in the megachiropteran fruit-bat *Rousettus aegyptiacus*. *Acta Chiropterologica*, 5, 209–219.
Wood, F. G. (1964). Discussion In W. Tavolga (Ed.), *Marine Bio-Acoustics Vol II* (pp. 321–334). Oxford: Pergamon Press.
Wyke, B. D., & Kirchner, J. A. (1976). Neurology of the larynx. In R. Hinchcliffe & D. Harrison (Eds.), *Scientific foundations of otolaryngology* (pp. 546–574). London: W. Heinemann.

Yovel, Y., Melcon, M. L., Franz, M. O., Denzinger, A., & Schnitzler, H-U. (2009). The voice of bats: How greater mouse-eared bats recognize individuals based on their echolocation calls. *PLoS Computational Biology*, 5, e1000400.
Yovel, Y., Falk, B., Moss, C. F., & Ulanovsky, N. (2010). Optimal localization by pointing off axis. *Science*, 327, 701–704.
Yuen, M. M. L., Nachtigall, P. E., Breese, M., & Supin, A. Y. (2005). Behavioral and auditory evoked potential audiograms of a false killer whale *Pseudorca crassidens. Journal of the Acoustical Society of America*, 118, 2688–2695.
Zhuang, Q., & Müller, R. (2006). Noseleaf furrows in a horseshoe bat act as resonance cavities shaping the biosonar beam. *Physical Review Letters*, 97, 218701.
Zhuang, Q., & Müller, R. (2007). Numerical study of the effect of the noseleaf on biosonar beamforming in a horseshoe bat. *Physical Review E*, 76, 051902.
Zimmer, W. M. X., Johnson, M. P., Madsen, P. T., & Tyack, P. L. (2005). Echolocation clicks of free-ranging Cuvier's beaked whales *Ziphius cavirostris. Journal of the Acoustical Society of America*, 117, 3919–3927.

Chapter 4
Sound Intensities of Biosonar Signals from Bats and Toothed Whales

Magnus Wahlberg and Annemarie Surlykke

Keywords Acoustic localization • Automatic gain control • Echolocation • Hearing sensitivity • Hydrophone array • Microchiroptera • Odontocetes • Source level • Transmission loss

4.1 Introduction

Echolocation is used by toothed whales and microchiropteran bats for navigation and prey detection. By emitting short, loud, and directional sound pulses, the animals can determine the range, size, and texture of targets, prey, obstacles, or background from the returning echoes of their own calls.

For both bats and toothed whales there is a huge diversity in animal size and behavior, as well as habitats in which they are echolocating. Concurrently, there is a large variation in the type of echolocation signals used by different bat species as well as by different toothed whales. Most bats emit a few milliseconds short frequency modulated signals, but a few species produce constant frequency signals longer than 100 ms. The frequency content varies by more than five octaves, from about 10 kHz up to more than 200 kHz (Fenton, Jensen, Kalko, & Tyack, Chap. 2; Au & Suthers, Chap. 3). The time-frequency modulation of bat signals varies between species, but also within species at different stages in the hunting sequence (Jakobsen and Surlykke 2010). For toothed whales, the echolocation signals fall

M. Wahlberg (✉)
Fjord & Bælt and Department of Biology, University of Southern Denmark,
Marine Biological Research Center, Hindsholmsvej 11, Kerteminde 5300, Denmark
e-mail: magnus@biology.sdu.dk

A. Surlykke
Department of Biology, University of Southern Denmark,
Odense M 5230, Denmark
e-mail: ams@biology.sdu.dk

A. Surlykke et al. (eds.), *Biosonar*, Springer Handbook of Auditory Research 51,
DOI 10.1007/978-1-4614-9146-0_4, © Springer-Verlag New York 2014

within four categories: broadband transients of high- or low-frequency emphasis, narrow band high-frequency signals, and frequency upsweeps of relatively long duration. The signals range in frequencies between 10 and 150 kHz and are short (20–200 μs; Fenton, Jensen, Kalko & Tyack, Chap. 2; Au & Suthers, Chap. 3).

In spite of the large differences in biosonar signals, both bats and toothed whales are under the same basic acoustic constraints. To obtain information through sonar it is important to (1) emit enough energy to receive a detectable echo (from prey or other objects) from a reasonable range, and (2) be able to obtain timing information between the outgoing signal and the returning echo for object localization. These simultaneous requirements can work against each other in signal design. For example, by increasing the signal duration more energy is emitted to facilitate the detection of the returning echoes. On the other hand, longer signals may impair precise timing measurements. Likewise, a large frequency content of the signals may be helpful to evaluate spectral interference patterns in the echoes of the target, but at the same time detectability may decrease as the energy is smeared over a larger range of frequencies. The sound intensity can be increased to improve the detection of small targets in open space, but high source levels may not be adequate in an environment where there are many unwanted echoes (clutter). Further, the echolocator may reveal its presence to both prey and potential predators by using intense signal emissions. Emitting intense signals may also affect the echolocator's own ability to hear the echoes (Nachtigall & Schuller, Chap. 4).

The relative importance of such constraints differs from animal to animal depending on ecological factors, for example, habitat and prey type (Madsen & Surlykke, Chap. 9). For example, the echo energy may be a constraint for bats or toothed whales operating in open habitats, where sonar range is the issue. For other species operating in a more complex acoustic environment, such as the extremely cluttered understory of the tropical jungle or the shallow waters of a river, discrimination may be the main problem, and a lower intensity sonar signal may in fact be advantageous.

The directionality and intensity of the emitted sound pulse define the ensonified space, and thus the intensity not only of target-echo, but also clutter. The echo-to-clutter ratio is crucial for target detection and recognition. The source level, frequency content, and directionality of the outgoing signal are readily under the control of the echolocating animal. Further, the receiving side also plays an important role. The hearing threshold determines how low sound levels can be detected, and the directionality of the hearing system affects the amount of noise collected while listening for the echo from the target.

4.2 Methodology

4.2.1 Transmission Loss

The intensity of an emitted signal is called the source level and is reported at a standard reference distance to the source, usually 1 m for toothed whales and 10 cm for bats, on the acoustic axis of the source, where the acoustic intensity is the highest.

The amount of energy or intensity lost from the source level to the received level at the microphone or hydrophone is termed the transmission loss (TL). To calculate the source level of a signal, the transmission loss has to be known. It depends critically on the range from the receiver to the source, and the frequency content of the signal. Transmission loss is due to geometric spreading loss and sound absorption. The *geometric spreading loss* is generally assumed to follow the so-called spherical spreading law. The sound intensity is spread out over the surface of an ever-increasing spherical "bubble" with its center at the location of the sound source, with a radius (corresponding to the distance from the source to the receiver) expanding with a velocity given by the sound speed of the medium (about 340 m s^{-1} in air and about 1,500 m s^{-1} in water). As the area of a sphere is $4\pi r^2$, the sound intensity will decrease by the squared range (r) from source to receiver. In decibel units this corresponds to a transmission loss of 20 log r.

Biosonar signals from toothed whales and bats are directional. For these signals, the sound energy is spread over a fraction of the surface of the expanding sphere. If this fraction stays constant as a function of range to the source, the geometric spreading will be spherical. There is some data confirming that this is true for both toothed whales (Au et al. 1978) and bats (Surlykke and Kalko 2008) up to the maximum ranges used in echolocation. As an example, consider a toothed whale, where the reference distance of source level measurements is 1 m. At 5 m distance the transmission loss experienced by the signal is 20 log (5) or 14 dB. A doubling of the range implies adding 6 dB to the transmission loss.

Besides geometric spreading, sound energy is also lost as heat. This is called *sound absorption*. In contrast to geometric spreading, absorption depends heavily on the frequency content of the signal and the medium through which the sound wave is propagating. Sound absorption is much more severe for high frequencies, and it is much larger in air than in water. For example, in air the excess attenuation as a result of absorption is about 0.1 dB m^{-1} at 10 kHz, about 3.5 dB m^{-1} at 100 kHz, and about 10 dB m^{-1} at 200 kHz at room temperature and 50 % relative humidity (Beranek 1996). In water the absorption at similar frequencies is only 10^{-3}, 0.03, and 0.04 dB m^{-1}, respectively (Medwin and Clay 1998; Madsen & Surlykke, Chap. 9).

Besides geometric spreading loss and sound absorption, there are other properties of the medium that attenuate the emitted sound on its path. For echolocation the most important of these is *scattering*, resulting from reflection by tiny water droplets or dust in air and by plankton and air bubbles in water (Beranek 1996; Medwin and Clay 1998). When working in dense foliage or close to a water surface the transmission loss may be severely affected by additional attenuation and by interference between the direct sound path and surface reflected paths. Therefore direct measurements are needed to describe the transmission loss properties for the habitat and signal type of interest and to estimate the source level of the recorded sound (Madsen and Wahlberg 2007; Surlykke and Kalko 2008).

4.2.2 *Acoustic Localization*

4.2.2.1 Different Types of Arrays

If one uses an array of spatially distributed receivers to record the vocalization from different angles it is possible to calculate the location of the animal by measuring the relative differences in arrival times of the same signal at the various receivers. If the transmission loss is known, one may back-calculate the received sound level to the source level, assuming the animal is pointing its acoustic axis toward one of the receivers. If the orientation of the animal to the receivers can be assessed (e.g., by video filming, by acoustically tracking the animal throughout several calls, or by calculating the recorded angles relative the assumed on-axis direction) the directionality of the signals may also be estimated.

The geometric shape of the receiver array, together with the number of receivers, determines the amount of information about the sound source location that can be derived with a specific localization system. They may be grouped into different source-array geometries. The simplest is the *one-dimensional linear array*, which in certain situations can be used for tracking animals in two dimensions (2D). In the *2D array* the receivers are spread out in one plane and two source coordinates can be obtained (x and y). 2D geometry is valuable when, for example, locating terrestrial animals in a restricted area or trawling bats flying over water surfaces (Dantzker et al. 1999; Surlykke et al. 2009). For 2D systems the practical array construction and handling, as well as the mathematics and the visualization of the localization process, are all relatively simple to implement. In many cases, it is practical to use a flat 2D array, even though the animal may fly, run, or swim off in a third independent direction (the z direction). In such systems, *2/3D arrays*, the animal cannot be localized completely unambiguously, but with the help of directional receivers, video recorders, or physical barriers the redundant locations can usually be efficiently ruled out. Finally, there are the *3D array* systems, where the receivers are extended in a volume. Such a system can ideally perform unambiguous three-dimensional localization of the animal, but the disadvantage is that it can be very complicated to understand the complete performance of the system (Wahlberg et al. 2001).

4.2.2.2 Theory of Acoustic Localization

Consider first a 2D source-receiver example using a linear array (Fig. 4.1). An animal (marked with a star, *) living in 'Flat Land' is calling, and its signals are recorded at four different receivers (marked with circles and denoted R1, R2, R3, and R4). The recorded signals from the receivers, sketched to the left in Fig. 4.1, show that the same call arrive at somewhat different times and with different amplitude in the four channels. By measuring the time-of-arrival differences (called TOADs) of the four signals in Fig. 4.1 we obtain three independent TOADs,

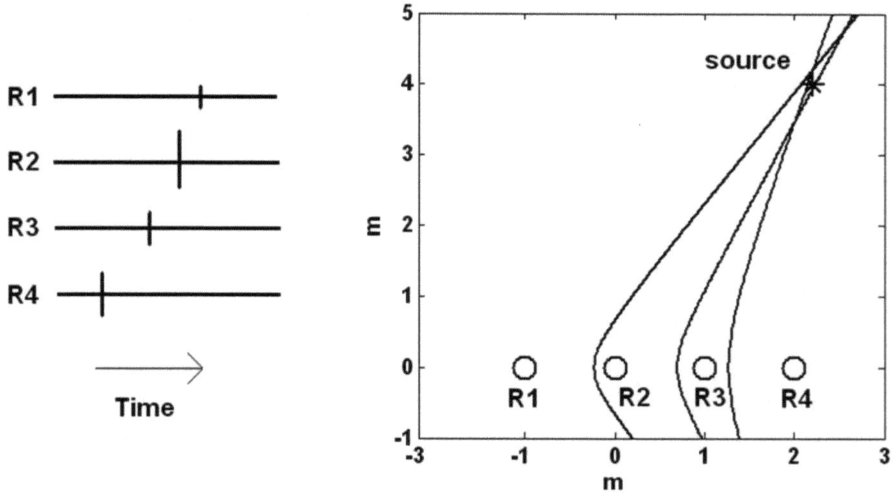

Fig. 4.1 Acoustic localization in two dimensions. The acoustic source is recorded with four receivers. To the *right* are shown the hyperbolas generated from the time-of-arrival differences of the signals recorded at the various receivers (shown to the *left*)

for example, T2–T1, T3–T1 and T4–T1. Any other TOAD between two receivers is a linear combination of these three and therefore does not contain any extra information.

Each TOAD restricts the possible location of the source to a hyperbolic curve, having its axis in the direction of the line connecting the two receivers. With one more receiver, another hyperbolic curve can be generated, and the source is ideally restricted to the intersection of the two curves. In some cases, the hyperbolas may cross in two points, making the source location ambiguous. With the help of additional receivers, additional hyperbolas can be plotted, both solving ambiguities and giving information on the precision in the derived source coordinates (Fig. 4.1). The hyperbolas in Fig. 4.1 should ideally all meet in one point, but measurement errors result in their intersections spreading out.

In three dimensions, each TOAD results in a hyperboloid surface, generated by a hyperbola rotated around its axis of symmetry (which is identical to the axis through the two receivers involved in the TOAD measurement). Two hyperboloids, generated by the signal recorded at three receivers, restrict the source to the intersection (a line) of the two surfaces. One more hyperboloid (i.e., all in all three surfaces or four receivers) intersecting this line is needed to determine the location of the sound source in 3D. Just as in the 2D case, the source location may be ambiguous in some cases. For example, if all four receivers are located on a straight line, the source is not restricted to a point but to a circle in space. Again, the solution is to use more receivers, generating more hyperboloids, solving ambiguities and giving more confidence in the derived source location.

The location of the source may be found either by graphic techniques or with linear algebra (Madsen and Wahlberg 2007). It can be shown that, just as inferred from the graphic solutions above, that for the 2D case at least three receivers are needed to algebraically solve for the two source coordinates, and in the 3D case that at least four receivers are needed. An array consisting of the smallest number of receivers needed to solve for the source location coordinates is denoted a MINNA ("minimum number of receivers array"); if more receivers are added the array is denoted an ODA ("over-determined array").

4.2.2.3 Precision in Source Localization

Adding receivers gives confidence to the derived source locations and can help assess the directionality of the signals. The quality of the assessed location of the acoustic source depends on several factors. The precision with which the TOADs between the different channels are measured is an important factor, as is the precision of the coordinates of the receivers. The TOADs may be measured by cross-correlating the receiver channels. This is prone to errors though, especially in situations where surface, ground, and other reflections are involved. The sound speed profile may have an additional effect on the localization performance, but at the ranges of interest here it will be only minor. Finally, the source-to-array geometry will have an extremely large effect on the localization precision. The intersecting hyperbolas or hyperboloids must intersect with angles as close to 90° as possible to give the optimal localization precision. In some areas around the array, the hyperbolas may become almost parallel, giving extremely erroneous source coordinates. There is a trade-off between making the array larger, which in general gives steeper intersects, and keeping the array sufficiently small to be able to handle and having control over the relative locations of the receivers, as well as allowing for a sufficient signal-to-noise ratio of the signals recorded on all the receivers. Placing receivers far apart may also affect the frequency content of the signals recorded on each channel. If, for example, a bat is flying parallel to the main axis of an array it may approach one receiver while flying away from another, resulting in different received signals due to directionality of the emitted signal as well as Doppler shifts caused by the bat's velocities relative to that of the receivers (Surlykke and Kalko 2008).

The optimal array geometry depends on which coordinates are desirable to obtain, how many receivers are available and which type of array system is practical to handle. In the case of recordings of source levels, linear or flat (2/3D) arrays are interesting choices (Table 4.1). The linear array will produce a range from the source to each receiver, thus making it possible to back-calculate to the source level. However, it is not possible to derive the exact 3D location of the bat or toothed whale with a linear array. Further, it can be difficult to ensure that the signals are recorded on the acoustic axis, which can be solved by adding an additional receiver off the array axis (Surlykke et al. 2009). With a flat 2/3D shaped array we obtain the 3D location of the animal, with an ambiguous location behind the array that can usually be ruled out. One drawback with such an array is that the localization precision is decreasing drastically when moving away from the axis perpendicular to the array plane.

Table 4.1 Different features of some common receiver array geometries for recording bat and toothed whale echolocation signals

Number of receivers	Geometry	Array name	Number of source coordinates	Examples of source coordinates
1	Point	Single	–	–
2	Pair	Stereo	1	(Bearing)[a]
3	Line	Linear	2	(*x*, *y*)[a], range
3	Plane	2D MINNA[b]	2	(*x*, *y*, bearing, range)[a]
>3	Plane	2D ODA[c] or 2/3D[d]	>2	x, *y* (*z*, bearing, range)[a]
4	Volume	3D MINNA[b]	3	(*x*, *y*, *z*, bearing, range)[a]
>4	Volume	3D ODA[c]	>3	*x*, *y*, *z*, bearing, range

[a]Parentheses in the right column indicate that coordinates are not unambiguous (i.e., for a pair of receivers, the bearing to the sound source can be rotated around the receiver axis)
[b]Minimum number of receivers array
[c]Overdetermined array
[d]Planar (2D) array geometry that can localize (at least partly) in 3D

4.3 Metrics

There are several ways to measure the sound pressure level of a transient signal (Fig. 4.2). The most direct and easy way is to measure the signal from its maximum to minimum level, the so-called peak-to-peak (pp) measure. However, this measure will usually drastically overestimate the actual sound pressure received and perceived by an animal. A standard averaging technique is the root mean square (RMS), which is the squared pressure averaged over a certain time window. The RMS measure is the time-averaged intensity of the signal and is related to the loudness perceived by the animal. The time window used for RMS averaging can be defined in many ways. A popular way is to use the absolute value of the analytical signal, called the envelope (Proakis and Manolakis 1991). The peak of the envelope is identified, and the interval between the points where the amplitude of the envelope has dropped with 3 or 10 dB relative the peak defines the "−3 dB" or "−10 dB" duration (Fig. 4.2). However, if the envelope has several peaks this measure may be ambiguous. Sometimes a more robust method is to calculate the "95 % energy window," also denoted τ_{95}. Here the signal is first squared, and the cumulative sum of the signal is calculated at each time sample. The interval between 2.5 and 97.5 % of the maximum of the cumulative sum is defined as the time window used to calculate the RMS (Fig. 4.2). Another well-defined measure is the RMS duration (τ_{RMS}), which is the standard deviation of the envelope function (Au 1993). The drawbacks of all time measurements are that they depend critically on good signal-to-noise ratios for the recordings. Also, both the τ_{95} and the τ_{RMS} measures depend on the size of the time window over which the cumulative energy function (for estimating τ_{95}) or the squared sample product (for estimating τ_{RMS}) is calculated. Especially for low signal-to-noise ratios this time window must therefore be accurately described to allow comparisons between different studies.

Presumably most or all animal ears are energy integrators. Thus, rather than only measuring the intensity of the signal its energy density should also be quantified.

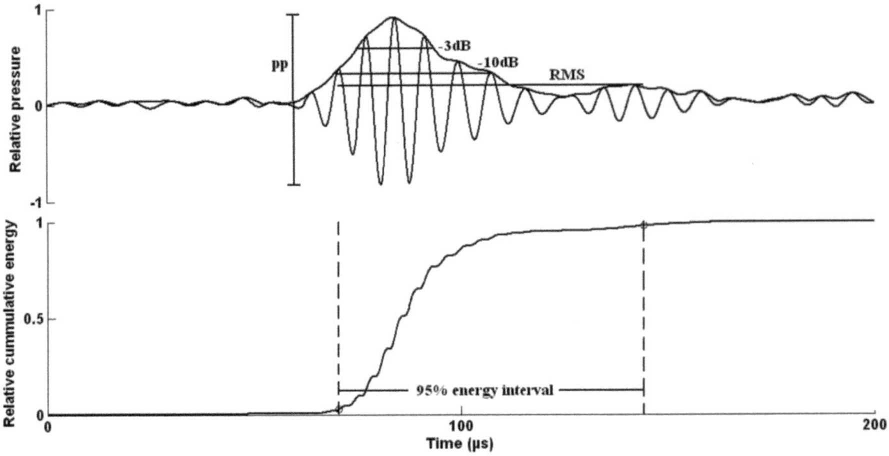

Fig. 4.2 Measuring the peak to peak (pp) and RMS levels of a harbor porpoise echolocation signal. The *solid line* right above the signal is the magnitude of the analytical signal (the envelope of the signal). Three duration measures are indicated, indicated by solid horizontal lines: the duration of the part of the signal with an envelope larger than –3 dB and –10 dB relative the peak of the envelope function, and the duration containing 95 % of the accumulated energy of the signal. The RMS level is calculated over the 95 % accumulated energy duration

Table 4.2 Examples of source level calculations from a bat and a dolphin, in various units and metrics

Unit	Bat (at 10 cm range)	Dolphin (at 1 m range)
Acoustic pressure	10 Pa RMS	10 kPa RMS
Acoustic intensity	0.24 W m^{-2}	67 W m^{-2}
RMS sound pressure level	114 dB re 20 μPa RMS	200 dB re 1 μPa RMS
pp sound pressure level	123 dB re 20 μPa pp	215 dB re 1 μPa pp
Sound energy	94 dB re 1 μPa^2 s	156 dB re 1 μPa^2 s
Duration	10 ms	40 μs

In decibel units, the energy can be found from adding the decibels derived with the RMS metric (in dB re 1 μPa in water and dB re 20 μPa in air) with 10 log (duration). The unit becomes dB re 1 $\mu Pa^2 s$ or dB re 20 $\mu Pa^2 s$. For example, a bat signal with a source level of 120 dB re 20 μPa RMS and a duration of 50 ms will have an energy content of $120 + 10 \log(0.05) = 107$ dB re 20 $\mu Pa^2 s$.

Table 4.2 illustrates that for a typical bat or dolphin signal, the various intensity measures may lead to variations in the reported sound levels of many tens of decibels. It is therefore important to make clear how the reported levels have been calculated, and in which units they are given. It is also clear from Table 4.2 that decibel units are, in general, much larger in water than in air. There are several reasons for this. First, water is less compressible than air, so for the same acoustic intensity, the acoustic pressure (and therefore also the sound pressure level) will be much larger in water. The difference in units of decibels is 36 dB (see Madsen & Surlykke,

Chap. 9, for details). Also, in water and air different reference units are usually used, that is, 1 and 20 μPa respectively, giving an additional difference of 26 dB. Thus, in total 36+26=62 dB should be added to the decibels values for intensities in-air when comparing with sound intensities in water. In some cases we want to compare the acoustic pressures rather than the intensities (e.g., Kastak and Schusterman 1999). In such cases the compensation should be only the 26 dB corresponding to the different reference pressures used in air and in water. In addition, when comparing source levels 20 dB has to be subtracted from bat source levels when comparing them to the ones from toothed whales, as the reference distance is only 10 cm for bat source levels but 1 m for toothed whale source levels. Thus, for a bat source level (at 10 cm) one should add 26 dB to compensate for the difference in reference pressures, add another 36 dB to compensate for the different acoustic impedances, and subtract 20 dB to compensate for the different reference distances. In all 26+36−20=42 dB should be added to the bat source level to compare its acoustic intensity to the one of a toothed whale.

Another important descriptor of the biosonar system of an animal is the signal directionality. When back-calculating from received level to source level using the assumption of spherical spreading and absorption, one may obtain very different levels in the various directions from the source to the receiver. This is usually because the bat or toothed whale emits signals in a directional beam. The width of the beam can be calculated in many ways. A common method is to measure the angle between the points where the intensity of the beam has fallen by 10 dB on either side of the direction of maximum intensity. In bats the −10 dB beam width is usually around 90°, whereas in dolphins it can be less than 8°. Thus, for each click, the back-calculated source level highly depends on the relative direction from the sound source to the observer. Usually one cannot be sure that the animal is recorded on its acoustic axis, especially when working in the field. One may assume that the loudest back-calculated sound level hitting the various receivers is the one most representative of the actual source level, but this may not be true as the animal usually modulates its emission within a click sequence. Beedholm and Møhl (2006), working on sperm whale data, and Guarato et al. (2011), working on bats, have tried to derive the clicks close to the acoustic axis using the spectral properties of the recorded signals. This approach may be the start of finding more robust methods to obtain information on whether or not a signal has been recorded on (or at least very close to) the acoustic axis (see also Mogensen and Møhl 1979).

Besides measuring the beam width one may calculate the directionality index, DI, which is the ratio (converted to decibels) between the total power radiated by an omnidirectional reference source that has the same on axis intensity as the source and the total power radiated from the sound source. Assume that we have measured the beam pattern of an animal and normalized it so that the intensity level is 1 at 0° in front of the animal. Then the DI can be estimated as:

$$DI = 10\log_{10}\frac{2}{\sum_i b_i \sin(v_i)\Delta v},$$

where b_i is the value of the ith bin of the beam pattern (defined as the normalized intensity on a linear scale in the direction v_i), the v_i is the angle between the on-axis direction and the direction to the ith measurement of the beam pattern, and Δv is the difference between two consecutive angular measurements of the beam pattern. The beam pattern is assumed to be rotational symmetric and should be sampled from 0 to 180° for the calculation of DI.

Both direct measurements of the beam width and the directionality index are important for the complete description of the beam pattern of a signal from an echolocating animal, and usually it is easy to convert from one to the other. The beam width is used when discussing how much off-axis information the animal receives while echolocating (Jakobsen and Surlykke 2010). The directionality index is essential when discussing the limitation of background noise to an animal's sonar (Møhl et al. 2003).

4.4 Source Levels and Directionality from Bats and Toothed Whales

The source level and directionality of bat and whale echolocation signals have been estimated both in the laboratory and in the field. For toothed whales, it seems that we have a good overall picture of what kinds of signals are emitted by many and perhaps most of the species, whereas our information of source levels and directionality is limited to a few species. For bats, we have only limiting data regarding the extreme variation in signal types, and we are still scratching the surface with a handful of species where source levels and directionality have been described among the more than 1,100 echolocating species.

Table 4.3 summarizes measurements of the source levels of bat sonar signals. Maximum source levels range from 96 to 144 dB re 20 μPa RMS at 10 cm. Even though a single bat can modulate the source level tremendously, there still seems to be a clear distinction between various bat species as to which levels of sound they are actually using. It is hypothesized that there is a correlation between habitat type and source level, with bats hunting in open air using higher levels (Brinkløv et al. 2010), as there are few unwanted echoes disturbing the bat's acoustic view. One may also assume that the source levels of various species are adapted to the size of the bat and the prey choice, but there are not yet enough data to demonstrate any clear pattern in this.

In Table 4.4, measurements of the signals from various species of toothed whales are grouped according to the four signal shapes described by Fenton, Jensen, Kalko, and Tyack (Chap. 2). A clear grouping in terms of source levels is evident. Some of the smaller animals, such as the porpoises and the dolphins of the genus *Cephalorhynchus*, emit relatively weak source levels of 153–205 dB re 1 μPa pp at 1 m. Larger animals, such as most delphinids, beaked whales, and the narwhal, emit signals with a source intensity of 160–228 dB re 1 μPa pp at 1 m. The sperm whale (*Physeter macrocephalus*), the largest of all toothed whales, emits clicks of

Table 4.3 Source properties of echolocation calls from microchiropteran bats, recorded in the field

Group	Species	Family	Max. ASL[a] dB SPL at 0.1 m	Peak frequency (kHz)	−10 dB BW (kHz)	Double −6 dB beamwidth	Reference
Gleaning bats "Whispering"	*Carollia perspicillata* Short-tailed leaf-nosed bat	Phyllostomidae	99^{RMS}	90	48	32° at 90 $\mathrm{kHz}^{\mathrm{L,Fly}}$	Brinkløv et al. (2011)
	Phyllostomus hastatus Greater spear-nosed bat	Phyllostomidae	112^{RMS}	55			Nørum et al. (2012)
	Phyllonycteris poeyi Cuban flower bat	Phyllostomidae	106^{RMS}	75			Nørum et al. (2012)
	Artibeus jamaicensis Jamaican fruit-bat	Phyllostomidae	110^{RMS}	79			Brinkløv et al. (2009)
Aerial insectivores	*Eptesicus bottae* Botta's serotine bat	Vespertilionidae	133^{pp}	33	9^{x}		Holderied et al. (2005)
	Eptesicus serotinus Serotine bat	Vespertilionidae	126^{RMS} 128^{pp}	26	7–17		Jensen and Miller (1999) Holderied and von Helversen (2003)
	Eptesicus nilsonii Northern bat	Vespertilionidae	$124^{\mathrm{c,pp}}$	26			Holderied and von Helversen (2003)
	Eptesicus fuscus Big brown bat	Vespertilionidae				60° at 30 $\mathrm{kHz}^{\mathrm{L,Fly}}$	Ghose and Moss (2003)
	Lasiurus ega Southern Yellow bat	Vespertilionidae	123^{RMS}	32			Surlykke and Kalko (2008)
	Pipistrellus pygmaeus Pygmy Pipistrelle	Vespertilionidae	$127^{\mathrm{c,pp}}$	55			Holderied and von Helversen (2003)

(continued)

Table 4.3 (continued)

Group	Species	Family	Max. ASL[a] dB SPL at 0.1 m	Peak frequency (kHz)	−10 dB BW (kHz)	Double −6 dB beamwidth	Reference
	Pipistrellus pipistrellus Common pipistrelle	Vespertilionidae	125[c,pp]	45			Holderied and von Helversen (2003)
	Pipistrellus kuhlii Kuhl's pipistrelle	Vespertilionidae	125[c,pp]	39			Holderied and von Helversen (2003)
	Pipistrellus nathusii Nathusius' pipistrelle	Vespertilionidae	128[c,pp]	39			Holderied and von Helversen (2003)
	Pipistrellus abramus Japanese house bat	Vespertilionidae	130[pp]	56			Hiryu et al. (2007)
	Nyctalus leisleri Lesser noctule bat	Vespertilionidae	124[c,pp]	26			Holderied and von Helversen (2003)
	Nyctalus noctula Noctule bat	Vespertilionidae	128[c,pp]	20			Holderied and von Helversen (2003)
	Nyctalus lasiopterus Greater noctule bat	Vespertilionidae	131[c,pp]	18			Holderied and von Helversen (2003)
	Miniopterus schreibersi Common bent-wing bat	Vespertilionidae	126[c,pp]	51			Holderied and von Helversen (2003)
	Hypsuga savii Savi's pipistrelle	Vespertilionidae	121[c,RMS]	35			
	Craesonycteris thonglongyai Bumblebee bat, Hognosed bat	Craseonycteridae	115[c,RMS]	73	12 (−20 dB)		Surlykke et al. (1993)

Molossus molossus Velvety free-tailed bat	Molossidae	122^{RMS}	36, 43, 46[d]			Surlykke and Kalko (2008)
Saccopteryx bilineata Greater sac-winged bat	Emballonuridae	134^{RMS}	45, 48[d]			Surlykke and Kalko (2008)
Saccopteryx leptura Lesser sac-winged bat	Emballonuridae	133^{RMS}	53, 56[d]			Surlykke and Kalko (2008)
Cormura brevirostris Wagner's sac-winged bat	Emballonuridae	127^{RMS}	25, 28, 31[d]			Surlykke and Kalko (2008)
Centronycteris centralis Thomas' shaggy bat	Emballonuridae	128^{RMS}	44			Surlykke and Kalko (2008)
Pteronotus gymnonotus Big naked-backed bat	Mormoopidae	131^{RMS}	55			Surlykke and Kalko (2008)
Pteronotus parnelli Mustache bat	Mormoopidae	96^{pp}	60		60° at 60 $kHz^{L,S}$	Henze and O'Neill (1991)
Hipposideros terasensis Taiwanese leaf-nosed bat	Hipposideridae		70		100–140° at 70 $kHz^{L,S}$	Hiryu et al. (2006)
Rhinolophus ferrumequinum Greater Horseshoe bat	Rhinolophidae		82		48° at 82 $kHz^{L,S}$	Schnitzler and Grinnell (1977)
Rhinolophus ferrumequinum Greater Horseshoe bat	Rhinolophidae	125^{RMS}	82			Tian and Schnitzler (1997)

(continued)

Table 4.3 (continued)

Group	Species	Family	Max. ASL[a] dB SPL at 0.1 m	Peak frequency (kHz)	−10 dB BW (kHz)	Double −6 dB beamwidth	Reference
	Rhinolophus mehelyi Mehely's horseshoe bat	Rhinolophidae	116[L]	108			Schuchmann and Siemers (2010)
	Rhinolophus blasii Blasius' horseshoe bat	Rhinolophidae	118[L]	95			Schuchmann and Siemers (2010)
	Rhinoplophus euryale Mediterranian horseshoe bat	Rhinolophidae	108[L]	106			Schuchmann and Siemers (2010)
	Rhinolophus hipposideros Lesser horseshoe bat	Rhinolophidae	119[L]	109			Schuchmann and Siemers (2010)
Trawling bats	*Macrophyllum macrophyllum* The long-legged bat	Phyllostomidae	111[RMS]	50	58		Brinkløv et al. (2010)
	Myotis daubentonii Daubenton's bat	Vespertilionidae	119[b,RMS]	45		20° at 55 kHz[F,Fly]	Surlykke et al. (2009a)
	Noctilio leporinus Greater bulldog bat	Noctilionidae	143[RMS]	56			Surlykke and Kalko (2008)
	Noctilio albiventris Lesser bulldog bat	Noctilionidae	144[RMS]	70			Surlykke and Kalko (2008)

ASL apparent source level (for a definition see text), *BW* bandwidth, *SPL* sound pressure level
[a]The table gives maximum ASL, when possible, but some references have only average value ([b]), and some average of upper 10 % ([c])
[d]Some bat species shift between two or three frequencies from call to call
[L]Recorded in the laboratory
[F]Recorded in the field
[Fly]Flying around during measurements
[S]Sitting restrained during measurements
[RMS]Measured using root-mean square
[pp]Measured using peak to peak
[x]−15 dB Bandwidth

Table 4.4 Some properties of signals from toothed whales recorded in the field

	Species	Latin name	ASL dB re 1 μPa pp at 1 m	Energy −10dB (dB re 1 μPa²s at 1 m)	Centroid frequency (kHz)	BW −3 dB (kHz)	Duration −10 dB (μs)	Beam width	DI (dB)	References
NBHF										
	Harbor porpoise	*Phocoena phocoena*	169–205	114–150	126–148	5–36	35–113	9.5–11°$^{-3\,dB}$	22.1–25.6	Au et al. (1999); Villadsgaard et al. (2007); Kyhn et al. (2013); Koblitz et al. (2012)
	Finless porpoise	*Neophocaena phocaenoides*			128	11	127[a]			Kamminga et al. (1988); Li et al. (2005, 2007)
	Dall's porpoise	*Phocoenoides dalli*	153–203	104–150	121–147	3–23	53–251	9.9°$^{-3\,dB}$	25.2	Kyhn et al. (2013)
	Hector's dolphin	*Cephalorhynchus hectorii*	161–187	110–126	125–132	12–26	41–65			Kyhn et al. (2009)
	Commerson's dolphin	*C. commersonii*	165–190	111–137	123–137	16–31	52–138		25	Kyhn et al. (2010)
	Heaviside's dolphin	*C. heavisidii*	161–186	108–135	121–130	6–21	53–115			Morisaka et al. (2011)
	Hourglass dolphin	*Lagenorhynchus cruciger*	161–187	140–152	124–132	5–11	179–176			Kyhn et al. (2009)
	Peale's dolphin	*Lagenorhynchus australis*	169–196	117–144	123–138	9–34	65–153		25	Kyhn et al. (2010)
	Pygmy sperm whale[b]	*Kogia breviceps*	175	136[c]	129	6	120[d]			Madsen et al. (2005a)

(continued)

Table 4.4 (continued)

	Species	Latin name	ASL dB re 1 μPa at 1 m pp	Energy density −10 dB (dB re 1 μPa^2 s)	Centroid frequency (kHz)	BW −3 dB (kHz)	Duration −10 dB (μs)	Beam width	DI (dB)	References
Broadband HF										
Delphinids	Bottlenose dolphin	*Tursiops truncatus and T. aduncus*	177–228	122–175	45–109	23–43[f]	35–45	10–21°[−10 dB]	26–29	Au (1974, 1993); Wahlberg et al. (2011a)
	Rissos dolphin	*Grampus griseus*	202–222	147–166	58–91	15–84	30–75			Madsen et al. (2004a)
	Whitebeaked dolphin	*Lagenorhynchus albirostris*	194–219		94		10–30	10°[−10 dB]	29	Rasmussen et al. (2002, 2004)
	Dusky dolphin	*Lagenorhynchus oscurus*	160–210		60–120	5–100	<70			Au and Wursig (2004)
	Long-snouted spinner dolphin	*Stenella longirostris*	222				31			Schotten et al. (2003)
	Pantropical spotted dolphin	*Stenella attenuata*	220				43			Schotten et al. (2003)
	Spotted dolphin	*Stenella frontalis*	180–223		25–150	0–80				Au and Herzing (2002)
	Killer whale	*Orcinus orca*	173–226	130–171	21–80	8–58	31–203			Au et al. (2004); Simon et al. (2007); Eskesen et al. (2011)
	False killer whale	*Pseudorca crassidens*	201–225	145–168	33–68	15–76	18–55			Madsen et al. (2004a)
	Melon-headed whale	*Feresa attenuata*	197–223	130–165	70–85		20–40			Madsen et al. (2004b)
	Long-finned pilot whale	*Globicephala melas*	189–202	132–145	37–73	24–71	18–40			Eskesen et al. (2011)
	Narwhal	*Monodon monoceros*	227			24–95[e]	29–45			Møhl et al. (1990); Miller et al. (1995)

	Beluga whale	*Delphinapterus leucas*	206–222	148–165		15–65		$6–10^{\circ -10\,dB}$	32	Au et al. (1985, 1988); Au (1993)
	Irrawaddy dolphin	*Orcaella brevirostris*	189–199	131–142	70–109	40–91	10–21			Jensen et al. (2013)
	Ganges river dolphin	*Platanista gangetica gangetica*	175–189	118–132	54–72	32–62	17–26	$14.5^{\circ -3\,dB}$		Jensen et al. (2013)
Upsweep										
	Blainville's beaked whale	*Mesoplodon densirostris*			35–42	19–33[e]	219–321			Johnson et al. (2006)
		Buzz click			45–56	50–58[e]	73–120[4]			Johnson et al. 2006
	Cuvier's beaked whale	*Ziphius cavirostris*	214	164	42	12	>25	$4^{\circ -3\,dB}$	25	Zimmer et al. (2005)
	Gervais' beaked whale	*Mesoplodon europaeus*			30–50	10	200			Gillespie et al. (2010)
	Bottlenose whale	*Hyperoodon ampullatus*	192–202	140–165	27–51	7–24	232–381			Wahlberg et al. (2011b)[h]
Broadband LF										
	Sperm whale	*Physeter catodon*[g]	230–240	186–196	13–20	2.6–12.5[f]	114	$4^{\circ -3\,dB}$	>27	Møhl et al. (2000, 2003); Møhl et al. (2006)

ASL apparent source level, *DI* Directionality Index, *NBHF* narrow B and high-frequency signals

[a]Duration measured as the interval in which the signal is discernable in the background noise

[b]Animal recorded in captivity

[c]Energy calculated over a 95 % energy time window

[d]Duration measured as 95 % energy content

[e]Bandwidth measured over 10 dB

[f]RMS bandwidth

[g]Also known as *Physeter macrocephalus*

[h]The energy levels were corrected in: Wahlberg 2012: Erratum, Journal of the Acoustical Society of America 131(4), 3181

extremely high intensities, up to 240 dB re 1 μPa pp at 1 m. Thus, there seems to be a clear correlation between source level and animal size in toothed whales.

Both bats and dolphins can emit calls of a whole range of intensities. It can be difficult to appreciate how loud these signals can be (Tables 4.3 and 4.4). At very close range the signals may have intensities close to or way over the threshold of pain in humans. As a matter of fact, if humans could hear ultrasound we would probably find it extremely annoying to walk around at dusk in areas where bats hunt, listening to their explosive vocalizations as they fly by. The same definitely holds true for dolphins. Dolphin source intensities are sometimes well beyond 200 dB re 1 μPa at 1 m range. Assuming spherical spreading, and transferring the decibels from water to air while preserving the energy content of the signal, allow us to estimate the sound level in air at 10 m distance. 200 dB re 1 μPa at 1 m range would correspond to $200-62-20=118$ dB re 20 μPa at 10 m in air, which is close to the threshold of pain for the hearing system in humans in air (see Madsen & Surlykke, Chap. 9). Humans do note hear these signals very well as they mainly contain ultrasonic frequencies, and also because in general we do not hear very well under water; in addition, comparing in-air and underwater decibels is never unproblematic. Still this example illustrates just how loud the source levels of dolphins are in the context of human hearing.

In contrast to the noisy echolocating species we usually encounter and record, there are also species or groups of both toothed whales and bats that can be extremely quiet, even while hunting. Killer whales hunting marine mammals rarely vocalize, as their prey have very sensitive underwater hearing (Barrett-Lennard et al. 1996). A quiet strategy may be useful for the echolocator to avoid being detected by either prey or predators around (Goerlitz et al. 2010). For other species, such as so-called whispering bats, emitting signals of low intensity may be used as an echolocation strategy to avoid receiving too many unwanted echoes. To record quieter animals can be very challenging, and it is quite possible that the present data are skewed toward results obtained from the louder animals.

Measurements of directionality of echolocation signals are given for bats in Table 4.3 and toothed whales in Table 4.4. For bats, measurements of DI are not included in Table 4.3, as there have only been a few measurements on Daubenton's bats (*Myotis daubentonii*), which has a half-power beam width of 20–40° and a DI between 16 and 11 dB, measured at 55 kHz in field and lab respectively (Surlykke et al. 2009). When comparing smaller and larger species of both bats and toothed whales, there are theoretical reasons to believe that the larger animals would have a more acute directionality. This does not seem to be the case in the studies made of bats so far. Directionality has been assessed in six bat species (Table 4.3). It seems that bats manipulate the signal type so that the bats share a somewhat similar beam pattern. It may therefore look as if it is extremely important for the bats across various species to achieve a certain transmission directionality in their signals, possibly to avoid echoes from unwanted clutter (Jakobsen et al. 2013).

For toothed whales, on the other hand, there seem to be a correlation between animal size and the directionality of the signals. The smallest animals, such as harbor porpoises (*Phocoena phocoena*), have a rather broad directionality with a directionality index of about 22–26 dB (Au 1993; Koblitz et al. 2012). The directionality of larger delphinids and the beluga ranges from 26 to 29 dB. The largest toothed

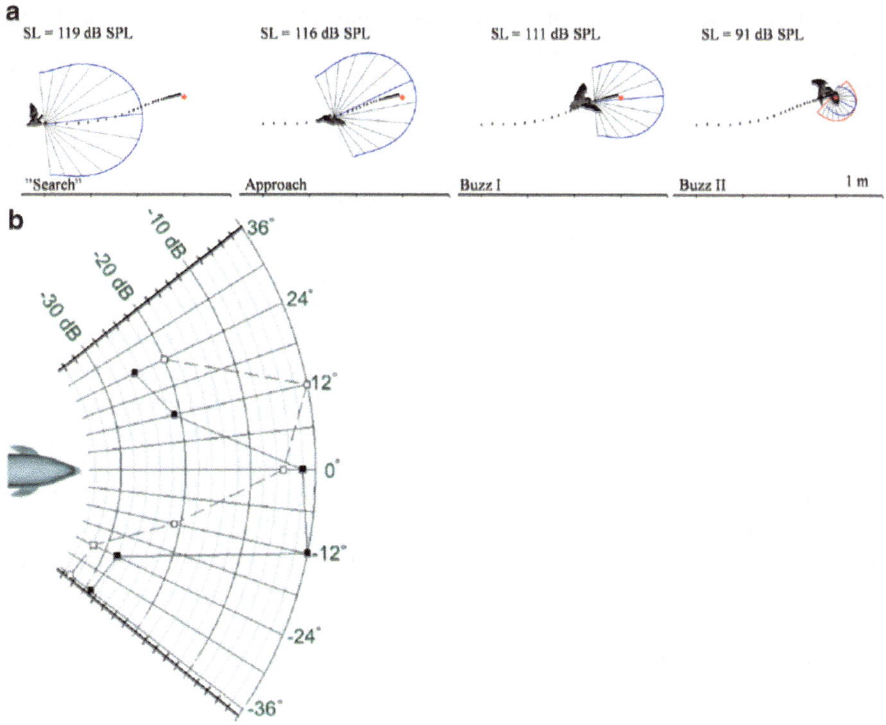

Fig. 4.3 Examples of the dynamics of signal directionality. (**a**) A bat homing in on a prey is expanding its acoustic beam (*blue* vs. *red* beam pattern; Jakobsen and Surlykke 2010). (**b**) A toothed whale is changing the direction of its sound beam during a target detection experiment with target presented off the acoustic axis (Moore et al. 2008)

whale, the sperm whale, seems to emit clicks with an extremely narrow beam pattern with a directionality index of at least 27 dB but possibly up to 39 dB (Møhl et al. 2003, 2006).

There are obvious advantages with using highly directional signals for echolocation. With a broader transmission beam the animal could expect receiving echoes from many directions around it. Therefore it would need to invest hugely in perceptual processing to discern the direction to the target. Knowing that all received echoes come from a certain direction probably also makes it perceptually easier for the animal to understand the auditory scene around it (Moss et al. 2011). In situations where the perception of echoes is hampered by clutter increasing the directionality will restrict the amount of received clutter. In addition, focusing the sound energy in a certain direction makes it possible to achieve higher source levels in that very direction. There are also drawbacks with using very directional sonar pulses. The animal's field of view is directly related to the beam width of the outgoing signal. Therefore, with a very directional signal the animal may fail to notice information of relevant targets slightly off-axis with respect to its sonar beam.

New evidence shows that both bats and dolphins can vary the directionality of the signals (Fig. 4.3). The beam width can be varied without changing the frequency content of the signals, for bats probably by changing the diameter of the emitter, that is, the degree of opening of the mouth (Surlykke et al. 2009), and for toothed whales by other, still unknown, means (Moore et al. 2008; Starkhammar et al. 2011). The beam can also be widened by lowering the frequency content of the signals as the animal is approaching the target (Jakobsen and Surlykke 2010), but at least for one species of nose-emitting rhinolophid bats thi scan also be accomplished without changing frequency by unknown mechanisms (Matsuta et al. 2013). The direction or aim of the beam can be changed by head movements (Ghose and Moss 2003; Ghose et al. 2007) or by other means (Moore et al. 2008). There are many reasons why this makes good sense. At close range, the animal's field of view would be more restricted by a narrow beam than at larger distances, and it would therefore be easier for the animal to miss the relevant target. Fanning out the beam comes at the price of receiving more clutter and being able to produce less intense signals.

More field data on directionality are needed to indicate "optimal" directionality for sonar signals. It probably depends critically on the context. It is likely that very directional signals are suitable for long-distance open range echolocation (Møhl et al. 2003; Surlykke et al. 2009). At close range a broad beam may be best (Ghose and Moss 2003; Jakobsen and Surlykke 2010), but in very cluttered surroundings a narrow beam may be a strategy to improve target-echo relative to the clutter level (Brinkløv et al. 2011; Vanderelst et al. 2010). The ability to vary the directionality of the signal has obvious advantages.

4.5 Modulation of the Source Level

The more intense signals an echolocating animal is producing, the more energy will return in the form of an echo, hence increasing the animal's probability of detecting the target. This would indicate that bats and toothed whales should use as loud signals as possible when echolocating. However, the optimal source level depends on a number of factors, such as the range to the target, the level of reverberation and clutter, and the hearing abilities of prey. There is a large variation in the actual source levels emitted, both for different species (Tables 4.3 and 4.4) and for the same individual animal, depending on the situation. In general, the animal drastically decreases the source level in the final stages of approaching the target (Fig. 4.4). There may be many reasons for this, such as improving target detectability and classification as well as to avoid the prey detecting the approaching predator. By modeling the influence of these factors insights can be gained into the advantages of choosing a specific source level under certain circumstances.

4.5.1 The Sonar Equations

To understand the intensity relationship between the outgoing signal and the returning echo, the sonar equation is very useful. In its simplest form:

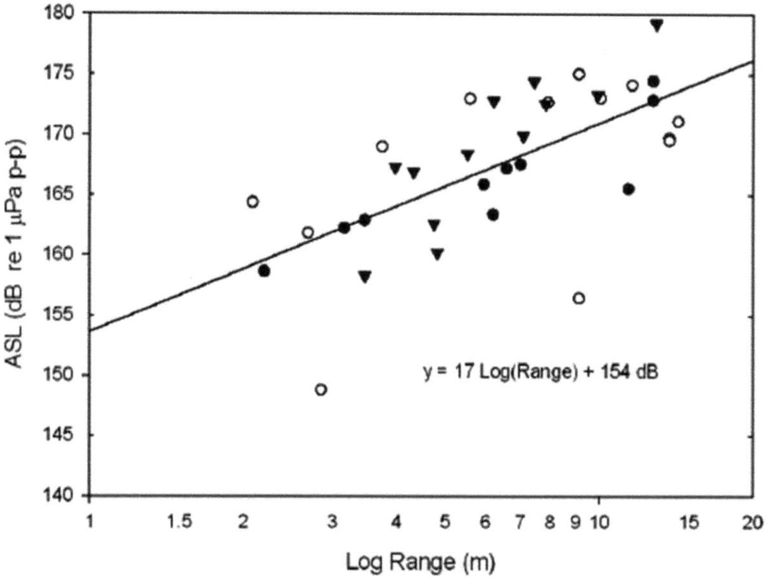

Fig. 4.4 The source level of echolocation clicks emitted by three harbour porpoises (each individual indicated by a different symbol) approaching a dead fish in a net pen (Atem et al. 2009)

$$\mathrm{RL} = \mathrm{SL} - \mathrm{TL},$$

where the received level (RL, in dB re 20 or 1 μPa) is the sound pressure level received at a location r meters away from the sound source, and SL is the source level in dB re 20 μPa at 0.1 m for bats and dB re 1 μPa at 1 m for toothed whales. TL is the transmission loss.

A sound pulse traveling from the source to the target is first reduced by an amount TL before it reaches the target. The amount of sound reflected from the target to the receiver is determined by the target strength, TS. The TS is ten times the logarithm of the fraction of the echo intensity (denoted I_{echo}) returning from distance of the target of 10 cm (for bats) or 1 m (for toothed whales), and the intensity impinging upon the target ($I_{\mathrm{impinging}}$, in decibel units denoted $\mathrm{RL}_{\mathrm{impinging}}$):

$$\mathrm{TS} = 10\log_{10}\frac{I_{\mathrm{echo}}}{I_{\mathrm{impinging}}} = 20\log_{10}\frac{p_{\mathrm{echo}}}{p_{\mathrm{impinging}}} = \mathrm{SL}_{\mathrm{echo}} - \mathrm{RL}_{\mathrm{impinging}} = \mathrm{SL}_{\mathrm{echo}} - (\mathrm{SL} - \mathrm{TL}).$$

The target strength is usually a complex function of the frequency content and duration of the impinging sound and is usually also varying in different directions to the target. Sometimes it is instructive to obtain some simple rules of thumb and work from more simple shapes to more complex ones. The target strength of spheres and cylinders for typical bat and toothed whale frequencies are given in Fig. 4.5. In this figure it can be seen that typical target strengths for prey sizes for bats are around −10 to −20 dB (re 10 cm; as has been confirmed by measurements in

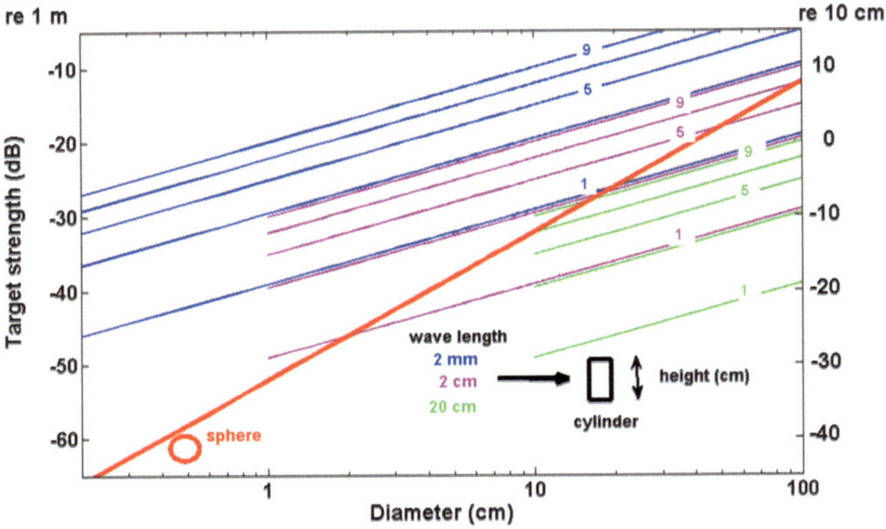

Fig. 4.5 Target strength (re 1 m and 10 cm) for simple geometric shapes, such as a sphere (*red line*, assuming a sphere diameter >>λ/π, where λ is the wavelength of the impinging signal) and cylinders of different lengths (1, 3, 5, 7 and 9 cm) for different wavelengths (λ=2 mm for *blue*, λ=2 cm for *magenta*, and 20 cm for *green lines*, respectively) and for diameters larger than λ/π

Surlykke et al. 1999), and for toothed whales they usually range from −50 to −30 dB (re 1 m, confirmed by measurements by, e.g., Madsen et al. 2007 and Au et al. 2010), except for mammal-eating toothed whales, such as some killer whales, whose prey may be in the −20 to +10 dB re 1 m range.

Thus, the echo intensity is reduced by the target strength, but in addition the echo is reduced on the way back by TL if the target is small enough to be a "point target." In this case, then the echo will be reradiated from the target as if the target were a small sound source. The radiation of energy will be efficient if the target's effective radius (a) is large compared to the wavelength. In mathematical terms this means that $2\pi fa/c > 1$, where f is frequency and c speed of sound in the medium. This is usually the case for both bat and toothed whale prey. Consider, for example, a bat emitting a 20-kHz tone and echolocating a 5 mm radius insect. Here one gets $2\pi fa/c = 2\pi 20 \cdot 10^3 \cdot 5 \cdot 10^{-3}/340 \approx 1.8$. As another example, take a sperm whale emitting 15-kHz clicks echolocating a fish with no swim bladder of 2 cm in radius. Now one gets $2\pi fa/c = 2\pi 15 \cdot 10^3 \cdot 2 \cdot 10^{-2}/1{,}500 \approx 1.3$. If the fish had a swim bladder, the effective size of the fish would be much greater, and therefore also the product would be much larger. The numbers chosen for both bats and toothed whales here are close to their lower limit: for higher frequencies and larger targets, these products will be larger and thereby there will be an efficient reradiation of echoes from the target.

The received echo level at the location of the echolocating animal may be estimated as:

$$\mathrm{RL}_{\text{small target}} = \mathrm{SL}_{\text{echo}} - \mathrm{TL} = \mathrm{SL} - 2\mathrm{TL} + \mathrm{TS}$$

With spherical geometric spreading (20 log r, or 6 dB/dd, where dd is distance doubling) the echo amplitude will decrease by 2TL, or 40 log r (equal to 12 dB/dd) as a function of the range between the source and the target.

Some targets are not point targets but consist of several small targets, or sometimes a whole wall or other surface structures. Let us first consider the easiest example, of the reflector being a plane wall, which is much larger than the cross section of the bat or dolphin transmission beam. In the case of bats this may be an even cliff surface or a dense foliage. For dolphins, it may be an even sea surface or bottom. Also assume the sound is impinging normal to the surface. Then the sound is reflected from the wall back to the animal very much as if it was an acoustic mirror, which means that the returning echo level corresponds to the received level a distance of $2r$ from the source. Assuming spherical spreading from the source to the receiver, the received level from such a reflecting surface is therefore:

$$\mathrm{RL}_{\text{planar surface}} = \mathrm{SL} - 20\log_{10}(2r) - 2\alpha r,$$

where α is the sound absorption. The transmission loss in this situation is extremely different from the transmission loss for small point targets. Take the example of a bat approaching an insect, assuming an absorption of 3 dB m^{-1}. When the bat is 3 m away from the insect, the returning echo has experienced a total transmission loss of $40 \log (3/0.1) + 2\cdot 3\cdot(3-0.1) = 76$ dB (the reference distance is 10 cm for bats). If the bat instead was emitting a cry 3 m away from an even wall, the transmission loss would be $20 \log (2\cdot 3/0.1) + 2\cdot 3\cdot(3-0.1) = 53$ dB. Thus, there is a $76-53=23$ dB difference in the transmission loss between the echo from the insect and the echo from the wall, at the site of the bat.

Another important issue is clutter, or unwanted echoes at a similar range as the target of interest. This may consist of small particles in the air, or air bubbles in water, or other prey items that have been deselected by the approaching bat or whale. Usually we may model each small item as a spherical target, for which the echo experiences spherical spreading. Assuming that they are evenly distributed in the medium, the number of the small targets contributing to the clutter will be the ones found within the cross section of the beam at the range of the target. The beam area is given by the product of the range and the beam width (in radians). The level of clutter is in this case the same independent of the range to the echolocating animal, so we may introduce a variable C (with the unit dB) denoting the summed target strength of all reflectors contributing to the clutter within the beam width of the animal at the reference range of 0.1 or 1 m in air and water, respectively. The total received clutter level (CL) is

$$\mathrm{CL} = \mathrm{SL} - 40\log_{10}(r) - 2\alpha r + C + 20\log_{10}(r) = \mathrm{SL} - 20\log_{10}(r) - 2\alpha r + C$$

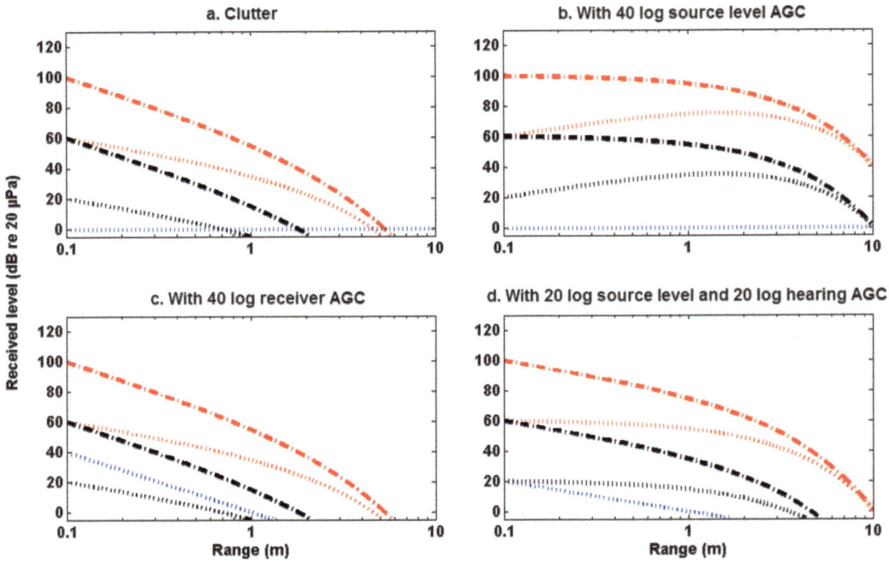

Fig. 4.6 The influence of unwanted echoes on biosonar performance. Bat–moth interaction example. The *red lines* are echoes returning from an echolocation signal of a high source level and the *black lines* echoes from signals of a lower source level. The *blue dotted line* is the hearing threshold of the bat. *Stippled line* is the echo returning from the insect, and the *dotted line* is the echo returning from the cluttered background. (**a**) Bat approaching insect in a cluttered environment. (**b**) Bat approaching insect in a cluttered environment, decreasing its source level as a function of range to the target by 40 log (range). (**c**) Bat approaching insect keeping source level constant, but reducing its hearing sensitivity by 40 log range. (**d**) Bat approaching insect, adjusting the source level by 20 log (range) and its hearing sensitivity by 20 log (range)

The received clutter is reduced by about 20 log r for short ranges. This is because the decrease in echo level due to geometric transmission loss (40 log r) from each small object in the clutter is counteracted by the fact that a larger and larger area, increasing by 20 log r, is ensonified, and therefore the number of contributing objects is also increased by 20 log r. This has been confirmed by measurements in air: using data from reflections of bat signals from vegetation by Jensen (2000), Ratcliffe et al. (2011) estimated that the echo from a clutter background decreases by 24 log r.

There is a wealth of extra reflectors in addition to the prey the bats or toothed whales are chasing. For bats these could be dense foliage, uneven surfaces on the earth, and there can also be small particles or animals in the air reflecting sound. In water, the bottom topography, the rugged or even surface, small air bubbles and particles in the water, will all generate echoes. Such echoes can be called reverberation or clutter depending on the situation and the magnitude of them, and the transmission loss will usually be somewhere between 20 log (r) and 40 log (r).

4.5.2 Modeling the Received Level from Echoes in Clutter

Using the sonar equation we can model the received level of echoes from small targets clutter. In Fig. 4.6a the consequences of reducing the source level in a cluttered situation is sketched. The model assumes that the clutter is situated at the same range as the target of interest, say an insect, and that it is evenly distributed. Further, it is assumed that the bat emits signals with certain directionality, so that the area of clutter ensonified is increasing with the range between the bat and the prey of interest, as described in Sect. 4.5.1. The signals are emitted at two source levels: black lines correspond to 60 dB and red lines to 100 dB re 20 μPa at 10 cm. The returning echoes from the prey are depicted with dash-dotted lines and the echoes from the clutter by the dotted lines.

The dashed-dotted lines show how the received echo level of the target varies with the range from the bat to the target. For short ranges these lines are almost straight when using a logarithmic x axis. This is the effect of geometric (spherical) spreading loss of 40 log r for a point target. At larger distances the received levels start to drop more than linearly, due to the effect of absorption. The approximate hearing threshold of a bat (assumed here to be 0 dB re 20 μPa; see Fay 1988) is indicated with a blue dotted line.

4.5.3 Automatic Gain Control

Even though bat and dolphin echolocation signals are extremely intense, the animals are actually far from always calling at their maximum level. As a matter of fact, there are many biotic and abiotic restrictions to production of intense sounds and there can be ecological, physiological, as well as behavioral reasons why very loud sounds are not ideal. Field observations of both bats and toothed whales have shown that animals normally adjust the emitted source level as a function of the range to the target when they are within a distance of a few body lengths (Johnson et al. 2006; Nørum et al. 2012). For many of the species so far measured, the source level adjustment quite closely follows a 20 log r curve (Au and Benoit-Bird 2003; Surlykke and Kalko 2008). The scatter is, however, huge and there are exemptions to this rule (e.g., the beaked whale *Mesoplodon densirostris*; Madsen et al. 2005b). Some measurements suggest that another function rather than the 20 log r function would more accurately describe the relationship between the source level and the range to the target (Nørum et al. 2012). Decreasing the source level with 20 dB per tenfold decrease of distance when approaching the target means that the sound level ensonifying the target remains constant, whereas at the site of the echolocator the returning echo will, assuming it is generated from a point target, *increase* by the amount of 20 dB per tenfold decrease of distance.

For technical sonars or echosounders, automatic gain control (AGC) is applied to the returning echoes, amplifying them by an amount of 40 log *T* or 20 log *T*, where *T* is the time since the emission of the sonar signal. The 40 log *T* function is proportional to the transmission loss functions for point targets (40 log *r*) and the 20 log *T* function is proportional to the transmission loss for clutter (20 log *r*) or oblique surfaces (20 log $(2r)=6+20$ log *r*). Using the relationship $r=T/(2c)$, where c is the speed of sound, we see that applying AGC helps to compensate the amplitude of the returning echoes for the transmission loss. For small single targets (such as single fish) the 40 log *T* AGC function is chosen, whereas for large targets (such as large fish schools) the 20 log *T* function is the appropriate one. "Small" and "large" targets are in this context defined in relation to the cross section of the transmission beam: if the target fits well within the beam, it is small; but if the target is only partly ensonified by the beam, the target is large. In this manner echoes can be displayed on the same scale, independent of the range from the transducer (Simmonds and MacLennan 2006).

Without range compensating the echo amplitudes, an echolocator (both machine and animal) needs to deal with a vast dynamic range of echoes. Consider an echo from a prey item of −40 dB target strength being investigated at 10 m distance by a dolphin. With a source level of 210 dB re 1 μPa, the echo will come back to the dolphin with a level of $210-40$ log $10-40=130$ dB re 1 μPa. If the dolphin investigates the same target at a range of 1 m with the same source level, the echo level is now $210-40$ log $1-40=170$ dB re 1 μPa, or 40 dB higher. Thus, the ratio between the intensities of the echo at 1 and 10 m is 10^4 for a tenfold change in distance. Not only man-made sonars, but also animals, have difficulties adjusting their representation of echoes to such vast changes in echo levels for the same target. Also, we know that for tracking and classifying targets with man-made sonars it is very valuable to maintain the echo amplitude as constant as possible, independent of target distance. It is reasonable that this would also be advantageous to animals, and therefore the idea of automatic gain control in bats and dolphins has been of keen interest to many scientists.

We can use the sonar equation to understand some of the consequences of AGC of the biosonar signals, and how these may be adjusted from other factors. Figure 4.6b illustrates a source level of the signal that is increased by 40 log (*r*) as a function of range to the target. The dashed-dotted lines, corresponding to the received echo level from a point target, are now almost flat with target range up to a few meters distance (where absorption starts to have an effect), because the 40 log *r* increase in signal level is compensating the geometric transmission loss of 40 log *r* for a point target. The clutter is, however, increasing in level with target distance at 20 log *r*, due to the lower spreading loss for clutter as compared to single-point targets. Thus, when the bat approaches the target it will experience the target echo as constant, and a drastic decrease in the clutter level. This is in contrast with the situation with no AGC (Fig. 4.6a), where both the target echo and clutter level increased, albeit with different rates, as the bat approached the target. Keeping the focal echo amplitude constant and the background amplitude varying may be favorable for the perception of the target while tracking and classifying it.

Instead of modulating the transmitted signal, we may assume the bat could vary the receiver sensitivity. This is the "normal" AGC in technical echosounders. In Fig. 4.6c the effect of applying a 40 log *r* AGC to the hearing threshold of the bat is displayed. Now the received echo and clutter levels are identical to in the non-AGC situation of Fig. 4.6a, but the hearing threshold is greatly reduced with range. The *perceived* echo and clutter level is the distance between the red curves and the blue line for the higher source level, and between the black curves and blue line for the lower source level. In this hypothetical example, a reduction in source level result in a clutter level below the hearing threshold. This may improve the target tracking and classification performance of the echolocating bat.

Finally, the echolocator could modulate both the transmission and receiver side of the biosonar system. It seems that many bats and dolphins employ approximately 20 log *r* gain on the transmitted signal. Hearing studies during echolocation indicate that the hearing sensitivity right after the outgoing click is reduced by forward masking, following an approximately 20 log *T* function in most cases, where *T* is the time from the emission of the sound pulse. Thus, it seems the hearing threshold of the animal can be varied during echolocation, presumably to compensate for the remaining 20 log *r* to obtain full transmission loss compensation (Hartley 1992a, b; Nachtigall and Supin 2008; Linnenschmidt et al. 2012). Such a hearing compensation could then explain the "missing 20 log *r*" in the biosonar system of both bats and dolphins (assuming they echolocate point targets). Therefore, the combined effect of decreased source level and decreased hearing sensitivity when approaching a target has been called AGC in the biosonar literature. In Fig. 4.6d this situation is modeled for the bat-insect-clutter situation discussed earlier in this section. The blue dotted line, which is the hearing threshold of the bat, is now varying as a function of the distance between the bat and the insect. The perceived echo level is given by the difference between the received echo level and the hearing threshold for a certain range. The dashed-dotted lines are following the hearing threshold slope closely, for shorter ranges. This corresponds to a constant perceived echo level independent of range, just as we saw in the 40 log *r*–compensated source level model in Fig. 4.6b. The perceived clutter level is again increasing with range, so that the target-to-clutter level will improve as the animal approaches the insect. By lowering the source level at close ranges to the insect, the bat may obtain a situation where the clutter level is close to or drops below the hearing threshold at short distances (such as for the black stippled-dotted line in Fig. 4.6d.

All in all, we may conclude from Fig. 4.6 that by using AGC on the transmitter and/or receiver side the signal detectability may be improved by reducing the source level when approaching the target. The reason why "whispering bats" emit lower source levels than open-area foragers is probably because hunting in cluttered space means hunting at short range, where prey echoes are relatively more intense than background echoes due to the difference in geometric spreading loss for point targets (prey) and background (extended reflectors) even if clutter walls do not reflect just like planar targets.

It should be emphasized that these models are great oversimplifications and care must be taken, as there is a risk of overinterpreting their biological relevance.

It is still unclear how bats and dolphins compensate for the variations in target ranges in the complex auditory scene of their natural habitats. However, the models are helpful in providing a framework for how echoes are perceived by the animal in different situations and may therefore help us understand various strategies employed by echolocating animals. The models would work just as well for toothed whale echolocation signals, the only difference being that the range scale would be expanded by a factor of about 10 (due to the larger source level and larger target strengths of target items of interest for the animals).

4.5.4 Acoustic Predator–Prey Interactions

Besides the physical reasons given in Sect. 4.5.3 there are also biological reasons for why bats and whales do not always maximize sound level while echolocating. Some of the prey that bats and toothed whales search for have acoustic detectors specialized to pick up the ultrasonic signals made during echolocation (Mann et al. 1997; Miller and Surlykke 2001; Goerlitz et al. 2010).

Many insects are sensitive to ultrasound and will react with evasive maneuvers. The type of reaction observed will usually depend on the received sound level (Miller and Surlykke 2001). Some insects will drop to the ground when exposed to ultrasound, whereas others will perform complicated evasive maneuvers. Still others will reply to the incoming clicks by emitting ultrasonic clicks back as an aposematic warning (for toxic moths) or to directly jam the sonar acoustically (Barber and Conner 2007; Corcoran et al. 2009). Recently, allis shad (herring species belonging to the subfamily Alosinae) has been shown to react with directional evasive maneuvers to dolphin-like ultrasound pulses (Mann et al. 1997; Plachta and Popper 2003; Wilson et al. 2011). It may therefore be advantageous to emit quieter sounds so as not to alert the prey before the predator has a chance to intercept with it (Surlykke 1988; Barrett-Lennard et al. 1996).

This is illustrated in Fig. 4.7, which is inspired by Surlykke (1988). Echolocators emitting intense signals for long-range sonar can be detected very far away by their hearing prey, whereas quieter echolocators have a reduced detection range, but the detection range of the prey is reduced even more dramatically. Thus, if the sonar-guided predator reduces the source level, the advantage is all of a sudden on the predator's side: the bat may now detect the insect at a range longer than the range at which the insect can detect the bat. This may explain the acoustic behavior of barbastelle bats (*Barbastella barbestellus*) reducing the source level abruptly when closing in on prey (Goerlitz et al. 2010).

Even though toothed whales constitute a heavy predation pressure on fish, as far as we know acoustic defense developed only very rarely under water, in spite of the fact that all fish already have ears. This suggests that toothed whales hunting fish do not gain much by lowering the source level of their signals to avoid being detected by their prey in many cases (see, however, Mann et al. 1997, and Wilson et al. 2011,

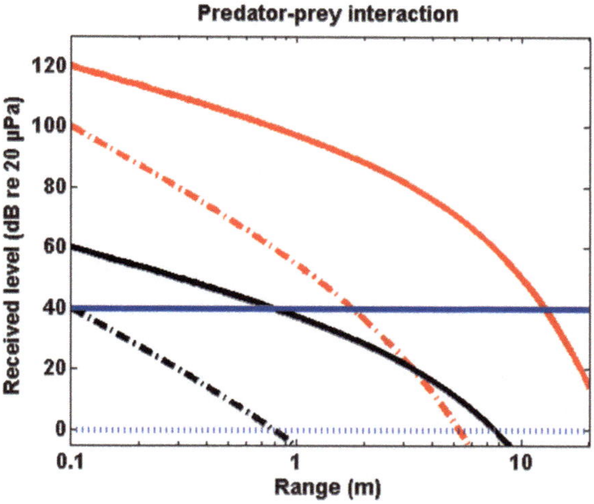

Fig. 4.7 Biosonar predator–prey interaction in air and water. The *solid red* and *black line* is the received level as a function of range of an echolocation click of two different source levels. The *stippled red* and *black lines* show the echo received level as a function of range to the target. *Blue lines* indicate hearing threshold of predator (*stippled*) and prey (*solid*). Source levels: 140 and 90 dB re 20 µPa at 10 cm; target strength: 0 dB at 10 cm; hearing threshold of moth: 40 dB re 20 µPa; hearing threshold of bat: 0 dB re 20 µPa. Lowering of the source level decreases the prey detection range (*intersection of solid blue line* with *solid red* and *black curves*) around 12 times from about 11 to 0.9 m. The decrease of the echolocation detection range for the same lowering of source level is around seven times, from 6 to 0.8 m (*intersection stippled blue line* with *stippled red* and *black curves*)

for examples where lowering the source level may have an effect). For toothed whales hunting for marine mammals the situation is different as illustrated by killer whales. Here, a great variation in the sound level of echolocation clicks is observed, such that killer whales hunting marine mammals are much quieter compared to those hunting fish (Barrett-Lennard et al. 1996).

Besides physiology may also play a role as to why bats and whales reduce the source level of their vocalizations. At very high click rates it may be difficult to sustain a high source intensity for every sonar emission. Data from Daubenton's bats show that laryngeal muscles are working at their maximum possible velocity in the terminal phases of the pursuit where the call rate is close to 200 Hz (Elemans et al. 2011). Also, evidence from Beedholm and Miller (2007) indicates that harbor porpoises cannot (or at least will not if they are not forced to) sustain a maximum source level when inter-click intervals (ICI) are below about 30 ms. Actually, the decrease in source level as a function of the ICIs in this experiment quite closely followed a 20 log (ICI) relationship. As the ICI is roughly linearly related to the range of the target during active echolocation at close ranges, the 20 log *r* function interpreted in Sect. 4.5.3 as AGC may therefore simply be a consequence of physical restrictions in the sound production apparatus.

4.6 Summary

Source levels of bats vary from 100 to 140 dB re 20 μPa at 10 cm RMS and for toothed whales from below 140 to beyond 235 dB re 1 μPa at 1 m pp. Taking into account the various measurement units as well as the differences between sound intensities in air and water when reported on a decibel scale with different pressure and range references, this means that the whale echolocation signals are up to 60 dB more intense than the bat signals. Owing to the differences in source level and transmission properties between air and water the sonar range for bats is some 1–10 m, whereas for dolphins it is up to hundreds of meters or even more. The intensity and directionality of the echolocation signals of bats and dolphins have been shaped to improve sonar performance in the two media. In general, toothed whale signals are shorter and are of higher intensity and directionality than bat signals. The reason for this is related to both the acoustic differences between the two media and the size difference between bats and toothed whales as well as between the prey of the two species.

Acknowledgments We thank Lee Miller, Paul Nachtigall, Arthur Popper, and Lutz Wiegrebe for comments on previous drafts of this chapter.

References

Atem, A. C., Rasmussen, M. H., Wahlberg, M. Petersen, H. C., & Miller, L. A. (2009). Changes in click source levels with distance to targets: Studies of free-ranging white-beaked dolphins (*Lagenorhynchus albirostris*) and captive harbour porpoises (*Phocoena phocoena*). *Bioacoustics*, 19(1), 49–65.

Au, W. W. L. (1974). Measurement of echolocation signals of the Atlantic bottlenose dolphin, *Tursiops truncatus* Montagu, in open waters. *Journal of the Acoustical Society of America*, 56(4), 1280–1290.

Au, W. W. L. (1993). *The sonar of dolphins*. New York: Springer-Verlag.

Au, W. W. L., & Herzing, D. L. (2002). Echolocation signals of wild Atlantic spotted dolphin (*Stenella frontalis*). *Journal of the Acoustical Society of America*, 113(1), 598–604.

Au, W. W. L., & Benoit-Bird, K. J. (2003). Automatic gain control in the echolocation system of dolphins. *Nature*, 423, 861–863.

Au, W. W. L., & Wursig, B. (2004). Echolocation signals of dusky dolphins (*Lagenorhynchus obscurus*) in Kaikoura, New Zealand. *Journal of the Acoustical Society of America*, 115(5), Pt. 1, 2307–2313.

Au, W. W. L., Floyd, R. W., & Haun, J. E. (1978). Propagation of Atlantic bottlenose dolphin echolocation signals. *Journal of the Acoustical Society of America*, 64(2), 411–422.

Au, W. W. L., Carder, D. A., Penner, R. H., & Scronce, B. L. (1985). Demonstration of adaptation in beluga whale echolocation signals. *Journal of the Acoustical Society of America*, 77(2),726–730.

Au, W. W. L., Penner, R. H., & Turl, C. H. (1988). Propagation of beluga echolocation signals. *Journal of the Acoustical Society of America*, 82(3), 807–813.

Au, W. W. L., Kastelein, R. A., Rippe, T., & Schooneman, N. M. (1999). Transmission beam pattern and echolocation signals of a harbor porpoise (*Phocoena phocoena*). *Journal of the Acoustical Society of America*, 106(6), 3699–3705.

Au, W. W. L., Ford, J. K. B., Horne, J. K., & Newman Allman, K. A. (2004). Echolocation signals of free-ranging killer whales (*Orcinus orca*) and modelling of foraging for Chinook salmon (*Oncorhynchus tshawytscha*). *Journal of the Acoustical Society of America*, 115(2), 901–909.
Au, W. W. L., Horne, J. K., & Jones, C. (2010). Basis of acoustic discrimination of chinook salmon from other salmons by echolocating *Orcinus orca*. *Journal of the Acoustical Society of America*, 128(4), 2225–2232.
Barber, J. R., & Conner, W. E. (2007). Acoustic mimicry in a predator–prey interaction. *Proceedings of the National Academy of Sciences of the USA*, 104, 9331–9334.
Barrett-Lennard, L., Ford, J. K. B., & Heise, K. A. (1996). The mixed blessing of echolocation: Differences in sonar use by fish-eating and mammal-eating killer whales. *Animal Behavior*, 51, 553–565.
Beedholm, K., & Møhl, B. (2006). Directionality of sperm whale sonar clicks and its relation to piston radiation theory. *Journal of the Acoustical Society of America*, 119(2), EL14–EL19.
Beedholm, K., & Miller, L. A. (2007). Automatic gain control in harbor porpoises (*Phocoena phocoe*na)? Central versus peripheral mechanisms. *Aquatic Mammals*, 33(1), 69–75.
Beranek, L. (1996). *Acoustics*. Melville, NY: American Institute of Physics.
Brinkløv, S., Kalko, E. K. V., & Surlykke, A. (2009). Intense echolocation calls from two 'whispering' bats, *Artibeus jamaicensis* and *Macrophyllum macrophyllum* (Phyllostomidae). *Journal of Experimental Biology*, 212, 11–20.
Brinkløv, S., Kalko, E. K. V., & Surlykke, A. (2010). Dynamic adjustment of biosonar intensity to habitat clutter in the bat *Macrophyllum macrophyllum* (Phyllosotomidae). *Behavioral Ecology and Sociobiology*, 64, 1867–1874.
Brinkløv, S., Jakobsen, L., Ratcliffe, J. M., Kalko, E. K. V., & Surlykke, A. (2011). Echolocation call intensity and directionality in flying short-tailed fruit bats, *Carollia perspicillata* (Phyllostomidae). *Journal of the Acoustical Society of America*, 129(1), 427–435.
Corcoran, A. J., Barber, J. R., & Conner, W. E. (2009). Tiger moth jams bat sonar. *Science* 325, 327.
Dantzker, M. S., Deane, G. B., & Bradbury, J. W. (1999). Directional acoustic radiation in the strut display of male sage grouns *Centrocercus urophasianus*. *Journal of Experimental Biology*, 202, 2893–2909.
Elemans, C. P. H., Mead, A. F., Jakobsen, L., & Ratcliffe, J. M. (2011). Superfast muscles set maximum call rate in echolocating bats. *Science*, 333, 1885–1888.
Eskesen, I. G., Wahlberg, M., Simon, M., & Larsen, O. N. (2011). Echolocation clicks from sympatric killer whales and long finned pilot whales. *Journal of the Acoustical Society of America*, 130(1), 9–12.
Fay, R. R. (1988). *Hearing in vertebrates: A psychophysics databook.* Winnetka, IL: Hill-Fay Associates.
Ghose, K., & Moss, C. F. (2003). The sonar beam pattern of a flying bat as it tracks tethered insects. *Journal of the Acoustical Society of America*, 114(2), 1120–1131.
Ghose, K., Moss, C. F., & Horiuchi, T. K. (2007). Flying big brown bats emit a beam with two lobes in the vertical plane. *Journal of the Acoustical Society of America*, 122, 3717–3724.
Gillespie, D., Dunn, C., Gordon, J. Claridge, D. Embling, C., & Boyd, I. (2010). Field recordings of Gervais' beaked whales *Mesoplodon europaeus* from the Bahamas. *Journal of the Acoustical Society of America*, 125(5), 3428–3433.
Goerlitz, H. R., der Hofstede, H. M., Zeale, M. R. K., Jones, G., & Holderied, M. W. (2010). An aerial-hawking bat uses stealth echolocation to counter moth hearing. *Current Biology*, 20(17), 1568–1572.
Guarato, F., Hallam, J., & Matsuo, I. (2011). Reconstruction of the signal produced by a directional sound source from remote multi-microphone recordings. *Journal of the Acoustical Society of America*, 130(3), 1689–1699.
Hartley, D. J. (1992a). Stabilization of perceived echo amplitudes in echolocating bats. I. Echo detection and automatic gain control in the big brown bat, *Eptesicus fuscus*, and the fishing bat, *Noctilio leporinus*. *Journal of the Acoustical Society of America*, 91, 1120–1132.
Hartley, D. J. (1992b). Stabilization of perceived echo amplitudes in echolocating bats. II. The acoustic behavior of the big brown bat, *Eptesicus fuscus*, when tracking moving prey. *Journal of the Acoustical Society of America*, 91, 1133–1149.

Henze, D., & O'Neill, W. E. (1991). The emission pattern of vocalizations and directionality of the sonar system in the echolocating bat, *Pteronotus parnelli*. *Journal of the Acoustical Society of America*, 89, 2430–2434.
Hiryu, S., Katsura, K., Lin, L.-K., Riquimaroux, H., & Watanabe, Y. (2006). Radiation pattern of echolocation pulse in Taiwanese leaf-nosed bat, *Hipposideros terasensis*. *Acoustical Science and Technology*, 27, 108–110.
Hiryu, S., Hagino, T., Riquimaroux, H., & Watanabe, Y. (2007). Echo-intensity compensation in echolocating bats (*Pipistrellus abramus*) during flight measured by a telemetry microphone. *Journal of the Acoustical Society of America*, 121, 1749–1757.
Holderied, M. W., & von Helversen, O. (2003). Echolocation range and wingbeat period match in aerial-hawking bats. *Proceedings of the Royal Society London B: Biological Sciences*, 270, 2293–2300.
Holderied, M. W., Korine, C., Fenton, M. B., Parsons, S., Robson, S., & Jones, G. (2005). Echolocation call intensity in the aerial hawking bat *Eptesicus bottae* (Vespertilionidae) studied using stereo videogrammetry. *Journal of Experimental Biology*, 208, 1321–1327.
Jakobsen, L., & Surlykke, A. (2010). Vespertilionid bats control the width of their biosonar sound beam dynamically during prey pursuit. *Proceedings of the National Academy of Sciences of the USA*, 107, 13930–13935.
Jakobsen, L., Ratcliffe, J. M., & Surlykke, A. (2013). Convergent acoustic field of view in echoloating bats. *Nature*, 493, 93–96.
Jensen, M. E. (2000). *The effect of acoustic interference and clutter objects on search signal design in echolocating bats*. Ph.D. thesis, University of Southern Denmark.
Jensen, M. E., & Miller, L. A. (1999). Echolocation signals of the bat *Eptesicus serotinus* recorded using a vertical microphone array: Effect of flight altitude on searching signals. *Behavioral Ecology and Sociobiology*, 47, 60–69.
Jensen, F. H., Rocco, A., Mansur, R. M., Smith, B. D., Janik, V. M., Madsen, P. T. (2013). Clicking in shallow rivers: short-range echolocation of Irrawaddy and ganges river dolphins in a shallow, acoustically complex habitat. *PLOS ONE*, 8(4), e59284.
Johnson M., Madsen, P. T., Zimmer, W. M. X., Aguilar de Soto, N., & Tyack, P. L. (2006). Foraging Blainville's beaked whales (*Mesoplodon densirostris*) produce distinct click types matched to different phases of echolocation. *Journal of Experimental Biology*, 209, 5038–5050.
Kamminga, C., Kataoka, T., & Engelsma, F. J. (1988). Investigations on cetacean sonar VII. Underwater sounds of *Neophocaena phcaenoides* of the Japanese coastal population. *Aquatic Mammals*, 12, 52–60.
Kastak, D., & Schusterman, R. J. (1999). In-air and underwater hearing sensitivity of a northern elephans seal (*Mirounga angustirostris*). *Canadian Journal of Zoology*, 77, 1751–1758.
Koblitz, J. C., Wahlberg, M., Stilz, P., Madsen, P., Beedholm, K., & Schnitzler, H. U. (2012). Asymmetry and dynamics of a narrow sonar beam in an echolocating harbor porpoise. *Journal of the Acoustical Society of America*, 131(3), 2315–2324.
Kyhn, L. A., Tougaard, J., Jensen, F., Wahlberg, M., Stone, G., Yoshinaga, A., Beedholm, K., & Madsen, P. (2009). Feeding at a high pitch: Source parameters of narrow band, high-frequency clicks from echolocating off-shore hourglass dolphins and coastal Hector's dolphins. *Journal of the Acoustical Society of America*, 125(3), 1783–1791.
Kyhn, L. A., Jensen, F. H., Beedholm, K., Tougaard, J., Hansen, M., & Madsen, P. T. (2010). Echolocation in sympatric Peale's dolphins (*Lagenothynchus australis*) and Commerson's dolphins (*Cephalorhynchus commersonii*) producing narrow-band high-frequency clicks. *Journal of Experimental Biology*, 213, 1940–1949.
Kyhn, L. A., Tougaard, J., Beedholm, K., Jensen, F. H., Ashe, E., Williams, R., & Madsen, P. T. (2013). Clicking in a killer whale habitat: narrow-band, high-frequency biosonar clicks of harbour porpoises (*Phocoena phocoena*) and Dall's poproise (*Phcoenoides dalli*). *PLoS ONE*, 8(5), e63763.
Li, S., Wang, K., Wang, D., & Akamatsu, T. (2005). Echolocation signals of the free-ranging Yangtze finless porpoise (*Neophocaena phocaenoides asiaeorientialis*). *Journal of the Acoustical Society of America*, 117(5), 3288–3296.

Li, S., Wang, D., Wang, K., Akamatsu, T., Ma, Z., & Han, J. (2007). Echolocation click sounds from wild inshore finless porpoise (*Neophocaena phocaenoides sunameri*) with comparisons to the sonar of riverine *N. p. asiaeorientalis*. *Journal of the Acoustical Society of America*, 121(6), 3938–3946.

Linnenschmidt, M., Beedholm, K., Wahlberg, M., Kristensen, J. H., & Nachtigall, P. E. (2012). Keeping returns optimal: Gain control elicited by dynamic hearing thresholds in a harbour porpoise. *Proceedings of the Royal Society of London B: Biological Sciences*, 279(1738), 2237–2245.

Madsen, P. T., & Wahlberg, M. (2007). Recording and quantification of ultrasonic echolocation clicks from free-ranging toothed whales. *Deep-Sea Research Pt. I: Oceanographic Research Papers*,, 54(8), 1421–1444.

Madsen, P. T., Kerr, I., & Payne, R. (2004a). Echolocation clicks of two free-ranging, oceanic delphinids with different food preferences: False killer whales *Pseudorca crassidens* and Risso's dolphins, *Grampus griseus*. *Journal of Experimental Biology*, 207, 1811–1823.

Madsen, P. T., Kerr, I., & Payne, R. (2004b). Source parameter estimates of echolocation clicks from wild pygmy killer whales (*Feresa attenuata*). *Journal of the Acoustical Society of America*, 116(4), 1909–1912.

Madsen, P. T., Carder, D. A., Beedholm, K., & Ridgway, S. H. (2005a). Porpoise clicks from a sperm whale nose: Convergent evolution of 130 kHz pulses in toothed whale sonars? *Bioacoustics*, 15, 195–206.

Madsen, P. T., Johnson, M., Aguilar de Soto, N., Zimmer, W. M. X., & Tyack, P. L. (2005b). Biosonar performance of foraging beaked whales (*Mesoplodon densirostris*). *Journal of Experimental Biology*, 208, 181–194.

Madsen, P. T., Wilson, M., Johnson, M., Hanlon, R. T., Bocconcelli, A., Aguilar de Soto, N., & Tyack, P. L. (2007). Clicking for calamari: toothed whales can echoloate squid *Loligo pealeii*. *Aquatic Biology*, 1, 141–150.

Mann, D. A., Lu, Z., & Popper, A. N. (1997). A clupeid fish can detect ultrasound. *Nature*, 389, 341.

Matsuta, N., Hiryu, S., Fojioka, E., Yamada, Y., Riquimaroux, H., & Watanabe, Y. (2013). Adaptive beamwidth control on echolocation sounds from CF-FM bats, Rhinolophus ferrumequinum nippon, during preycapture flight. *Journal of Experimental Biology*, 216, 1210–1218.

Medwin, H., & Clay, C. S. (1998). *Fundamentals of acoustical oceanography*. New York: Academic Press.

Miller, L.A., Pristed, J., Møhl, B., & Surlykke, A. (1995). The click-sounds of narwhals (*Monodon monoceros*) in Inglefield Bay, Northwest Greenland. *Marine Mammal Science*, 11(4), 491–502.

Miller, L. A., & Surlykke, A. M. (2001). How some insects detect and avoid being eaten by bats: Tactics and countertactics of prey and predator. *Bioscience*, 51, 570–581.

Mogensen, F., & Møhl, B. (1979). Sound radiation patterns in the frequency domain of cries from a vespertilionid bat. *Journal of Comparative Physiology A*, 134, 165–171.

Møhl, B., Surlykke, A., & Miller, L. A. (1990). High intensity narwhal clicks. In J. Thomas & R. Kastelein (Eds.), *Sensory abilities of cetaceans* (pp. 295–304). New York: Plenum Press.

Møhl, B., Wahlberg, M., Madsen, P. T., Miller, L. A., & Surlykke, A. (2000). Sperm whale clicks: Directionality and source levels revisited. *Journal of the Acoustical Society of America*, 107 (1), 638–648.

Møhl, B., Wahlberg, M., Madsen, P. T., Heerfordt, A., & Lund, A. (2003). The monopulsed nature of sperm whale clicks. *Journal of the Acoustical Society of America*, 114(2), 1143–1154.

Møhl, B., Wahlberg, M., & Heerfordt. A. (2006). Hyper-directionality in clicks from the sperm whale (*Physeter macrocephalus*). *Journal of the Acoustical Society of America*, 119(5) Pt. 2, 3333.

Moore, P. W. B., Dankiewicz, L. A., & Houser, D. A. (2008). Beamwidth control and angular target detection in an echolocating bottlenose dolphin (*Tursiops truncatus*). *Journal of the Acoustical Society of America*, 124, 3324–3332.

Morisaka, T., Karczmarski, L., Akamatsu, T., Sakai, M., Dawson, S., & Thornton, M. (2011). Echolocation signals of Heaviside's dolphins (*Cephalorhynchus heavisidii*). *Journal of the Acoustical Society of America*, 129(1), 449–457.

Moss, C. F., Chiu, C., & Surlykke, A. (2011). Adaptive vocal behavior drives perception by echolocation in bats. *Current Opinion in Neurobiology*, 21, 1–8.

Nachtigall, P. E., & Supin, A. Y. (2008). A false killer whale adjusts its hearing when it echolocates. *Journal of Experimental Biology*, 211, 1714–1718.

Nørum, U., Brinkløv, S., & Surlykke, A. (2012). New model for gain control of signal intensity to object distance in echolocating bats. *Journal of Experimental Biology*, 215, 3045–3054.

Plachta, D. T., & Popper, A. N. (2003). Evasive responses of American shad (*Alosa sapidissima*) to ultrasonic stimli. Acoustic. *Research Letters Online*, 5, 25–30.

Proakis, J. G., & Manolakis, D. G. (1991). *Digital signal processing*, 3rd ed. Upper Saddle River, NJ: Prentice-Hall.

Rasmussen, M., Miller, L. A., & Au, W. W. L. (2002). Sounds and calculated source levels from whitebeaked dolphins (*Lagenorhynchus albirostris* Gray 1846) recorded in Icelandic waters. *Journal of the Acoustical Society of America*, 111, 1122–1125.

Rasmussen, M. H., Wahlberg, M., & Miller, L. A. (2004). Estimated transmission beam pattern of clicks recorded from free-ranging white-beaked dolphins *Lagenorhynchus albirostris*. *Journal of the Acoustical Society of America*, 116, 1826–1831.

Ratcliffe, J. M., Jakobsen, L., Kalko, E. K. V., & Surlykke, A. (2011). Frequency alternation and an offbeat rhythm indicate foraging behavior in the echolocating bat, *Saccopteryx bilineata*. *Journal of Comparative Physiology A*, 197(5), 413–423.

Schnitzler, H.-U., & Grinnell, A. D. (1977). Directional sensitivity of echolocation in the horseshoe bat, *Rhinolophus ferrumequinum*. I Directionality of sound emission. *Journal of Comparative Physiology A*, 116, 51–61.

Schotten, M., Au, W. W. L., Lammers, M. O., & Aubauer, R. (2003). Echolocation recordings and localization of wild spinner dolphins (*Stenella longirostris*) and pantropical spotted dolpins (*Stenella attenuata*) using a four-hydrophone array. In J. Thomas, C. Moss, & M. Vater (Eds.), *Echolocation in bats and dolphins* (pp. 393–400). Chicago: University of Chicago Press.

Schuchmann, M., & Siemers, B. (2010). Variability in echolocation call intensity in a community of horseshoe bats: A role for resource partitioning or communication? *PLoS ONE*, 5(9): e1 2842. doi:10.1371/journal.pone.0012842.

Simmonds, J., & MacLennan, D. (2006). *Fisheries acoustics*. Oxford: Blackwell.

Simon, M., Wahlberg, M., & Miller, L. A. (2007). Echolocation clicks from killer whales (*Orcinus orca*) feeding on herring (*Clupea harengus*) in Norwegian waters. *Journal of the Acoustical Society of America*, 121(2), 749–752.

Starkhammar, J., Moore, P. W., Talmadge, L. & Houser, D. S. (2011). Frequency-dependent variation in the two-dimensional beam pattern of an echolocating dolphin. *Biology Letters*, doi: 10.1098/rsbl.2011.0396.

Surlykke, A. (1988). Interaction between echolocating bats and their prey. In P. E. Nachtigall & P. W. B. Moore (Eds.), *Animal sonar* (pp. 551–556). New York: Plenum Press.

Surlykke, A., & Kalko, E. K. V. (2008). Echolocating bats cry out loud to detect their prey. *PLoS ONE*, 3(4), e2036(1)-e2036(10).

Surlykke, A., Miller, L. A., Møhl, B., Andersen, B. B., Christensen-Dalsgaard, J., & Jørgensen, M. B. (1993). Echolocation in two very small bats from Thailand: *Craseonycteris thonglongyai* and *Myotis siligorensis*. *Behavioral Ecology and Sociobiology*, 33, 1–12.

Surlykke, A., Filskov, M., Fullard, J. H., & Forrest, E. (1999). Auditory relationships to size in noctuid moths: Bigger is better. *Naturwissenschaften*, 86, 238–241.

Surlykke, A., Pedersen, S. B., & Jakobsen, L. (2009). Echolocating bats emit a highly directional sonar sound beam in the field. *Proceedings of the Royal Society B: Biological Sciences*, 276, 853–860.

Tian, B. & Schnitzler, H.-U. (1997). Echolocation signals of the greater horseshoe bat (*Rhinolophus ferrumequinum*) in transfer flight and during landing. *Journal of the Acoustical Society of America* 101, 2347–2364.

Vanderelst, D., De Mey, F., Peremans, H., Geipel, I., Kalko, E. K. V., & Firzlaff, U. (2010). What noseleaves do for FM bats depends on their degree of sensorial specialization. *PLoS ONE*, *5*, e11893 (1)–e11893 (13).

Villadsgaard, A., Wahlberg, M., & Tougaard, J. (2007). Echolocation clicks of wild harbour porpoises, *Phocoena phocoena*. *Journal of Experimental Biology*, 210, 56–64.

Wahlberg, M., B. Møhl, & P. T. Madsen (2001). Estimating source position accuracy of a large aperture hydrophone array for bioacoustics. *Journal of the Acoustical Society of America*, 109 (1): 397–406.

Wahlberg, M., F. H. Jensen, N. A. Soto, K. Beedholm, L. Bejder, C. Oliveira, M. Rasmussen M. Simon, A. Villadsgaard, P. T. Madsen (2011a). Source parameters of echolocation clicks from wild bottlenose dolphins (*Tursiops truncatus* and *T. aduncus*). *Journal of the Acoustical Society of America*, 130 (4), 2263–2274.

Wahlberg, M., Beedholm, K., Heerfordt, A., & Møhl, B. (2011b). Characteristics of biosonar signals from the Northern bottlenose whale, *Hyperoodon ampullatus*. *Journal of the Acoustical Society of America*, 130(5), 3077–3084.

Wilson, M., Schack, H. B., Madsen, P. T., Surlykke, A., & Wahlberg, M. (2011). Directional escape behavior in allis shad (*Alosa alosa*) exposed to ultrasonic clicks mimicking an approaching toothed whale. *Journal of Experimental Biology*, 214, 22–29.

Zimmer, W. M. X., Johnson, M., Madsen, P. T., & Tyack, P. L. (2005). Echolocation clicks of free-ranging Cuvier's beaked whales (*Ziphius cavirostris*). *Journal of the Acoustical Society of America*, 117(6), 3919–3927.

Chapter 5
Hearing During Echolocation in Whales and Bats

Paul E. Nachtigall and Gerd Schuller

Keywords Animal acoustic reflex • Auditory evoked potential hearing measurement • Automatic gain control • Bat echolocation hearing • Click hearing • Echo • Echolocation calls • Echolocation vocalization • Evoked potential • FM sweep • Forward masking • Hearing sensation control • Micro-stimulation • Self-hearing • Self-stimulation • Stapedial reflex • Whale echolocation hearing

5.1 Introduction

Echolocating animals, both toothed whales and bats, generate loud chirps, clicks, or cries that are broadcast so that they bounce off prey and produce echoes. Loud outgoing sounds are efficient because they produce strong echoes (Wahlberg & Surlykke, Chap. 4). When targets are at a considerable distance, increasing the outgoing sound may help in prey detection (Rasmussen et al. 2002; Au and Benoit-Bird 2003). The problem for both bats and toothed whales, however, is that the louder the outgoing sound the more it will affect the animal's own ability to hear the much quieter echoes that contain the essential information. Loud sounds cause a temporary forward masking of the sounds heard shortly afterwards, and echolocation requires exactly that—a loud sound followed by listening to a much quieter echo. The information that is required for the animal to identify and catch its prey is

P.E. Nachtigall (✉)
Marine Mammal Research Program, Hawaii Institute of Marine Biology,
University of Hawaii, P.O. Box 1106, Kailua, HI 96734, USA
e-mail: nachtiga@hawaii.edu

G. Schuller
Biocenter LMU, Department Biologie II, Neurobiology, Ludwig-Maximilians-University
of Munich, Planegg-Martinsried 82152, Germany
e-mail: gschuller@lmu.de

A. Surlykke et al. (eds.), *Biosonar*, Springer Handbook of Auditory Research 51,
DOI 10.1007/978-1-4614-9146-0_5, © Springer-Verlag New York 2014

in the echo and the echoes must be heard well and interpreted rapidly to allow the animal to adjust its behavior and catch the prey.

Both bats and toothed whales (including dolphins and porpoises) produce loud echolocation signals (Wahlberg & Surlykke, Chap. 4). Peak-to-peak sound pressures of 100–110 dB re 20 μPa and more have been measured directly in front of echolocating bats (Griffin 1958; Kick and Simmons 1984) while wild sperm whale (*Physeter macrocephalus*) echolocation clicks may be as high as 236 dB re 1 μPa RMS (Madsen et al. 2002), and laboratory bottlenose dolphin clicks, measured while echolocating a small sphere more than 100 m away in a noisy bay, averaged as high as 227 dB re 1 μPa peak to peak during a single click train (Au 1980). The louder the sound, the more likely it is to forward mask the quiet echoes that follow, and the longer it will take for the hearing system to recover from the masking produced by the loud sound (Supin et al. 2001).

5.2 Hearing Sensation Level Changes

Echoes from prey return very rapidly. In air, echoes from 5 m away return in 30 ms or less and, given that the speed of sound is five times as fast under water, echoes from 5 m away underwater return within 6 ms. Prey 5 m away certainly might have the opportunity to avoid an approaching bat or toothed whale and echolocation must remain functional as a sense to influence behavior of the echolocating animal in order to allow adjustments to catch the prey in these very short time frames. Thus, it is important to consider how auditory systems handle this forward masking problem. The analysis of this problem has taken different paths in the study of bats and whales. Those who study bats can examine the neural structures and neurophysiological mechanisms underlying behavior and auditory mechanisms while those studying dolphins and whales must primarily deal with whole animals and behavior.

Taking advantage of a very rich neurophysiological background (Henson 1965; Suga and Schlegel 1972; Suga and Jen 1975) investigations had shown that during the production of the outgoing echolocation cries the bat's stapedial muscles contracted within their middle ears to dampen and protect their hearing (Wever and Vernon 1961). Kick and Simmons (1984) conducted a series of behavioral experiments that showed that the big brown bat (*Eptesicus fuscus*) controlled its hearing during echolocation in order to protect its ears from the loud outgoing cries. Whereas most of the outgoing cries of the bats were between 100 and 110 dB re 20 μPa, the hearing thresholds for the simulated echo returns varied greatly depending on how long it had been since the animal had produced the cry. The longer the time since the cry, the better was the hearing. At a delay of 6.4 ms the bat's threshold for echo detection at a criterion of 75 % correct responses was approximately 6 dB sound pressure level (SPL) peak to peak. At a shorter delay of 1.9 ms the thresholds were about 31 dB SPL showing an increase in sensitivity as time after the cry increased. Taking these and other data and fitting a regression line, Kick and Simmons (1984) found a slope of −11 dB for each increase in echo delay by a factor of 2. These data fit very nicely with Suga and Jen's (1975) data showing that an interval of 5–8 ms being required to relax the stapedial muscles of the middle ear.

Most importantly for hearing during echolocation, however, the hearing of echoes remained the same independent of distance because of a cancellation of distance causing reduction in echo signal and time causing an increase in hearing response. The hearing of sonar echoes became more sensitive (by around 11 dB) for each doubling of target distance due to time passing and hearing recovery, while the strength of the echoes themselves declined by 12 dB for every doubling of range due to signal attenuation (Kick and Simmons 1984). So within the range of 17 cm out to beyond 1 m the bat essentially heard the echoes at the same level no matter if the echo was very high intensity because it came from a high-intensity cry and was nearby or was of lower intensity because it was farther away and had attenuated because of distance. This range of 17 cm to 1 m is an important range for bats tracking insects, and it is critical for successful gathering of prey by bats.

Given societal mores there has been considerably less of a neurophysiological background underlying the study of hearing and echolocation in whales and dolphins than there has been with bats. The forward masking problem, though, is even more pronounced if one considers both intensity and time of the outgoing signals. The 227 dB peak to peak re 1 μPa dolphin underwater echolocation clicks are much more intense than signals occurring with bats and the echoes return much faster. A distance of 5 m, and therefore 6 ms, is a reasonable range to consider if fish are being detected and chased. Perhaps the first question to reasonably ask is: Does the stapedial reflex work in whales the way it does in bats? The answer, unfortunately, is not clear. Though there is apparently a great diversity in cetacean middle ear structure there is "no clear consensus on how cetacean middle ears function" (Wartzog and Ketten 1999). There has, however, been an examination of how well the false killer whale (*Pseudorca crassidens*) hears its own outgoing echolocation clicks in comparison to similar clicks played back to it (Supin et al. 2006). But that study did not explore the protective mechanisms. Before explaining how that experiment was accomplished, and its results, it is important to provide some general background on recent advances in the method of studying whale and dolphin hearing. Though a considerable effort has been devoted to the neurophysiology of the bat brain and echolocation system, little has been dedicated to the exploration of toothed whale neurophysiology. Toothed whales and dolphins are large, expensive, and very protected, and little laboratory neurophysiological work has been completed since the passage of the U.S. Marine Mammal Protection Act of 1972. Overt measurements of toothed whale brain function have been examined through the use of auditory evoked potentials recorded from the surface of the skin.

5.3 Auditory Evoked Potential Thresholds

The use of auditory evoked potential (AEP) threshold measurements in dolphins and whales has recently allowed the examination of high-frequency hearing in a number of toothed whales and dolphins, and the AEP method has been shown to produce auditory thresholds comparable to behaviorally obtained audiometrics (Yuen et al. 2005; Houser and Finneran 2006a; Nachtigall et al. 2007a). Because the animals have large brains, the potentials are available on the skin surface. Small human skin surface electrodes can be imbedded in soft latex suction cups and placed on the surface of the animal's skin over the brain. The evoked potentials can be used to measure hearing by examination of the response to the patterns of sound presented to the animal. The direct comparability of behavioral and the AEP procedures (Yuen et al. 2005) was made possible through the use of the envelope following response (EFR) (sometimes termed the ASSR or auditory steady-state response; Finneran et al. 2007) method of measuring toothed-whale hearing thresholds. Because of echolocation and the fast processing of echolocation pulses, the toothed whales are especially adapted for following fast modulation rates and make the technique particularly useful for that group of animals (Dolphin et al. 1995; Supin and Popov 1995a; Mooney et al. 2009). Hearing thresholds can be obtained by modulating the carrier frequency of interest at rates near one thousand times per second and filling the sound "envelopes" with a particular carrier frequency for which the investigator desires to know the threshold. The AEPs are examined in response to the envelopes and then transformed using Fast Fourier Transforms. The levels of the peaks at the modulation rates are noted, and a linear regression is performed on the peaks to estimate thresholds where the regression line crosses zero (Supin et al. 2001; Nachtigall et al. 2007a).

Since the development of the EFR procedure there have been attempts to measure hearing in gray whales (Ridgway and Carder 2001), sperm whales (Ridgway and Carder 2001; Nachtigall et al. 2007a), beaked whales (Cook et al. 2006; Pacini et al 2011), as well as the hearing of a stranded infant Risso's dolphin (Nachtigall et al. 2005). The use of AEP has allowed a rapid increase in the number of marine mammal hearing measurements. Rather than having to rely on a single animal's audiogram to represent the species, there are now good measures of population variability in the audiograms of bottlenosed dolphins from newly captured groups (Popov et al. 2007) as well as captive populations (Houser and Finneran 2006b). The AEP measurement in response to tone pips has been used to examine the hearing of anesthetized polar bears (Nachtigall et al. 2007b).

Though there has been a good history of study on the characteristics of the outgoing signals of toothed whales and dolphins during echolocation (Busnel and Fish 1980; Nachtigall and Moore 1988; Au 1993; Thomas et al. 2004), the direct measurement of evoked potentially measured *hearing* during actual echolocation has only recently been accomplished (Supin et al. 2003). The quantification of hearing while the whales and dolphins echolocated required the development of a technique to measure the AEPs in response to the outgoing clicks and returning echoes during an active echolocation task. This technique requires that the hearing be measured to individual events of outgoing and return pulses. So, unlike the envelope following response procedure for measuring hearing, where a thousand AEP responses to 1,000 ms envelopes of sound can be quickly averaged to examine a response level to a pure tone, single hearing responses to each outgoing click and its corresponding echo during echolocation must be gathered and averaged to produce thresholds to outgoing clicks and matched corresponding echoes (Supin et al. 2003).This newly developed ability to measure what a whole free-swimming active captive whale hears of its outgoing echolocation clicks, and its returning echoes, has enabled investigators to ask very basic questions about the processes underlying toothed whale echolocation, including questions about automatic gain control and forward masking.

5.4 Hearing Loud Signals and Quiet Returns

The discussion now returns to the question of loud outgoing signals and listening for quiet echo returns during whale echolocation. Supin et al. (2006) examined the hearing of a female false killer whale while the animal echolocated. The hearing of the false killer whale was examined with evoked potentials: (1) to her own outgoing clicks while she was actually echolocating aluminum cylinders presented 3 m away and (2) to simulated false killer whale clicks presented directly in front of the animal at a variety of comparable levels similar to her own clicks. Interestingly, as can be seen in Fig. 5.1, when targets were present the whale heard her own clicks 40 dB lower than she heard simulated false killer whale clicks of equal intensity presented directly in front of her.

So, at least a part of the issue of echolocation and outgoing signals causing problems with the hearing of quiet quickly returning echoes has been solved by the whale's hearing system by an overall dampening of the whale's hearing of its own

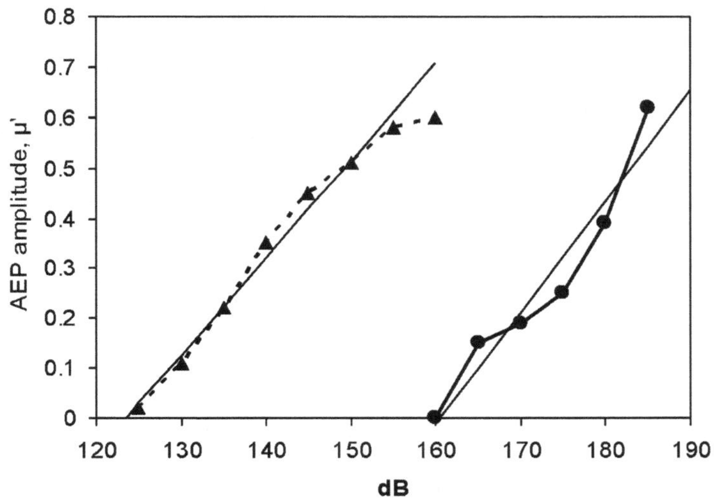

Fig. 5.1 *Triangle line* represents AEP passive listening response levels to external clicks while the *filled circle line* represents similar responses to clicks that the animal produced herself. Note a nearly 40-dB difference in "hearing" level between whether a click was self produced or presented from the outside. (Adapted from Supin et al. 2006)

outgoing signal. This has likely been accomplished by anatomical isolation of the ears from the skull (e.g., Ketten 2000) and air-filled peribullar and pterygoid sinuses along with the possibility of regulated hearing sensitivity by middle ear activation or potentially other mechanisms.

Thus the false killer whale has some protection from its very intense outgoing echolocation clicks. The whale heard its own outgoing clicks 40 dB less well than similar artificial whale clicks presented directly in front of her. Does the whale's hearing dampen, then immediately recover, in response to self-produced loud sounds, as was shown by Kick and Simmons (1984) in bats? There are several ways that a hearing system might accomplish a dampening and recovery of hearing in response to a loud short sound. One way is to anticipate the loud sound and dampen the hearing as the sound occurs, which is the way that the bat's stapedial muscle contracts at the same time the bat's cry is made to protect its ear. The other way is that the auditory system may be simply forward-masked by the loud sound. A very simple, and extreme, explanation of forward masking might be that when a person shoots a rifle and the sound is heard next to the ear. Sounds are then not heard very well with that ear for a while. The hearing of subsequent sounds in that ear are said to be forward-masked by the sound of the rifle shot.

Supin and Popov (1995b) played two pulses to Atlantic bottlenosed dolphins (*Tursiops truncatus*) while measuring their hearing response via auditory brain stem responses (ABRs) in order to see how the hearing of the second sound was affected by the first. If the two pulses were of the same amplitude, the hearing of the second one was reduced by two variables: (1) time since the first pulse and (2) the overall amplitude of the pulses. The first pulse forward masked the hearing of the second pulse. The longer the time between pulses the better the recovery from forward masking and the better the second pulse was heard, and the louder the pulses were the longer it took to recover from the effects of the first pulse. Hearing of the second pulse with two pulses at 80 dB, for example, appeared to be fully recovered after 2 ms, while hearing was not fully recovered at 10 ms if both pulses were 120 dB. Similar work with the false killer whale (Supin et al. 2007) showed similar results.

The work of Supin et al. (2007) provided some idea of what may be going on with the hearing system when two pulses are presented to the animal from the outside, but what of actual hearing during the echolocation task? The two variables that mattered in the forward-masking work were the amplitude of the signals and the time between them. In actual echolocation tasks those two can be varied by changing the distance to the target. A target farther away produces a weaker echo and it also produces an echo with a greater time between the click production and the echo return. Supin et al. (2004) examined the AEP hearing responses of the false killer whale while it echolocated an aluminum cylinder target at distances varying from 1 to 8 m. As can be seen by examining the AEP traces shown in Fig. 5.2, although there was very little difference in the outgoing echolocation pulses produced by the whale, the response to the echo returns also did not differ.

The AEP hearing responses to the quieter echoes were as large as they were to the much louder outgoing clicks (as much as 64 dB difference), and the responses

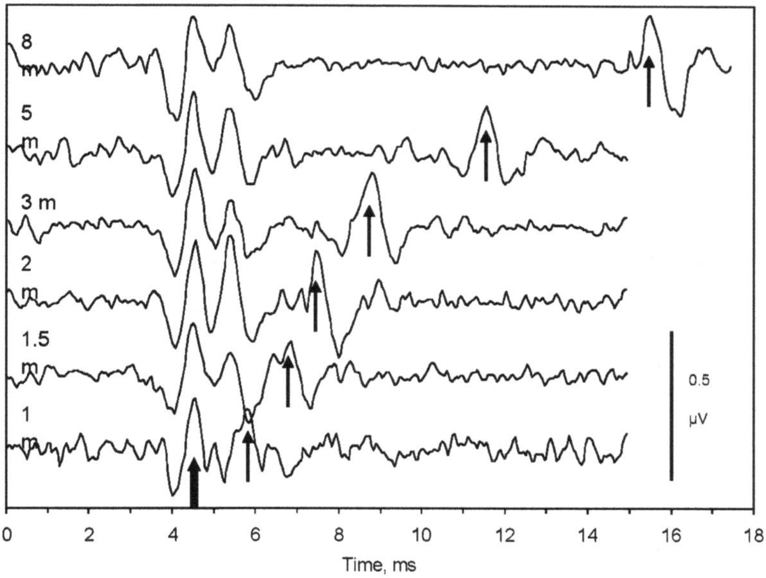

Fig. 5.2 AEP response level to echolocation targets presented at distances of 1, 1.5, 2, 3, 5, and 8 m from the false killer whale. Note that AEP response level is the same for quiet echoes and for loud outgoing clicks. AEP level does not change with target distance even though the echo return level changes by 36 dB. (Adapted from Supin et al. 2004)

to the target echoes from targets 8 m away were the same amplitude as AEP responses to targets 1 m away even though there was a 36 dB difference in the intensity of the different echoes. How could the animal hear 64 or 36 dB differences in intensity at the same level? How could the animal's auditory system avoid the forward masking "rifle shot" effect of the loud outgoing forward masking that should occur with those loud echolocation pulses? How could the target echoes differ by 36 dB yet the animal's auditory system respond as if they were all the same level? Three possible explanations were offered. The first was that the toothed whales have a very high temporal auditory resolution as had been shown in previous experiments by Moore et al. (1984) and Dubrovskiy (1990); the second was that the animal has a "significant muffling" of the outgoing signal so that it is not heard at full strength; and the third was that there were automatic gain control mechanisms in the auditory system based on release from forward-masking. As the time between the outgoing pulse and the received return echo increased, the hearing recovered and compensated. The longer time due to distance allowed the hearing to recover, enabling the distant quieter echoes to be heard at the same level as closer louder echoes in a manner similar to that been previously demonstrated in the dolphin two pulse forward-masking experiments (Supin and Popov 1995b).

Supin et al. (2005) then performed an experiment in which not only distance to the target but also the magnitude of the echo return was varied by using multiple cylinders of varying target strengths. These targets were presented at three distances

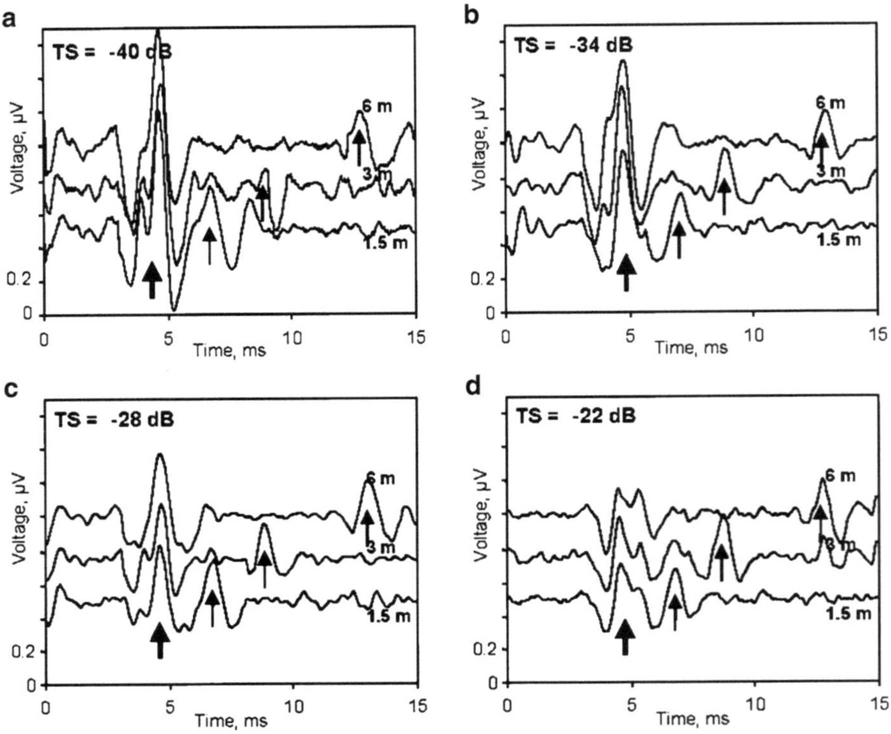

Fig. 5.3 Echolocation echo return AEPs recorded for outgoing clicks and echo returns at three distances (6, 3, and 1.5 m) and four different sized cylinders (**a**–**d**). Distance increased time between click and return and reduced echo strength. Decrease in cylinder size decreased target strength of cylinder. The smallest cylinder (**a**) –40 dB at 6 m and the largest cylinder (**d**) –22 dB at 1.5 m show little difference in AEP response size. Overall hearing, as demonstrated by size of AEP to outgoing clicks, apparently increased as the target size decreased. (Redrawn from Fig. 3 in Supin et al. 2005)

to further alter the strength of the returning echoes: 1.5, 3, and 6 m. The AEP averaged AEP brain wave records are presented in Fig. 5.3.

Looking first at the echo returns from targets at the various distances, it can be seen that the false killer whale heard most of the echoes at about the same level. The size of the AEP response was relatively equal across targets independent of distance to the target. The AEP responses to the target echoes were also relatively the same independent of the size of the target. The smallest target echo AEP response at the greatest distance seen in the top line of Fig. 5.3d to the right side is at about the same level as the largest target at the closest distance seen in Fig. 5.3a nearest the click echo. While the hearing of the echoes was measured at a near constant level, the *hearing* of the outgoing pulse changed dramatically. Despite the fact that there was negligible difference in the amplitude of the outgoing pulses, the hearing of those outgoing pulses changed as the targets changed. The differences in the relative sizes of the AEPs may be seen presented numerically graphed in Fig. 5.4.

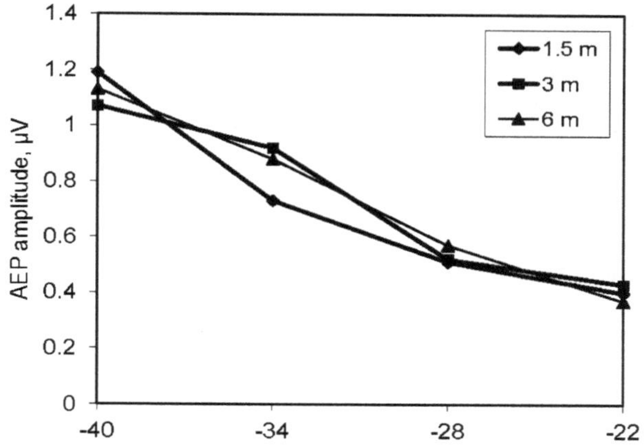

Fig. 5.4 AEP response level to the outgoing click depending on the size of the target echolocated. (Redrawn from Fig. 5a in Supin et al. 2005)

The AEPs to the outgoing pulses around 1.2 µV to the smallest targets were *three times* the size of the AEPs to the largest targets around 0.4 µV.

The only logical explanation for these data was that there was some other sort of active hearing control process going on for the whale. Whereas the bats control vocal self stimulation by attenuating the hearing of the call (Suga and Jen 1975), it seems as though perhaps a very different process is going on with the whale. The whale's ability to hear the outgoing pulse is obviously modulated by the size of the target echo. If the whale's echolocation were like that of the bats, all outgoing pulses would be essentially heard at about the same level. The whale would simply be protecting its ears from its own outgoing signal. However, the whale hears the same outgoing pulses at very different levels depending on the size of the echo return. It is as if the whale needs to adjust its hearing by "opening its ears" to keep the *echo level* high enough to hear it, and when the whale "opens its ears," it also must necessarily hear more of the outgoing signal. So, when the animal had a very small echolocation target at distance the AEPs to the outgoing signals were as high as 1.2 µV, but when the targets were larger and nearby giving larger echoes, the hearing of the outgoing signals was lessened to the 0.4 µV level, meanwhile keeping the echoes from the targets at a near constant level independent of their distance or target size. But, was there any other evidence for active control of hearing? During the Supin et al. (2006) work comparing the AEP responses to internal outgoing clicks and external clicks of similar amplitudes, AEPs were collected in both target present and target absent conditions. As may be seen in Fig. 5.5, the AEP responses to the outgoing clicks of similar levels were different depending on the target condition.

If targets were not present the animals heard outgoing clicks at a level about 15 dB lower than if the targets were there. Perhaps in the absence of stronger echoes, the increase in hearing sensitivity may be a way to search for weaker echoes. But, once again, outgoing clicks were heard at differing levels depending on the level of the echo return independent of level of the outgoing clicks and any forward-masking effect. If there were a stapedial control of hearing, it would likely function in a manner very different than the bat's assumed protective mechanism.

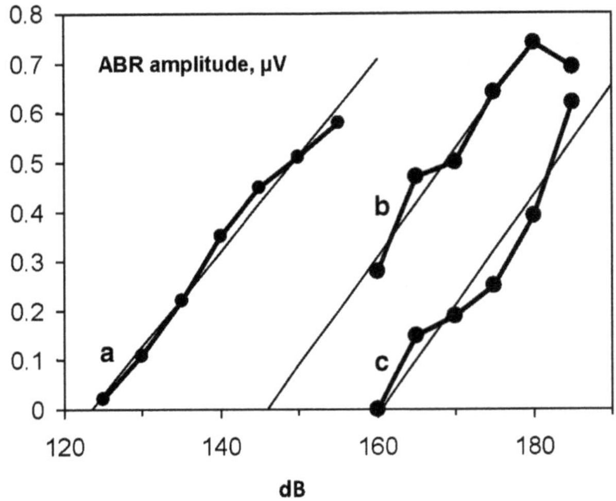

Fig. 5.5 AEP passive response levels to external clicks, clicks produced by the animal when echolocation targets are present, and when targets are absent. Note that the animal appears to be more sensitive and hears outgoing clicks at about a 15-dB lower level when targets are absent. (Adapted from Supin et al. 2006)

The question then arose as to the generality of the hearing change and difference when echolocation targets present and absent. Was the change in hearing limited to only echolocation clicks or was it a more general phenomenon? Supin et al. (2008) continued work with the false killer whale presenting two different sized cylinders: −22 and −34 dB. But they also presented a modulated 22.5-kHz tone to test the animal's hearing simultaneous with the echolocation performance. When a screen was lowered allowing the animal to echolocate, the amplitude-modulated 22.5-kHz tone was also presented for the 2 s that it took for the animal to echolocate the cylinder's presence or absence. The main finding was a "remarkable difference between the estimates of hearing sensitivity of the (whale) in the target present and target absent conditions." When the targets were absent, overall hearing of the 22.5-kHz tone, estimated with the EFR AEP, was almost 20 dB better than when targets were present. Apparently the strategy for echolocation searching at relatively close range in this whale is that in the absence of an echo that is loud enough an increased sensitivity should help to pick up fainter echoes, and adjustments in sensitivity are accomplished via a gain control of the auditory system.

Though the initial work measuring cetacean hearing during echolocation was carried out on the false killer whale, recently similar studies examining the hearing of the bottlenose dolphin (Li et al. 2011, 2012) and the harbor porpoise (*Phocoena phocoena*) (Linnenschmidt et al. 2012) while the animals echolocated have shown similar hearing changes.

5.5 Neural Mechanisms for Hearing in Echolocating Bats

The study of bat echolocation and hearing has proceeded differently than the study of toothed whale and dolphin hearing during echolocation. Echolocation has been considered as involving self-produced vocalizations that represent acoustical input to the hearing system. Thus, hearing in echolocating bats has been considered in a variety of conditions including (1) simple passive hearing (resting state, attentive state); (2) passive hearing during echolocating behavior (e.g., within groups, eavesdropping); and (3) listening to their own echoes after active emission of echolocation calls.

During active vocalization there are different possibilities as to how vocal activity and hearing can interact, and consequently different neural mechanisms will be involved. The physiological study of bat hearing and echolocation processes has allowed an objective look at those processes and their underlying mechanisms. Intrinsically effective corollary signals from the descending vocal motor system to the auditory processing pathway may eventually influence the processing of the echoes. This influence could consist in priming the hearing system and inducing a certain expectancy concerning the spectral properties and the temporal occurrence of the echoes.

Extrinsically, vocal utterings make for an acoustic stimulation prior to the return of the associated echoes and establish a typical two-tone situation involving forward masking effects. The acoustical self-stimulation is mediated by both internal and external sound transmission, and it is important to know the signal strength stimulating the cochlea during ongoing vocalization when attempting to estimate the masking or other auditory effects of the vocal utterings.

In addition to provoking masking effects, which would have to be considered as counteracting the reception of weak echoes, the vocalization most probably is an important constituent in active echolocation behavior, as it marks the reference parameters in time, intensity, and spectral content, thus narrowing down and specifying the expected parameter space in the echoes. This sort of priming can take effect not only intrinsically but also by acoustic stimulation, such as listening to one's own vocalization.

5.5.1 Self-Stimulation

An objective direct measurement of vocal self-stimulation in horseshoe bats (*Rhinolophus rouxi*, a bat species emitting an echolocation call consisting of a long constant frequency [CF] portion preceded and terminated by frequency modulated [FM] parts) has been performed only by Pietsch and Schuller (1987) using cochlear microphonic (CM) potential responses to vocalizations in comparison to the characteristics of CM responses to artificial stimuli at the CF resting frequencies of the bats. The equivalent stimulation level at resting frequency was on average −26.3 dB (range: −22 to −34 dB) lower than the emitted sound pressure level, and corresponded to that measured with a microphone near the ear canal entrance under the best receiving angle of the pinna (Fig. 5.6).

This level of acoustic self-stimulation is below the threshold for eliciting an acoustic middle ear reflex. However, the middle ear contraction concurrent to laryngeal activity during vocalization is active throughout the duration of the call and

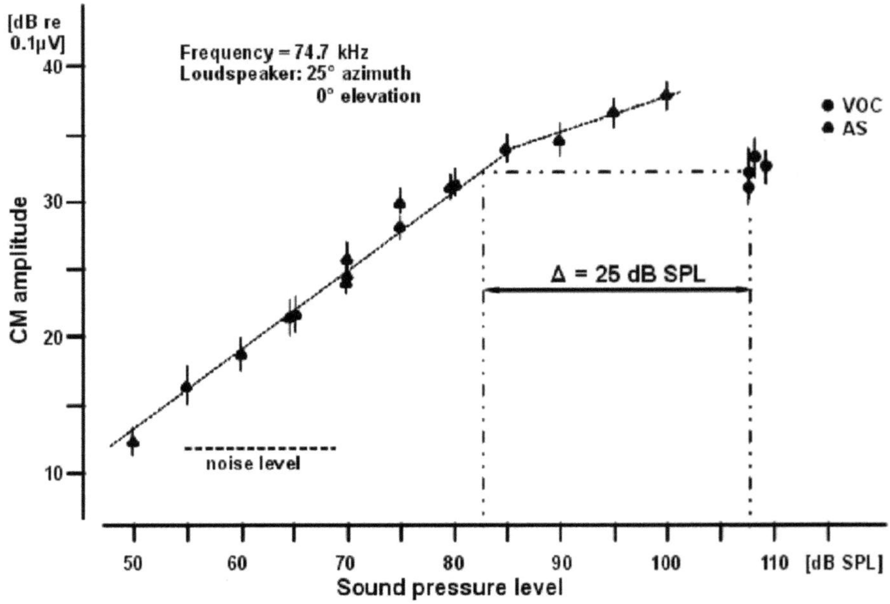

Fig. 5.6 Calibrations of auditory self-stimulation by echolocation call emission in the horseshoe bat, *Rhinolophus rouxi*. *Triangles* (AS) show microphonic potential responses to increasing sound pressure levels of artificial constant frequency signals at the bat's resting frequency (74.7 kHz). The *dots* (VOC) give the amplitude of microphonic potentials elicited by active echolocation call emission at around 108 dB SPL, which corresponds to a stimulation level of approximately 83 dB SPL. (Adapted from Figure 3 in Pietsch and Schuller 1987)

reaches a maximum of approximately −10 dB attenuation measured for a 30-kHz probe tone. This middle ear contraction seems to counteract the contribution of internal sound conductance as measured self-stimulation by the CM reflected the acoustic sound pressure near the ear canal entrance. No determination of the internal sound conduction from the vocal apparatus to the cochlea is currently available. Equivalent measurements of vocal self-stimulation in the large group of non-CF-FM bats, the FM-bats, which generally use much shorter mostly broad-band echolocation signals, are currently not available. The laryngeally bound middle-ear contraction is reported to be activated even before the onset of the vocalization in FM-bats such as the Mexican free-tailed bat (*Tadarida brasilensis*; Henson 1965) and little brown bat (*Myotis lucifugus*; Suga and Jen 1975), and contributes to attenuation of self-stimulation.

5.5.2 Masking

Does the self-stimulation constitute an effective masker to impair the detection and processing of fainter echoes in bats? This question has to be discussed separately for the different groups of bats because the masking affects hearing in bats differentially depending on the different call types and distinct response properties of the respective hearing systems.

The so-called CF-FM bats use long-lasting (several tens of milliseconds) calls constituted mainly of a narrow band CF-portion, so that self-stimulation is also long lasting and would mask echoes over several tens of milliseconds, thus deafening the bat for its near distance range. Doppler shift compensating CF-FM bats such as horseshoe bats (*Rhinolophus*) and mustached bats (*Pteronotus*) use two different main strategies for echolocating and prey detection: (1) the so-called fly-catcher style, in which the roosting bat is not flying but scanning the surroundings for moving prey (Neuweiler et al. 1987), or (2) the flying bat lowering its emitted frequency when compensating for Doppler shifts to stabilize the carrier echo frequency at a reference CF-frequency, where the hearing threshold is lowest (Schuller and Pollack 1979). As the frequency tuning of neurons at the reference frequency is extremely narrow (quality factors $Q_{10\text{dB}}$ of several 100) and as threshold toward lower frequencies shows an enormous rise, the slightest difference between the emitted vocal frequency and the frequency of the returning echo leads to a drastic decrease of forward masking effect of the vocalization (Schuller and Pollack 1979). In the fly-catcher situation the vocal signal lies at the increasing threshold slope at reduced self-stimulation level, whereas moving targets of interest for the bat produce frequency components at or above the reference frequency, and are little affected by forward masking. In the flying bat the lowered vocal frequency leads to even less self-stimulation in a high-threshold frequency band (reducing self-stimulation strength), and therefore loses forward masking impact on the response of narrowly tuned neurons in the low-threshold reference frequency band. Thus the separation of vocal and echo frequencies into different processing channels, as well as the effective reduction of self-stimulation due to higher thresholds at vocal frequencies, is an effective mechanism to minimize the problem of vocal self-masking in CF-FM bats (Neumann and Schuller 1991).

The echolocation situation in FM-bats is somewhat different in two aspects: the emitted calls are generally shorter (in the milliseconds range) and the signals cover a broader spectral range, thus spreading the call energy over a larger range of frequency channels. The only estimations of vocal self-stimulation in an FM-bat drawn from neurophysiological measurements stem from the Mexican free-tailed bat (Henson 1965). Henson's measurements yielded levels between 25 and 35 dB below the sound pressure level of the emitted sound, which corresponds well to the measures obtained for CF-FM bats.

The masking efficiency of short sounds depends, in addition to the intensity, largely on the duration in relation to the integration times encountered in the hearing system. If the hearing processing is conceived as frequency channel-oriented in the periphery, the effective stimulation during a FM-sweep (e.g., 10 kHz/ms) is short. As a function of the tuning characteristics of the system, the transient stimulation duration can be below the typical integration times of 100–200 μs after Wiegrebe (2008) or up to 350 μs after Saillant et al. (1993) or Peremans and Hallam (1998), and thus would exert only reduced forward masking effects at the peripheral level of processing.

This does not exclude that vocal self-stimulation has suppressive effects on advanced levels of processing, which would, however, signify more a context-dependent effect produced in the auditory processing network (see also Pecka et al. 2007; Nelson et al. 2009).

5.6 Vocal Influence on Auditory Processing and Facilitation

Besides suppressive effects of the vocal self-stimulation, there is experimental evidence that processing of acoustic stimulus features are processed differently depending on whether they are preceded by vocalization or presented in a passive listening situation. The available information from recordings in vocalizing bats is extremely scarce (Schuller 1979; Henson et al. 1982; Pietsch and Schuller 1987; Behrend and Schuller 2000), whereas neurophysiological experiments in awake, as well as anesthetized, but non-vocalizing bats have been numerous. A large variety of stimulus configurations mimicking a wide range of echolocation situation have been tested in single-unit and multi-unit recordings, covering the encoding of spectral, temporal, and spatial parameters.

The value of neurophysiologic results from stationary, non-behaving animals is generally under scrutiny, and there are strong indications that, for example, the tuning properties of neurons are largely dependent on what the animal is doing at the moment of acoustic stimulation (e.g., Fritz et al. 2007). Anesthetics can have marked influence on auditory responsiveness affecting differentially various auditory processing characteristics (e.g., Gaese and Ostwald 2001). Generally the dependence of the response characteristics in the hearing system on the behavioral state, for example, active behavior, wakefulness, inactivity, or under anesthesia is the more compelling the higher the processing level in the ascending auditory system is. In particular, results from auditory cortex have to be carefully considered when interpreting them in a behavioral context.

In bats, the true echolocation situation, as composed of a vocal signal and a delayed modified version of it as an echo, has been simulated in many neurophysiological recordings. The combination of two temporally delayed components, that is, in CF-FM-bats the combination of the first harmonic and the second harmonic of the FM-portion of the call, and in FM bats the combination of two FM sweep of the same harmonic, proved to be responded to by a strongly facilitated response when distinct temporal conditions were met, that is, the neural responses showed a distinct maximum of spike discharges around a so-called best delay between the components (auditory cortex: O'Neill and Suga 1982; Dear et al. 1993; medial geniculate body: Olsen and Suga 1991; inferior colliculus: Dear and Suga 1995; Mittmann and Wenstrup 1995).

A large number of experiments have been performed mostly in awake bats, but only few recordings have been taken in actively vocalizing bats (Kawasaki et al. 1988). In these experiments, the responses to an echo signal preceded by an active vocalization showed the same facilitation at a similar best delay as was found for the artificial combination tones. Thus the simulated vocal signal in the combination stimulus proved equivalent to an active vocalization as first "priming" signal of the facilitatory combination. The correspondence of the stimulus level optimal for the facilitated responses as found in the horseshoe bat *Rhinolophus rouxi* (Schuller et al. 1991) also corresponds well to the level of self-stimulation by the first harmonic as heard by the bat during vocalization (Pietsch and Schuller 1987). Although these experiments apparently show that vocalization can be simulated by an artificial replica, caution should be taken in making a straightforward interpretation of the existing data on facilitation that all such neurons would behave the same way in echolocation as they do in the recording situation. That prior ongoing vocalization can have marked influence on the processing of subsequent delayed stimulation was shown for inferior colliculus neurons in the horseshoe bat (Schuller 1979). Besides neurons that were inhibited due to the preceding vocalization, neurons of the inferior colliculus (IC) were also found that displayed clearly enhanced response to an echo stimulus when active vocalization, and not an artificial replica, preceded it. Some neurons encoded features of a stimulus (sinusoidal frequency modulation) only when presented in a distinct time period following active vocalization and the neuron stayed silent if the vocalization was replaced by a replica of the vocalization under otherwise identical conditions (Fig. 5.7).

Very recently, a new approach to record from neuronal clusters or single neurons in the IC of the phyllostomid bat *Phyllostomus discolor*, implanted permanently with stimulation electrodes for eliciting species-specific vocalizations by mild electrical micro-stimulation in the paralemniscal area (Schuller and Radtke-Schuller 1990) was presented by Hoffmann and Firzlaff (2009). The elicited vocalizations were picked up electronically, modified in real time by convolution with impulse responses characterizing reflecting virtual objects, and played back to the bat in a method similar to that used by Aubauer et al. (2000) for echolocating dolphins in behavioral echolocation discrimination experiments. The results of the dolphin experiments indicated that dolphins would accept computer-generated echoes as real targets only if the echoes were generated from their own outgoing clicks and then convoluted with the impulse response of the real targets. Recorded replicas of echoes were not accepted as real targets to be discriminated. Hoffmann and Firzlaff (2009) found that collicular responses to vocalizations and echoes in an actively vocalizing bat could markedly differ from the responses in a mimicked passive listening situation, where the vocalization was replaced by a recorded replica. Some neurons were evidently not purely driven by the acoustical input, but were also influenced intrinsically by active vocal behavior similar to what was found in the dolphin work. These examples demonstrate clearly two important aspects for hearing processing during vocalization and probably echolocation: (1) Responses to the same echo stimulus can differ considerably depending on whether the signal trails behind an active vocalization or an acoustically identical replica of it, and (2) the process of vocalization by itself may intrinsically influence how and within which time window the echo signal is processed. As a consequence, the simulation of a typical echolocation configuration portrays passive listening to a two-tone stimulation, but does not obligatorily reflect the processing of echoes as going on in an animal actively engaged in echolocation.

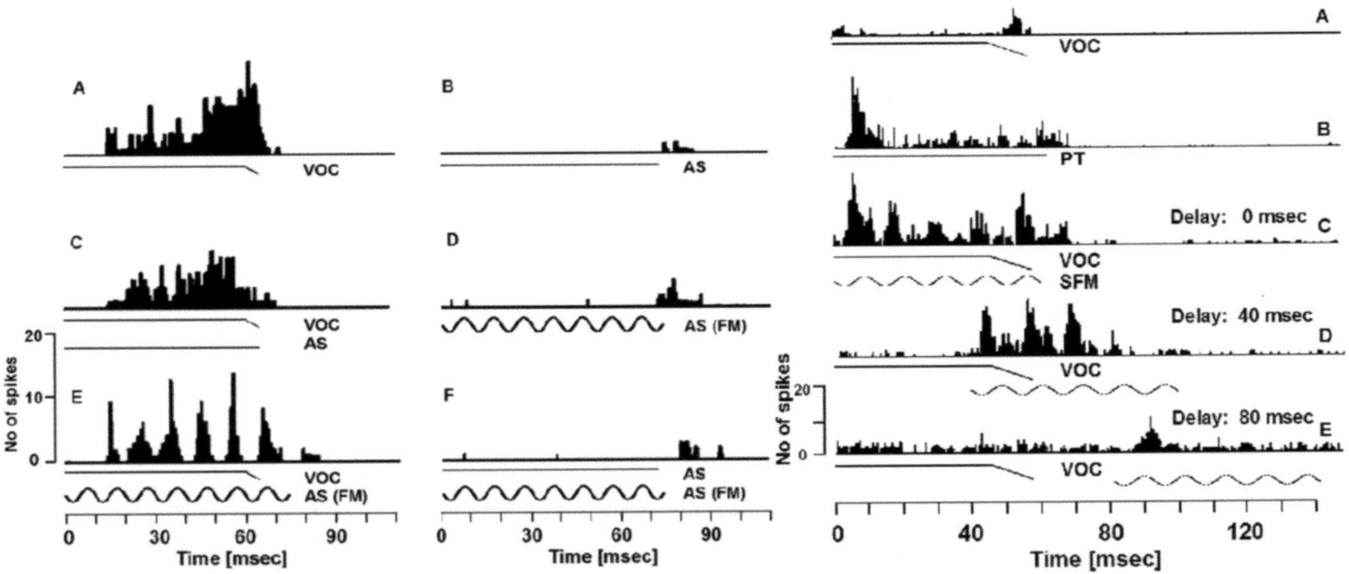

Fig. 5.7 Responses of inferior colliculus neurons in the horseshoe bat, *Rhinolophus ferrumequinum*, to emitted vocalizations (VOC) (*left* **a** and *right* **a**), simulations of the constant frequency portion (AS or PT) (*left* **c** and *right* **b**), sinusoidally frequency modulated stimuli (AS(FM) or SFM), and combination of these stimuli. *Left*: Responses of neurons to modulated signals in the actively vocalizing bat (**e**) differ distinctly from those to a mimicked, acoustically comparable, situation (**f**) and the responses to the different stimuli alone (**b, d**). *Right*: Synchronized responses to modulated echoes in the actively vocalizing bat show time-dependent responses (**c–e**) with enhancement in specific delay ranges (e.g., **d**). (Adapted from Schuller 1979, Figures 3 and 4)

The scarceness of recordings in echolocating bats is most probably due to the methodological difficulties in recording from vocalizing animals. Each vocalization leads to slight shocks transmitted from the moving larynx over the spinal cord to the brain, limiting considerably the recording probability and stability, at least when recording with highly selective electrodes. There is no realistic workaround to this problem other than the use of more flexible or floating electrodes in a fixed preparation in which vocalization can be elicited with electrical micro-stimulation, and to make do with multi-unit recordings combined with post-processing with offline spike sorter software. Freely moving behaving bats at a body mass of above 15 g may be implanted with a micro-electrode drive carrying four tetrodes as long as the behavioral experiment can be run under tethered conditions (Ulanovsky and Moss 2007). Small and light enough telemetry systems with proper adaptation for the transmission of neural signals over several parallel channels should technically be in reach for laboratory use (e.g., Schregardus et al. 2006).

5.7 Corollary Discharges and Efferent Influences on Auditory Processing

The auditory system in echolocating bats represents, as in other animals, only one sensory branch of an integrated sensory-motor ensemble with numerous interactions in either direction and subjected to influences from general modulating systems such as those found in arousal or attention. The processing of auditory signals is influenced on various levels by efferent inputs from sources within the auditory pathway or from nonauditory brain regions (e.g., vocal, motor; Schuller and Radtke-Schuller 1988; Huffman and Henson 1990; for general review: Crapse and Sommer 2009). Which of the acoustical features are extracted and how they are processed, such as from echoes in bat echolocation, may be largely influenced by these efferent, corollary, and modulatory activities. Whereas efferent auditory feedback mechanisms are efficient in inactive passively listening bats, corollary discharges that can influence afferent processing as well as efferent motor feedback are operating only in behaving bats, either vocalizing or flying.

Efferent influences emanating from auditory cortex and targeting the inferior colliculus have been studied in the bat most extensively by Suga and co-workers (for a review see Suga 2008). Electrical micro-stimulation in auditory cortex at a specific best frequency region (BF_{Stim}) results in an enhancement of the responses of collicular neurons with the same best frequency together with a sharpening of their frequency tuning. The best frequencies of collicular neurons slightly deviant from the best frequency of the cortical stimulation site's BF are either shifted toward or away from this best frequency. This signifies a shift of the tuning area nearer to the stimulated BF_{Stim}, meaning that more neurons are activated by the stimulus at this frequency, which together with the enhanced center response leads to an overall increase of cluster activity. In the alternative case the response areas are pushed away by the efferent cortical influence, and the activation of neurons with BF slightly off the enhanced frequency BF_{Stim} is reduced. This corresponds to a contrast enhancement along the frequency axis around the center frequency. These efferent effects of cortical stimulation have been demonstrated in neurons of the central nucleus of the IC, which is not considered to be a direct target of descending cortical efferents (Herbert et al. 1991), and must therefore result from more complicated interaction between neurons of the dorsal cortex or the external nucleus of the IC with those of the central nucleus. The dorsal cortex and external nucleus of the IC are in addition targets of non-auditory inputs from locations such as the somatosensory and the vocal system (Prechtl 1995). Electrical micro-stimulation in the external nucleus and in the pericentral rostral pole of the IC elicits emission of echolocation calls and points to a connection with the descending vocalization system (Prechtl 1995). The central nucleus of the IC is only the final target on the midbrain level of control mechanisms mediated by the dorsal collicular cortex, the rostral pole, and the external nucleus of the IC. Studies in the context of hearing during echolocation should therefore concentrate on corollary discharges and efferent auditory feedback to these sub-nuclei of the IC, in order to get more insight into the mechanisms at work at auditory midbrain level.

Short-term plasticity of cortical neurons induced by electrical micro-stimulation in the medial geniculate body (Ma and Suga 2009), or by fear conditioning or unspecific conditioning in the big brown bat (Suga 2008, for review), has again demonstrated the crucial dependency of auditory processing on the animal's behavioral state. The essential question remains whether and to which extent such induced plasticity of the auditory processing similarly occurs during active vocalization and echolocation behavior.

5.8 Echolocation and Passive Listening in Groups

Hearing during echolocation in a bat orienting and hunting alone at safe distance from other bats is fundamentally different from the situation of bats hunting concertedly in groups. All echoes in the former situation are causally related to the own emitted signal and give a coherent image of the echo reflecting scene, which is earmarked by compatible echo parameters (delay, sound pressure level, spectral peculiarities). Such a referencing relationship between echoes and emitted vocalization does not occur undisturbed by a bat flying and hunting in conjunction with other conspecifics or other species. Besides a large body of investigations dealing with jamming effects and interference and possible avoidance strategies (Ulanovsky et al. 2004; Gillam et al. 2007; Chiu et al. 2009), some studies have demonstrated that bats can also benefit from unrelated echoes originating from calls of other bats (Chiu et al. 2008). These authors found that bats flying in company of another bat reduced their own vocalizations by interspersing long silent periods in dependence on the distance to the other bat, mainly below 1 m. Besides the explanation that this behavior is mainly a strategy to avoid jamming, it could also mean that a bat can possibly locate prey by passive listening to vocalizations and echoes from other bats under specific echolocation conditions, for example, flying in known surroundings. Although the silent bat does not have its own vocal reference, the loudest signal when flying near to each other will definitely be the emitted vocalization of the active bat and could thus serve as a spatial reference to locate the passively received echoes in relation to the bat's own spatial coordinates. Owing to a certain spectral uniformity of the echolocation calls used by individuals of the same species, the auditory spectral filtering mechanisms would be rather compatible and provide valuable information to the non-echolocating bat during passive hearing. Echolocation in conjunctively orienting and hunting bats could thus provide realistic advantages of such a group strategy. This same strategy may also apply to swimming hunting cetaceans but remains to be investigated (Gregg et al. 2007).

Because of non-correspondence of the spectral call patterns in conjunctively echolocating bats of different species, the useful information content extractable by passive hearing would be reduced to the detection of the spatial presence of echo reflectors rather than yielding information on the structure of the reflectors. There is, however, no experimental evidence that passive listening is used at all in encounters of bats of different species during echolocating behavior.

5.9 Comparisons of Whale and Bat Hearing Measured During Echolocation

The precise and objective measures of bat hearing, like that shown by monitoring a bat's hearing of its own calls by measuring cochlear microphonics at the round window of the cochlea (Schuller 1979), will unlikely be duplicated in the whales and dolphins. Cetaceans are voluntary breathers and anesthesia requires extraordinary efforts (Ridgway and McCormick 1967). That, along with Society's current respect for cetaceans, makes it very unlikely that underlying physiological hearing processes and mechanisms will be examined. Measures of the evoked auditory potentials taken from the skin surface of the false killer whale do, however, show some interesting comparative values to those obtained from bats. Suga and Shimozawa (1974) found that bats of the genus *Myotis* attenuate the sound of self-stimulation during echolocation type calls by 20–25 dB by activating the stapedial muscles of the middle ear and an additional 15 dB by neural processes for a total attenuation of 35–40 dB. The false killer whale hears its own outgoing clicks 40 dB down from those presented directly in front of it (Supin et al. 2006). While the mechanisms may differ, the initial level of protection from the outgoing echolocation signals appears to be similar.

There is still actually little known about echolocation in whales and dolphins and even less known about hearing while they echolocate. Though there are more than 800 species of echolocating bats there are only 72 species of described toothed whales that are likely to echolocate. The jury is still out on whether the larger mysticete whales, such as the humpback whale, might have some sort of rudimentary echolocation (Stimpert et al. 2007). Echolocation outgoing signals of only a few of those 72 species have been adequately described and basic audiograms of only 17 of those species are currently available. There are obvious differences among the described toothed whales: Sperm whales produce intense low frequency clicks (Madsen et al. 2002; Møhl et al. 2003), harbor porpoises produce frequency modulated narrow-band high frequency signals (Kamminga and Wiersma 1981), and beaked whales produce relative narrow band lower frequency, frequency modulated signals (Madsen et al. 2005). However, we know little about how those different signals or different hearing among the animals might produce different mechanisms or processes of hearing during echolocation.

5.10 Summary

In both whales and bats, behavioral studies have yielded a large body of information on performance in different echolocation tasks and on the use of sound parameters. The study of hearing and of neuronal processing mechanisms in whales and bats has taken different paths because of profound differences in feasibility. Much neurophysiological bat work has allowed objective measures of processes occurring within the brain during passive hearing while most all of the work with whales and dolphins has been behavioral. Neither behavioral investigations nor neural characteristics of passive hearing provide for an unequivocal clarification of auditory processing going on when the animals are actively performing echolocation tasks. Only recently have larger brain events been examined in whales through measurement of AEPs from the surface of the skin through suction cup electrodes. Also in bats, the investigation of hearing in actively vocalizing bats has found new interest. Both animal groups have evidently evolved mechanisms to actively manipulate hearing in order to overcome the forward masking effect of loud outgoing signals that could overshadow the essential information found within the quieter returning echoes.

Because hearing changes can now be experimentally monitored, the effects of learning and voluntary control of hearing processes can be explored. Interesting questions such as "Can a whale change its hearing thresholds to avoid anticipated passive loud noise events?" can be answered (Nachtigall and Supin 2013). Besides further recording neural activity in actively behaving animals confined to auditory areas, experimenters should also consider that auditory processing is not an isolated process and investigators should pay more attention to influences of the inputs from various non-auditory systems such as motor system (vocal, outer ear) or general systems modulating the level of arousal, attention and motivation. Although the experimental approaches will inherently be different in the two animal groups, a comparison of modulatory non-auditory influences might mutually complement the understanding of "hearing while echolocating" in these animals.

References

Au, W. W. L. (1980). Echolocation signals in open waters. In R. G. Busnel & J. F. Fish (Eds.), *Animal sonar systems* (pp. 251–282). New York: Plenum Press.

Au, W. W. L. (1993). *The sonar of dolphins*. New York: Springer-Verlag.

Au, W. W. L., & Benoit-Bird, K. J. (2003). Automatic gain control in the echolocation system of dolphins. *Nature*, 423, 861–863.

Aubauer, R., Au, W. W. L., Nachtigall, P. E., Pawloski, D. A., & DeLong, C. M. (2000). Classification of an electronically generated phantom target by an Atlantic bottlenose dolphin (*Tursiops truncatus*). *Journal of the Acoustical Society of America*, 107, 2750–2754.

Behrend, O., & Schuller, G. (2000). The central acoustic tract and audio-vocal coupling in the horseshoe bat, *Rhinolophus rouxi. European Journal of Neuroscience*, 12, 4268–4280.

Busnel, R. G., & Fish, J. F. (1980). *Animal sonar systems*. New York: Plenum Press.

Chiu, C. H., Xian, W., & Moss, C. F. (2008). Flying in silence: Echolocating bats cease vocalizing to avoid sonar jamming. *Proceedings of the National Academy of Sciences of the USA*, 105(35), 13116–13121.

Chiu, C., Xian, W., & Moss, C. F. (2009). Adaptive echolocation behavior in bats for the analysis of auditory scenes. *Journal of Experimental Biology*, 212(9), 1392–1404.

Cook, M. L. H., Varcia, R. A., Goldstein, J. D., McCullock, S. D., Bossart, G. D., Finneran, J. J., Houser, D., & Mann, D. (2006). Beaked whale auditory evoked potential hearing measurements. *Journal of Comparative Physiology A*, 192, 489–495.

Crapse, T. B., & Sommer, M. A. (2009). Corollary discharge across the animal kingdom. *Nature Reviews in Neurosciences*, 9, 587–600.

Dear, S. P., & Suga, N. (1995). Delay-tuned neurons in the midbrain of the big brown bat. *Neurophysiology*, 73(3), 1084–1100.

Dear, S. P., Fritz, J., Haresign, T., Ferragamo, M., & Simmons, J. A. (1993). Tonotopic and functional organization in the auditory cortex of the big brown bat, *Eptesicus fuscus. Journal of Neurophysiology*, 70(5), 1988–2009.

Dolphin, W. F., Au, W. W. L., Nachtigall, P. E., & Pawloski J. L. (1995). Modulation rate transfer functions to low frequency carriers by three species of cetaceans. *Journal of Comparative Physiology A*, 177, 235–245.

Dubrovskiy, N. A. (1990). On the two auditory systems in dolphins. In J. Thomas & R. Kastelein (Eds.), *Sensory abilities of cetaceans: Laboratory and field evidence* (pp. 233–254). New York: Plenum Press.

Finneran, J. J., Schlundt, C. E., Branstetter, B., & Dear, R. L. (2007). Assessing temporary threshold shift in a bottlenose dolphin (*Tursiops truncatus*) using multiple simultaneous auditory evoked potentials. *Journal of the Acoustical Society of America*, 122, 1249–1264.

Fritz, J. B., Elhilali, M., David, S. V., & Shamma, S. A. (2007). Does attention play a role in dynamic receptive field adaptation to changing acoustic salience in AI? *Hearing Research*, 229, 186–203

Gaese, B. H., & Ostwald, J. (2001). Anesthesia changes frequency tuning of neurons in the rat primary auditory cortex. *Journal of Neurophysiology*, 86, 1062–1066.

Gillam, E. H., Ulanovsky, N., & McCracken, G. F. (2007). Rapid jamming avoidance in biosonar. *Proceedings of the Royal Society B: Biological Sciences*, 274, 651–660.

Gregg, J. D., Dudzinski, K. M., & Smith H. V. (2007). Do dolphins eavesdrop on the echolocation signals of conspecifics? *International Journal of Comparative Psychology*, 20, 65–88.

Griffin, D. R. (1958). *Listening in the dark*. New Haven CT: Yale University Press.

Henson, O. W. (1965). The activity and function of the middle ear muscles in echolocating bats. *Journal of Physiology*, 180, 871–887.

Henson, O. W., Pollak, G. D., Kobler, J. B., Henson, M. M., & Goldman, L. J. (1982). Cochlear microphonic potentials elicited by biosonar signals in flying bats, *Pteronotus p parnellii. Hearing Research*, 7, 127–147.

Herbert, H., Aschoff, A., & Ostwald, J. (1991). Topography of projections from the auditory cortex to the inferior colliculus in the rat. *Journal of Comparative Neurology*, 304(1), 103–122.

Hoffmann, S., & Firzlaff, U. (2009). Vocalization influences neural processing of acoustic echoes. In *5th Animal Sonar Symposium*, September 14 –18, 2009 Doshisha University, Kyoto, Japan.

Houser, D. S., & Finneran, J. J. (2006a). A comparison of underwater hearing sensitivity in bottlenose dolphins (*Tursiops truncatus*) determined by electrophysiological and behavioral methods. *Journal of the Acoustical Society of America*, 120, 1713–1722.

Houser, D. S., & Finneran, J. J. (2006b). Variation in the hearing sensitivity of a dolphin population determined through the use of evoked potential audiometry. *Journal of the Acoustical Society of America*, 120, 4090–4099.

Huffman, R. F., & Henson, O. W., Jr. (1990). The descending auditory pathway and acousticomotor systems: Connections with the inferior colliculus. *Brain Research Review*, 15(3), 295–323.

Kamminga, C., & Wiersma, H. (1981). Investigations on cetacean sonar II. Acoustical Similarities and differences in odontocete sonar signals. *Aquatic Mammals*, 8, 41–62.

Kawasaki, M., Margoliash, D., & Suga, N. (1988). Delay-tuned combination-sensitive neurons in the auditory cortex of the vocalizing mustached bat. *Journal of Neurophysiology*, 59, 623–635.

Ketten, D. R. (2000). Cetacean ears. In W. W. L. Au, A. N. Popper, & R. R. Fay (Eds), *Hearing by whales and dolphins* (pp. 43–108). New York: Springer-Verlag.

Kick, S. A., & Simmons, J. A. (1984). Automatic gain control in the bat's sonar receiver and the neuroethology of echolocation. *Journal of Neuroscience*, 4(11), 2725–2737.

Li, S., Nachtigall, P. E., & Breese, M. (2011). Dolphin echolocation: Evoked potential responses in an Atlantic bottlenose dolphin (*Tursiops truncatus*). *The Journal of Experimental Biology*, 214, 2027–2035.

Li, S., Nachtigall, P. E., & Breese, M. (2012). Auditory brain stem responses associated with echolocation in an Atlantic bottlenose dolphin (*Tursiops truncatus*) In A. N. Popper & A. Hawkins (Eds.), *The effects of noise on aquatic life* (pp. 45–47). New York: Springer Science+Business Media.

Linnenschmidt, M., Beedholm, K., Wahlberg, M., Hojer-Kristensen, J., & Nachtigall, P. E. (2012). Keeping returns optimal: Gain control exerted through sensitivity adjustments in the harbour porpoise auditory system. *Proceedings of the Royal Society of London B: Biological Sciences*, 279, 2237–2245.

Ma, X., & Suga, N. (2009). Specific and nonspecific plasticity of the primary auditory cortex elicited by thalamic auditory neurons. *Journal of Neuroscience*, 29(15), 4888–4896.

Madsen, P. T., Wahlberg, M., & Møhl, B. (2002). Male sperm whale (*Physeter macrocephalus*) acoustics in a high-latitude habitat: Implications for echolocation and communication. *Behavioral Ecology and Sociobiology*, 53, 31–41.

Madsen, P. T., Johnson, M., Aguilar de Soto, N., Zimmer, W. M. X., & Tyack, P. (2005). Biosonar performance of foraging beaked whales (*Mesoplodon densirostris*). *Journal of Experimental Biology*, 208, 181–194.

Mittmann, D., & Wenstrup, J. J. (1995). Combination-sensitive neurons in the inferior colliculus. *Hearing Research*, 90(1–2), 185–191.

Møhl, B., Wahlberg, M., Madsen, P. T., Heerfordt, A., & Lund, A. (2003). The monopulsed nature of sperm whale clicks. *Journal of the Acoustical Society of America*, 114, 1143–1154.

Mooney, T. A., Nachtigall, P. E., Taylor, K. A., Rasmussen, M., & Miller, L. (2009). Auditory temporal resolution of a wild white-beaked dolphin (*Lagenorhynchus albirostris*). *Journal of Comparative Physiology A*, 195, 375–384.

Moore, P. W. B., Hall, R. W., Friedl, W. A., & Nachtigall, P. E.(1984). The critical interval in dolphin echolocation: What is it? *Journal of the Acoustical Society of America*, 76, 314–317.

Nachtigall, P. E., & Moore, P. W. B. (1988). *Animal sonar: Processes and performance*. New York: Plenum Press.

Nachtigall, P. E., Yuen, M. E., Mooney, T. A., & Taylor, K. A. (2005). Hearing measurements from a stranded infant Risso's dolphin (*Grampus griseus*). *Journal of Experimental Biology*, 208, 4181–4188.

Nachtigall, P. E., Mooney, T. A., Taylor, K. A., & Yuen, M. L. (2007a). Hearing and auditory evoked potential methods applied to odontocete cetaceans. *Aquatic Mammals*, 33(1), 6–13.

Nachtigall, P. E., Supin, A. Y., Amundin, M., Röken, B., Møller, T., Mooney, T. A., Taylor, K. A., & Yuen, M. E. (2007b). Polar bear (*Ursus maritimus*) hearing measured with auditory evoked potentials. *Journal of Experimental Biology*, 210, 1116–1122.

Nachtigall, P.E. and Supin, A. Ya. (2013) False killer whales reduce their hearing sensitivity if a loud sound is preceded by a warning. *Journal of Experimental Biology*, 216, 3062–3070

Nelson, P. C., Smith, Z. M., & Young, E. D. (2009). Wide-dynamic-range forward suppression in marmoset inferior colliculus neurons is generated centrally and accounts for perceptual masking. *Journal of Neuroscience*, 29(8), 2553–2562.

Neumann, I., & Schuller, G. (1991). Spectral and temporal gating mechanisms enhance the clutter rejection in the echolocating bat, *Rhinolophus rouxi. Journal of Comparative Physiology A*, 169(1), 109–116.

Neuweiler, G., Metzner, W., Heilmann, U., Rübsamen, R., Eckrich, M., & Costa, H. H. (1987). Foraging behaviour and echolocation in the rufous horseshoe bat (*Rhinolophus rouxi*) of Sri Lanka. *Journal of Behavioral Ecology and Sociobiology*, 20, 53–67.

Olsen, J. F., & Suga, N. (1991). Combination-sensitive neurons in the medial geniculate body of the mustached bat: Encoding of target range information. *Journal of Neurophysiology*, 65(6), 1275–1296.

O'Neill, W. E., & Suga, N. (1982). Encoding of target range and its representation in the auditory cortex of the mustached bat. *Journal of Neuroscience*, 2(1), 17–31.

Pacini, A. F., Nachtigall, P. E., Quintos, C., Schofield, D., Look, D. A., Levine, G., & Turner, J. (2011). Audiogram of a stranded Blainville's beaked whale (*Mesoplodon densirostris*) measured using auditory evoked potentials. *Journal of Experimental Biology*, 214, 2409–2415.

Pecka, M., Zahn, T. P., Saunier-Rebori, B., Siveke, I., Felmy, F., Wiegrebe, L., Klug, A., Pollak, G. D., & Grothe, B. (2007). Inhibiting the inhibition: A neuronal network for sound localization in reverberant environments. *Journal of Neuroscience*, 27(7), 1782–1790.

Peremans, H., & Hallam, J. (1998). The spectrogram correlation and transformation receiver, revisited. *Journal of the Acoustical Society of America*, 104(2), 1101–1110.

Pietsch, G., & Schuller, G. (1987). Auditory self-stimulation by vocalization in the CF-FM bat, *Rhinolophus rouxi. Journal of Comparative Physiology A*, 160, 635–644.

Popov, V. V., Supin, A. Ya., Pletenko, M. G., Tarakanov, M. B., Klishin, V. O., Bulgakavo, T. N., & Rosanova, E. I. (2007). Audiogram variability in normal bottlenose dolphins (*Tursiops truncatus*). *Aquatic Mammals*, 33, 14–24.

Prechtl, H. (1995). Senso-motorische Wechselwirkung im auditorischen Mittelhirn der Hufeisennasen-Fledermaus, *Rhinolophus rouxi*. Doctoral dissertation, University of Munich, Germany.

Rasmussen, M. H., Miller, L. A., & Au, W. W. L. (2002). Source levels of clicks from free-ranging white beaked dolphins (*Lagenorhinchus albirostris* Gray 1846) recorded in Icelandic waters. *Journal of the Acoustical Society of America*, 111, 1122–1125.

Ridgway, S. H., & McCormick, J. G. (1967). Anesthetization of porpoises for major surgery. *Science*, 158, 510–512.

Ridgway, S. H., & Carder, D. A. (2001). Assessing hearing and sound production in cetacean not available for behavior audiograms: Experiences with sperm, pygmy sperm, and gray whales. *Aquatic Mammals*, 27(3), 267–276.

Saillant, P. A., Simmons, J. A., Dear, S. P., & McMullen, T. A. (1993). A computational model of echo processing and acoustic imaging in frequency-modulated echolocating bats: The spectrogram correlation and transformation receiver. *Journal of the Acoustical Society of America*, 94(5), 2691–2712.

Schregardus, D. S., Pieneman, A. W., Ter Maat, A., Jansen, R. F., Brouwer, T. G. F., & Gahr, M. L. (2006). A lightweight telemetry system for recording neuronal activity in freely behaving small animals. *Journal of Neuroscience Methods*, 155(1), 62–71.

Schuller, G. (1979). Vocalization influences auditory processing in collicular neurons of the CF-FM bat, *Rhinolophus ferrumequinum. Journal of Comparative Physiology A*, 132, 39–46.

Schuller, G., & Pollack, G. (1979). Disproportionate frequency representation in the inferior colliculus of Doppler-compensating greater horseshoe bats: Evidence for an acoustic fovea. *Journal of Comparative Physiology A*, 132(1), 47–54.
Schuller, G., & Radtke-Schuller, S. (1988). Midbrain areas as candidates for audio-vocal interface in echolocating bats. In P. E. Nachtigall & P. W. B. Moore (Eds.), *Animal sonar: Processes and performance*. New York: Plenum Press.
Schuller, G., & Radtke-Schuller, S. (1990). Neural control of vocalization in bats: Mapping of brainstem areas with electrical microstimulation eliciting species-specific echolocation calls in the rufous horseshoe bat. *Experimental Brain Research*, 79(1), 192–206.
Schuller, G., O'Neill, W. E. & Radtke-Schuller, S. (1991) Facilitation and delay sensitivity of auditory cortex neurons in CF-FM bats, *Rhinolophus rouxi* and *Pteronotus p. parnellii*. *European Journal of Neuroscience*, 3, 1165–1181.
Stimpert, A. K., Wiley, D. N., Au, W. W. L., Johnson, M. P., & Arsenault, R. (2007). Megaclicks: Acoustic click trains and buzzes produced during nighttime foraging of humpback whales (*Megaptera novaeangliae*). *Biological Letters*, 3, 467–470.
Suga, N. (2008). Role of corticofugal feedback in hearing. *Journal of Comparative Physiology A*, 194, 169–183.
Suga, N., & Schlegel, P. (1972). Neural attenuation of responses to emitted sounds in echolocating bats. *Science*, 177, 82–84.
Suga, N., & Shimozawa, T. (1974). Site of neural attenuation of responses to self-vocalized sounds in echolocating bats. *Science*, 183, 1211–1212.
Suga, N., & Jen, P. H. S. (1975). Peripheral control of acoustic signals in the auditory system of echolocating bats. *Journal of Experimental Biology*, 62, 277–311.
Supin, A. Ya., & Popov, V. V. (1995a). Envelope-following response and modulation rate transfer function in the dolphin's auditory system. *Hearing Research*, 92, 38–45.
Supin, A. Ya., & Popov, V. V. (1995b). Temporal resolution in the dolphin's auditory system revealed by double-click evoked potential study. *Journal of the Acoustical Society of America*, 97, 2586–2593.
Supin, A. Ya. , Popov, V. V., & Mass, A. M. (2001). *The sensory physiology of aquatic mammals*. Boston: Kluwer Academic.
Supin, A. Ya., Nachtigall, P. E., & Pawloski, J. L. (2003). Evoked potential recording during echolocation in a false killer whale (*Pseudorca crassidens*). *Journal of the Acoustical Society of America*, 113(5), 2408–2411.
Supin, A. Ya., Nachtigall, P. E., Au, W. W. L., & Breese, M. (2004). The interaction of outgoing echolocation pulses and echoes in the false killer whale's auditory system: Evoked potential study. *Journal of the Acoustical Society of America*, 115(6), 3218–3225.
Supin, A. Ya., Nachtigall, P. E., Au, W. W. L., & Breese, M. (2005). Invariance of echo-responses to target strength and distance in an echolocating false killer whale: Evoked potential study. *Journal of the Acoustical Society of America*, 117(6), 3928–3935.
Supin, A. Ya., Nachtigall, P. E., & Breese, M. (2006). Source to sensation level ratio of transmitted biosonar pulses in an echolocating false killer whale. *Journal of the Acoustical Society of America*, 120, 518–526.
Supin, A. Ya., Nachtigall, P. E., & Breese, M. (2007). Evoked-potential recovery during double click stimulation in a whale: A possibility of biosonar automatic gain control. *Journal of the Acoustical Society of America*, 121, 618–625.
Supin, A. Ya., Nachtigall, P. E., & Breese, M. (2008). Hearing sensitivity during target presence and absence while a whale echolocates. *Journal of the Acoustical Society of America*, 123, 534–541.
Thomas, J. A., Moss, C. F., & Vater, M. (2004). *Echolocation in bats and dolphins*. Chicago: University of Chicago Press.
Ulanovsky, N., & Moss, C. F. (2007). Hippocampal cellular and network activity in freely moving echolocating bats. *Nature Neuroscience*, 10, 224–233.

Ulanovsky, N., Fenton, M. B., Tsoar, A., & Korine, C. (2004). Dynamics of jamming avoidance in echolocating bats. *Proceedings of the Royal Society of London B: Biological Sciences*, 27, 1467–1475.

Wartzog, D., & Ketten, D. R. (1999). Marine mammal sensory systems. In J. E. Reynolds III & S. A. Rommel (Eds.), *Biology of marine mammals.* Washington, DC: Smithsonian Institution Press.

Wever, E. G., & Vernon, J. A. (1961). The protective mechanism of the bat's ear. *Annals of Otology, Rhinology and Laryngology*, 70(1), 5–17.

Wiegrebe, L. (2008). An autocorrelation model of bat sonar. *Biological Cybernetics*, 98, 587–595.

Yuen, M. E., Nachtigall, P. E., Supin, A. Ya., & Breese, M. (2005). Behavioral and auditory evoked potential audiograms of a false killer whale (*Pseudorca crassidens*). *Journal of the Acoustical Society of America*, 118 (4), 2688–2695.

Chapter 6
Localization and Classification of Targets by Echolocating Bats and Dolphins

James A. Simmons, Dorian Houser, and Laura Kloepper

Keywords Biosonar • Echo delay • Echo spectrum • *Eptesicus fuscus* • Glint delay • Pulse-echo ambiguity • *Tursiops truncatus*

6.1 Introduction

Echolocating bats and dolphins project sounds into their surroundings and listen to the returning echoes to detect and identify objects. These animals must deal with a potentially wide range of acoustic interference that is dependent on the amount of clutter, or the distribution of extraneous objects, in the environment. For both dolphins and bats, the ability to detect and resolve targets of interest is due to the intricacies of the sound projection and echo reception systems in association with sophisticated neural processing. The present chapter offers an integrated view of selected research findings regarding the principal purpose of wideband biosonar: the localization and classification of targets based on accurate determination of the delay and spectrum of echoes.

J.A. Simmons (✉) • L. Kloepper
Department of Neuroscience, Brown University, Providence, RI 02912, USA
e-mail: james_simmons@brown.edu; laura_kloepper@brown.edu

D. Houser
National Marine Mammal Foundation, 2240 Shelter Island Drive, San Diego, CA 92106, USA
e-mail: dorian.houser@nmmfoundation.org

A. Surlykke et al. (eds.), *Biosonar*, Springer Handbook of Auditory Research 51,
DOI 10.1007/978-1-4614-9146-0_6,

6.1.1 Limitations on Comparisons Between Dolphins and Bats

It is difficult to articulate a single, comprehensive account, or even a reasonably specific unifying hypothesis, about how echolocation "works." One reason is that there are a number of different systems used by different species of bats and odontocetes. Few species of either bat or odontocete have been sufficiently studied to yield a realistic, mechanistic assessment of even one type of echolocation system. The inevitable consequence of different kinds of experiments being carried out by different investigators using different species leads to a description of a fictional sonar system in a fictional, composite animal. Extracting broadly applicable principles from this diversity is not feasible until individual systems have been explored to unambiguously reveal their inner mechanisms and their relation to environmental demands and prior evolutionary pressures. Here, we highlight information derived from new research on echolocation by bottlenose dolphins (*Tursiops truncatus*) and big brown bats (*Eptesicus fuscus*). Both species transmit wideband biosonar sounds, with most of their energy in the range of 20 kHz to about 110–130 kHz. Studies of echolocation in bats have developed along somewhat different lines than research on cetaceans, largely from practical considerations—the difficulty inherent in observing these animals using their sonar in natural conditions plus the availability, suitability, and expense of equipment and facilities required for laboratory studies. Nevertheless, a relatively wide range of experiments has been conducted on both species to further the goal of understanding the mechanisms of echolocation.

6.2 Target Detection and the Operating Range of Echolocation in Relation to the Emission Patterns of Broadcast Signals

Identifying relations between the structure and pattern of biosonar emissions and performance in behavioral tests of target localization and perception is necessary to understand how echoes are processed in bats and dolphins. Figure 6.1 illustrates the spectrogram, spectrum, and autocorrelation function for a typical FM biosonar sound emitted by a big brown bat and an echolocation click emitted by a bottlenose dolphin. For big brown bats in most conditions, the duration of FM pulses ranges from a maximum of 20–25 ms when flying in open spaces to about 1–3 ms when flying in vegetation or clutter (Petrites et al. 2009; Aytekin et al. 2010; Hiryu et al. 2010; Moss and Surlykke 2010). When the bat is about to land on a surface or capture a flying insect, broadcast durations shorten further to 0.3–0.5 ms. The most salient characteristics of these sounds are their wide frequency band (extending from roughly 20 kHz at the low end to 100–110 kHz at the high end, to achieve a total bandwidth of 80–90 kHz), their downward FM sweeps, and their multiple harmonics (FM1, FM2, FM3). In contrast, bottlenose dolphins produce signals much shorter than big brown bats referred to as clicks. Echolocation clicks have

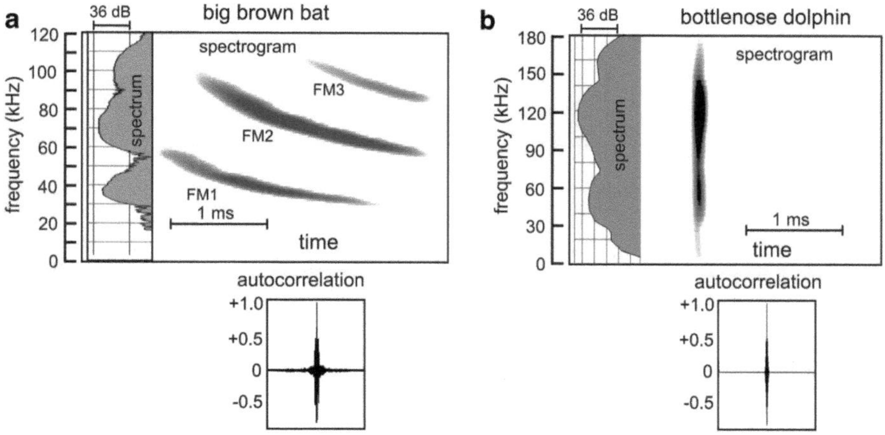

Fig. 6.1 Three different signal representations for (**a**) big brown bat and (**b**) bottlenose dolphin wideband echolocation sounds. Plots show the spectrogram (time–frequency), spectrum (frequency), and autocorrelation (time) representations for a representative biosonar signal. The very compressed autocorrelation functions illustrate how sharply echoes originating from ensonification with these sounds can be localized in delay, or target range (Simmons and Stein 1980)

durations between 40 and 70 μs for laboratory conditions (Au et al. 1974; Capus et al. 2007) and 10–23 μs for free-ranging conditions (Akamatsu et al. 1998; Wahlberg et al. 2011), although differences in data collection efforts and means of calculating durations may explain some of the observed differences between laboratory and free-ranging conditions. Dolphin echolocation clicks can have a wide frequency bandwidth (>85 kHz) with energy commonly between 25 and 130 kHz, but the dolphin can also control the spectral content of the click and capitalize on narrower band signals (Houser et al. 1999; Muller et al. 2008). Because of the transient nature of the dolphin echolocation signal, significantly less control in duration is observed relative to that of the big brown bat. Nevertheless, both big brown bats and dolphins change characteristics of their broadcasts in a multidimensional adaptation to the prevailing acoustic conditions.

Bats and dolphins produce FM pulses or clicks, respectively, in temporal sequences that result in a stream of echoes returning to the animal's ears (Fig. 6.2). The sequence of clicks or pulses, often called "trains," is important as each successive echo received from a target increases the amount of information about the object available to the animal. The sensitivity or probability of detection gets worse as the number of received echoes decreases in both species (Altes et al. 2003; Surlykke 2004). In the dolphin, each successive echo received reduces the signal-to-noise ratio (SNR) of the detection threshold, suggesting that the dolphin potentially conforms to a summation or integration receiver model (Au 1993). In the bat, a very different pattern emerges—as the number of echoes increases from one to three, sensitivity remains poor, but when a minimum of four to seven echoes are available, sensitivity abruptly improves (Surlykke 2004).

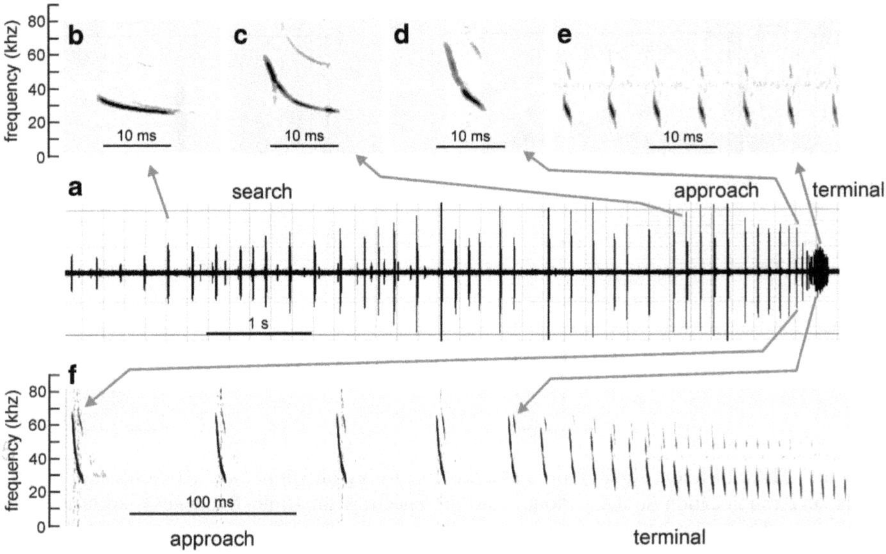

Fig. 6.2 Natural acoustic behavior of a big brown bat during aerial interception. (**a**) Time series of biosonar broadcasts emitted by a big brown bat during an interception maneuver (Simmons 2005). Labels indicate the approximate segments of the search, approach, and terminal stages. Spectrograms of (**b**) shallow-sweeping "quasi-CF" search-stage sound, (**c**) first broad-sweeping FM sound following initial detection, (**d**) broad-sweeping FM sound emitted during active approach, (**e**) series of seven successive FM sounds emitted during terminal approach ("buzz"), and (**f**) sequence of sounds emitted during the transition into the terminal stage. Note the progression of shortened pulse durations and the decrease in interpulse intervals as the bat moves nearer to the insect

During flight or swimming, the bat or dolphin aims its head or steers the broadcast beam to ensonify the target of immediate interest. The receiving beams are such that the focus of reception is from echoes arriving from straight ahead. However, objects off to the sides still are ensonified relatively strongly and the echoes they return are picked up by the ears with considerable sensitivity (although the ear on the same side as one of these objects does receive a stronger version of the echoes). The raw transmitted and received beams thus do not directly segregate the target's echoes from surrounding clutter.

The target strength is a representation of how strongly an object reflects sound. Because target strength is calculated as the logarithm of the ratio of the intensity of the reflected echo to the intensity of the incident sound, it is often presented in decibels (dB). For the bat, small objects such as flying insects have target strengths of roughly −10 to −40 dB depending on their size. In contrast, objects comprising clutter, such as leaves, have target strengths that depend more on their perpendicular orientation to the incident sound path than size; they reflect specular echoes with little intrinsic target-related attenuation other than that due to distance or direction because their dimensions are so much larger than the incident wavelengths. Thus, when bats fly through vegetation or close to the ground, they receive numerous, very strong echoes from the parts of the scene located at close range, even from parts that are off the beam axis because the transmitted beam is quite broad. Moreover, as bats move relative to their surroundings (typically at velocities of several meters per second), most of the clutter fluctuates in strength because different elements of the scene shift in and out of the necessary perpendicular, specular orientation. Surprisingly, more distant parts of the scene still can return strong echoes because the array of reflecting elements acts as an extended surface—at greater distances more and more reflections are recruited into the incident beam and add together to create an overall backscatter that declines only very gradually with range.

Bottlenose dolphins feed on prey much larger than those of bats, with dimensions often larger than the incident wavelength, and with a more complicated echo structure that fluctuates depending on aspect angle for tonal signals (Au et al. 2007). Despite these challenges, dolphins often forage in highly reverberant shallow waters and locate prey hiding among sea grass and beneath sand or mud. The broadband structure of the bottlenose dolphin click reduces the inherent fluctuations in echo target strength (based on energy), allowing the dolphin to maintain a relatively consistent echo target strength regardless of angle (Au et al. 2007). This allows for high levels of accuracy, with dolphins detecting targets with echoes less than 1 dB greater than the background clutter echoes (Au and Turl 1983).

Both big brown bats and bottlenose dolphins change the interval between sonar broadcasts (interpulse intervals [IPI for bats] or interclick interval [ICI for dolphins]) in response to prevailing acoustic conditions and variation in target distance. The echo stream is the series of echoes that returns to a bat or dolphin after they have ensonified the environment. The IPI or ICI defines the portion of the echo stream that returns to the bat or dolphin before the production of another pulse or click. The duration of the echo stream (ESD) following the production of a pulse or click is critical to resolving target echoes from clutter both for the bat and the dolphin. If the bat or dolphin emits a second sound before all of the first sound's echoes have arrived, uncertainty arises about which broadcast is responsible for which echoes. This pulse-echo ambiguity occurs whenever successive ESDs overlap.

Echoes from targets at different distances return at different delays (5.8 ms/m in air and 1.3 ms/m in sea water), creating an equivalence between objects arrayed in space and echoes arrayed in time. The outer limit for this equivalence—the operating range of biosonar—is determined by propagation losses, the time delays themselves, and the degree of sensitivity of the hearing apparatus. For a point target, such as an insect, at all frequencies, spreading loss combined across the outward-bound and inward-bound paths as a function of range is $1/d^2$, where d is the distance. For a planar target, such as leaf clutter, it is $1/d$. As distance increases, echo strength declines according to these terms, but added to geometric spreading is absorption due to the medium through which the sound travels, which is frequency dependent. Absorption discriminates sharply against higher frequencies in both air and water, but is greatly diminished relative to the propagation loss in water. For higher-frequency in-air broadcasts, detection ranges will be shorter than 5–10 m in the best conditions of very intense broadcasts and large reflecting surfaces (Fig. 6.3).

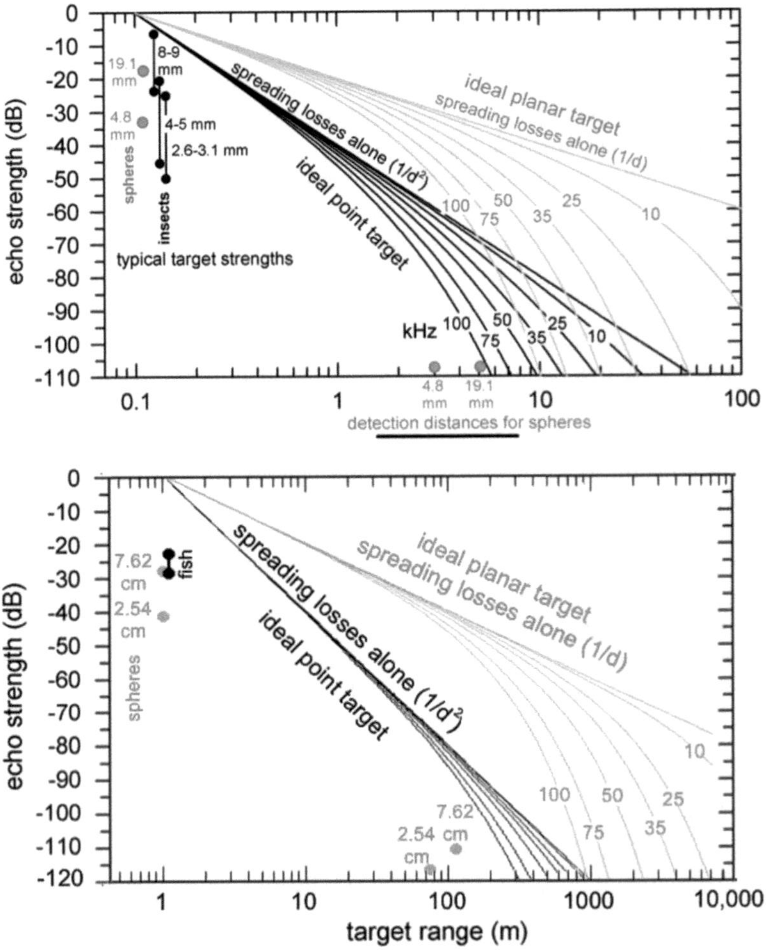

Fig. 6.3 Operating range of echolocation for bats (*top*) and dolphins (*bottom*). *Solid black curves* trace the reduction in strength of echoes returned by an ideal point target (i.e., 0 dB target strength) assuming spherical spreading losses separately for the broadcast and the echo (accumulating as $1/d^2$), plus absorption at selected frequencies. *Solid gray curves* trace the reduction in strength incurred by echoes returned by an ideal planar target (i.e., 0 dB target strength) assuming spherical spreading losses together for the broadcast and the echo (accumulating as $1/d$ overall), plus atmospheric absorption at selected frequencies. *Top* (bats): Data on echo strength at different distances are presented as echo attenuation relative to the broadcast (0 dB) recorded at a distance 10 cm from the bat's open mouth. The *gray curves* give the likely maximum operating range for the strongest possible reflecting source, while the *black curves* give the likely operating range for an insect-like small reflector. *Upper left*: *Gray circles* show target strengths for 4.8 and 19.1 mm diameter spheres (Simmons and Chen 1989). *Black circles* and *vertical lines* show the spread of target strengths for insects having wing lengths of 2.6–3.1 mm, 5–5 mm, and 8–9 mm (Houston et al. 2004). *Lower right*: *Gray circles* show measured big brown bat distances of detection for spheres (Kick 1982). *Horizontal black bar* shows approximate spread of big brown bat distances of detection for different-sized insects. *Bottom* (dolphins): Data on echo strength are presented as echo attenuation relative to the broadcast (0 dB) recorded at 1.0 m from the dolphin's rostrum. *Upper left*: *Gray circles* show target strengths for a 2.54 and 7.62 cm sphere (Au and Snyder 1980; Murchison 1980). *Black circles* show the spread of target strengths for fish of lengths 20–26 cm (Au et al. 2007). *Lower right*: *Gray circles* show the measured bottlenose dolphin distances of detection for spheres (Au and Snyder 1980; Murchison 1980)

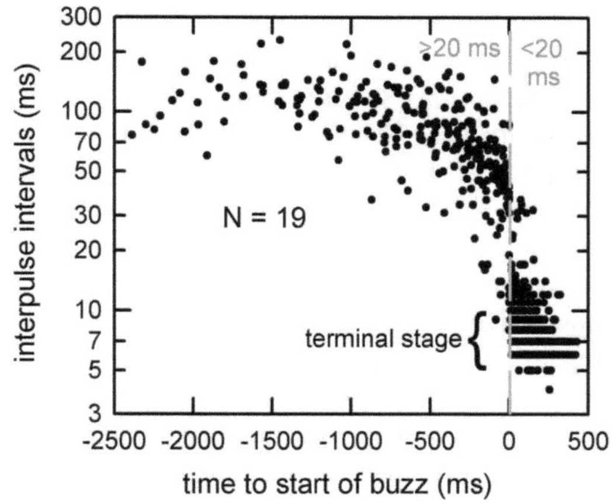

Fig. 6.4 Interpulse intervals (IPIs) for big brown bats during open-space interceptions. Data points show IPIs for biosonar sounds emitted during 19 separate aerial captures of June beetles (see Fig. 6.2 and Simmons 2005). The data are plotted with respect to the transition into the terminal stage, indicated by the first IPI consistently shorter than 20 ms (0 origin of horizontal axis)

Bats progress through three stages during foraging: the search stage; the approach stage, where the bat has found and approaches its prey; and the terminal stage, which is the short period starting just before prey capture that is typified by increased rates of pulse production (i.e., "buzzing"). In the big brown bat, the IPIs recorded in the field seem surprisingly long (90–180 ms; Fig. 6.4) during the search stage (before about 1,000 ms relative to the onset of the terminal stage buzz; Moss and Surlykke 2010). If the bat is waiting for reception of all audible echoes from various parts of the whole scene (e.g., trees in the distance or the ground) before emitting the next sound, the IPIs indicate that large sources of reflections that define the boundaries of the space must be detectable at long ranges of 15–30 m (see Fig. 6.3). For big brown bats using frequencies of 25–35 kHz in FM1 of shallow-sweeping sounds or the tail-end frequencies of broader-sweeping sounds, echoes of 0–10 dB SPL from point targets are detectable at distances of 3–5 m for small spheres (Kick 1982). The corresponding echo delays are about 18–30 ms. At these same broadcast levels, planar targets could plausibly be detected at distances of 20–25 m. The corresponding echo delays are 120–150 ms. It thus appears that the long IPIs used in open-area searches seem designed to accommodate the return of echoes from large background surfaces at long ranges (Schnitzler et al. 2003; Moss and Surlykke 2010).

To avoid pulse-echo ambiguity in open-area searches, big brown bats wait until all echoes have returned before emitting the next sound. However, when flying in dense clutter where the demands of obstacle avoidance become paramount, big brown bats shorten their IPIs by emitting sounds in strobe-group pairs or triplets (two or three sounds emitted close together followed by a longer interval between pairs) so that streams of echoes from successive sounds overlap and pulse-echo ambiguity does occur (Petrites et al. 2009). In this difficult situation, the bats make subtle changes in the frequencies of their broadcasts—alternately increasing and decreasing the tail-end frequencies of the sounds comprising the strobe groups to allow their echoes to be distinguished (Hiryu et al. 2010).

Another interesting feature of the IPIs shown in Fig. 6.4 is that the bats make an unexpectedly abrupt transition from the approach stage to the terminal buzz. In the field recordings, the bats appear to jump from emitting sounds at intervals mostly longer than 25–30 ms to mostly shorter than 15 ms in the terminal buzz, with practically no IPIs in the intervening region. This unexpected gap in the IPI distributions suggests that, even when the bat is actively engaged in tracking the target, it keeps emitting sounds at long enough intervals to preserve some space for echoes that arrive from background objects farther away than the insect. Broadcast durations, however, do continue to track the steadily declining distance to the insect, confirming that a true, acoustically specified approach stage is in progress. Laboratory experiments have confirmed that big brown bats distinguish between regulation of broadcast duration by the distance to targets of interest and regulation of IPIs by the larger scale of the surrounding space (Saillant et al. 2007; Petrites et al. 2009; Aytekin et al. 2010).

Bottlenose dolphins in both field and laboratory settings produce clicks that typically vary according to target range and with ICIs long enough to prevent overlap of successive ESDs (Au et al. 1974; Jensen et al. 2009). Work by Penner (1988) demonstrated that the generation of ICIs was conscious and that dolphins produce ICIs consistent with the expectation of target location, that is, randomization of target distances showed an influence of prior target location on ICIs and an increase in detection error rates relative to targets presented at a constant distance. For bottlenose dolphins producing clicks with a center frequency of 75 kHz, echoes from point targets with target strengths of –28 to –41 dB are detectable at distances of 74–113 m for small spheres (Au and Snyder 1980; Murchison 1980). The corresponding echo delays are 96–147 ms, and fall within the maximum ICI value of 462 ms recorded in the field (Jensen et al. 2009). At these same broadcast levels, planar targets could plausibly be detected at distances of 1,147–1,260 m with corresponding echo delays of 1,491–1,638 ms. Unlike the field recordings of foraging bats, most field recordings of bottlenose dolphins are obtained by encouraging the animal to focus its attention on an array of hydrophones that is hanging perpendicular to the ocean floor. Further, on-axis clicks are typically recorded when the animal is oriented perpendicular to the array and thus parallel to the ocean floor. In such a configuration, the dolphin is essentially swimming in a clutter-free zone and would have no need to produce echolocation signals with ICIs long enough to process echoes from large, distant surfaces.

Dolphins, like the big brown bat, significantly decrease the ICI in the final approaches to a target or prey item, that is, they also demonstrate a terminal buzz. However, more interesting may be the behavior that occurs when targets are at long range. At target distances greater than approximately 100–200 m, dolphins have been observed to produce packets of clicks and have been shown capable of detecting targets and changes in target echoes at distances of up to 800 m, provided the echo level was sufficiently high for detection (Ivanov 2004; Finneran 2013). Because the number of clicks per packet (approximately 4–10) does not significantly vary according to changes in target strength or echo level, the production of packets appears to be related mostly to the target distance (Finneran 2013). The production of packets should result in pulse-echo ambiguity, but the duration between packets follows the same pattern as that of individual clicks at shorter ranges; mainly, the packets are not produced until the echoes from all clicks within a packet are received by the dolphin. Thus, the duration between packets is likely used to resolve the pulse-echo ambiguity. Why packets are produced is unknown, although its occurrence may signify a limitation in the delay that can occur between echoes if the dolphins indeed utilize multi-echo processing (Au et al. 1988; Altes et al. 2003).

6.3 Perception of Target Range from Echo Delay

Having developed a means for estimating the overall operating range of echolocation, it is of interest to determine how effectively bats and dolphins can locate targets along this range dimension. For measuring the delay of echoes, a sonar system uses the time of the broadcast as a reference, or a trigger signal, and reception of an echo culminates in registration of the elapsed time since the trigger occurred. Assuming that the receiver has arbitrarily precise knowledge of this reference time, the accuracy of delay determination depends on the nature of the broadcast signal, in particular, its bandwidth. The availability of more frequencies equals sharper determination of delay, which means that wideband sonar signals are especially suited to precisely determining target range from echo delay. Indeed, it has been suggested that there is no purpose for bats to emit wideband FM sonar sounds unless they exploit the bandwidth by internally dechirping the echoes—removing the frequency modulation to minimize the duration of the sound, which then maximizes the accuracy of echo delay estimates (Glaser 1974).

The theoretical influence of the broadcast signal's composition on the accuracy of target ranging is portrayed by the cross correlation function between echoes and broadcasts. (The autocorrelation function in Fig. 6.1 is equivalent to the cross correlation function of a delayed, attenuated replica of the broadcast, as is the case for the reflection from a point target, which provides one reflective surface, or glint, at close range. The shape of this function displays the intrinsic timing accuracy of the signal.) The example of a big brown bat sound in Fig. 6.1 has a broad bandwidth (approximately 80 kHz), which gives it a tightly compressed autocorrelation function, with little spread in time around its central peak (i.e., few side peaks and those of lower amplitude than the central peak). The total time span, including the main peak and the most prominent side peaks, is about 150 μs, which corresponds to about 2–3 cm in target range. The central peak alone is 7–8 μs wide, corresponding to slightly more than 1 mm in target range, and its very tip is even narrower, 1 μs or less, corresponding to a fraction of a millimeter. Autocorrelation functions of bottlenose dolphin signals demonstrate similar characteristics of bats (Fig. 6.1). The dolphin's broad bandwidth signal (approximately 85 kHz) produces even tighter autocorrelation functions of total duration around 100 μs, which corresponds to 3–4 cm in target range. The central peak is even narrower, around 6–7 μs, corresponding to

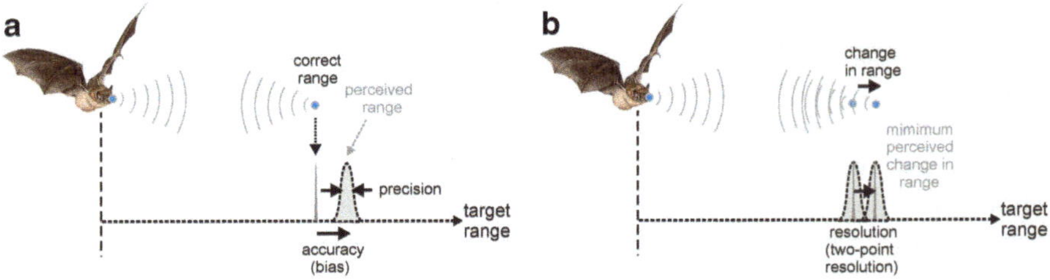

Fig. 6.5 The measurement process for target ranging. (**a**) Accuracy, bias, and precision of range estimation using echo delay. See text for details. (**b**) Psychophysical experiments on the bat's perception of range assess accuracy and precision indirectly, by determining resolution. This can be done by giving the bat two separate objects at the same time situated in different directions at different ranges and training it to chose one (the rewarded stimulus object at the correct range, S+) over the other (the unrewarded object at an incorrect range, S−), or by presenting the bat with one object at a time and training it to respond to the object presented at the rewarded range (S+) and not to the object presented at other ranges (S−). The bats limit of resolution is determined by reducing the difference in range between S+ and S−

slightly more than 2 mm target range. To make a crude but useful first approximation, the core question about how echolocation works revolves around which of these time scales embodied in the cross correlation function best describes the animal's acuity for perceiving echo delay.

The accuracy, bias, and precision of echo delay perception is visualized in Fig. 6.5a. The *accuracy* of a range estimate refers to the agreement between the estimated value of target range in relation to the target's true, or objective distance. Any error in range accuracy reflects a *bias* in the delay-estimation process. If the bat creates a separate delay estimate for each broadcast-echo pair, and then emits a series of sounds so that multiple estimates accumulate, the width of the distribution of these estimates is the *precision* of echo delay estimation. Note that precision refers to the variability of delay estimates around a mean estimate for all of the echoes, not a statement about whether the bat experiences a bias away from the objective delay. If precision is high enough (i.e., the measurement distribution is narrow enough), the presence of a bias can be detected in the measurements, but if precision is low, any bias would go unnoticed owing to the excessive variability.

Psychophysical experiments allow for controlled measurements of echolocation delay precision by training animals to respond to echoes that arrive at different delays and measuring the smallest change in delay the animal can perceive. Figure 6.5b illustrates what is meant by delay resolution, which is what psychophysical experiments measure. This procedural alternative to the sensorimotor approach seems straightforward enough, but it turns out not to be so simple in practice. Numerous psychophysical experiments have been carried out to estimate the bat's target range or echo delay resolution (Simmons and Grinnell 1988; Moss and Schnitzler 1995). Such experiments often use targets that return one reflection for each incident broadcast; that is, each target has one glint. However, some of these experiments even estimate the bat's two-point resolution (Fig. 6.5b), which is a special case in which the subject is presented two different ranges within the same target so that the object returns two reflections for every incident broadcast. This object, which is a two-glint object, is used to determine the smallest range separation between the glints for which the object still is perceived as containing two separate glints.

Psychophysical experiments investigating echolocation delay fall to two different types—two-choice discrimination of echoes delivered at different delays for a series of broadcasts emitted by the bat over a time span of several seconds, and detection of echoes that jitter in their delay from one broadcast to the next. Their methods differ most obviously in the time scale for the presentation of the stimuli—the two-choice method gives the bat several seconds to examine the echoes at each of two delays to determine which echoes arrive at the correct delay; the jitter method delivers the bat with alternating examples of both delays in just a few tens of milliseconds.

Figure 6.6 illustrates the two-alternative forced-choice method (the “two-choice” discrimination procedure) most commonly employed to measure the bat’s echo delay resolution with single-glint echoes (Moss and Schnitzler 1995). In this example, the bat sits on an elevated Y-shaped platform while its broadcasts are picked up by microphones (m), delayed electronically, and then returned from loudspeakers (s) as echoes at a particular delay that simulates a target at a particular range. The bat is trained to respond by moving forward toward the loudspeaker that returns the rewarded stimuli (S+), here shown as being presented at a fixed delay of 3.2 ms to simulate a target at a range of 55 cm. The bat should not approach the loudspeaker that returns the unrewarded stimuli (S–), which is presented at a series of different delays (illustrated schematically as proceeding from 1 to 10 in the diagram) that bracket the fixed delay of S+ (e.g., 3.5 ms down to 3.1 ms compared to 3.2 ms). The presentation of S+ and S– on the left or right is alternated randomly, and the bat’s task is to locate S+ at the rewarded delay in the presence of S– at a series of different delays. Insets show spectrograms for the FM broadcast followed by either S+ (upper) or by S– (lower). The example shows S– at a longer delay than S+ (dashed curves in S– spectrogram show delay of S+). As the delay of S– is changed from one value to the next, the hypothetical performance curve (S+ to S– delay difference vs. % errors) traces the masking effect of S– on S+, or the region where the delays of S+ and S– appear indistinguishable to the bat. The peak in the error curve marks the delay of S– that equals the perceived delay of S+, and the shape of the curve traces the bat’s representation of the delay—ideally, its width should be the bat’s delay accuracy.

The two-choice experiment arrives at an estimate for the bat’s resolution for echo delay by gradually reducing the difference in delay between S+ and S–. Figure 6.6 shows schematically how the difference in delay changes (in steps 1–10) so that, across many trials, S- is delivered at a series of delays that bracket the delay of S+. The bat’s performance changes according to the size of the delay difference—it makes more errors in its choices (chance performance is 50 % errors) when S+ and S– have similar delays. At this performance level, the bat cannot distinguish between the delays. For the simplest two-choice experiment, the stimuli both consist of a single echo for each broadcast (i.e., a simulated single-glint target)—one echo from the

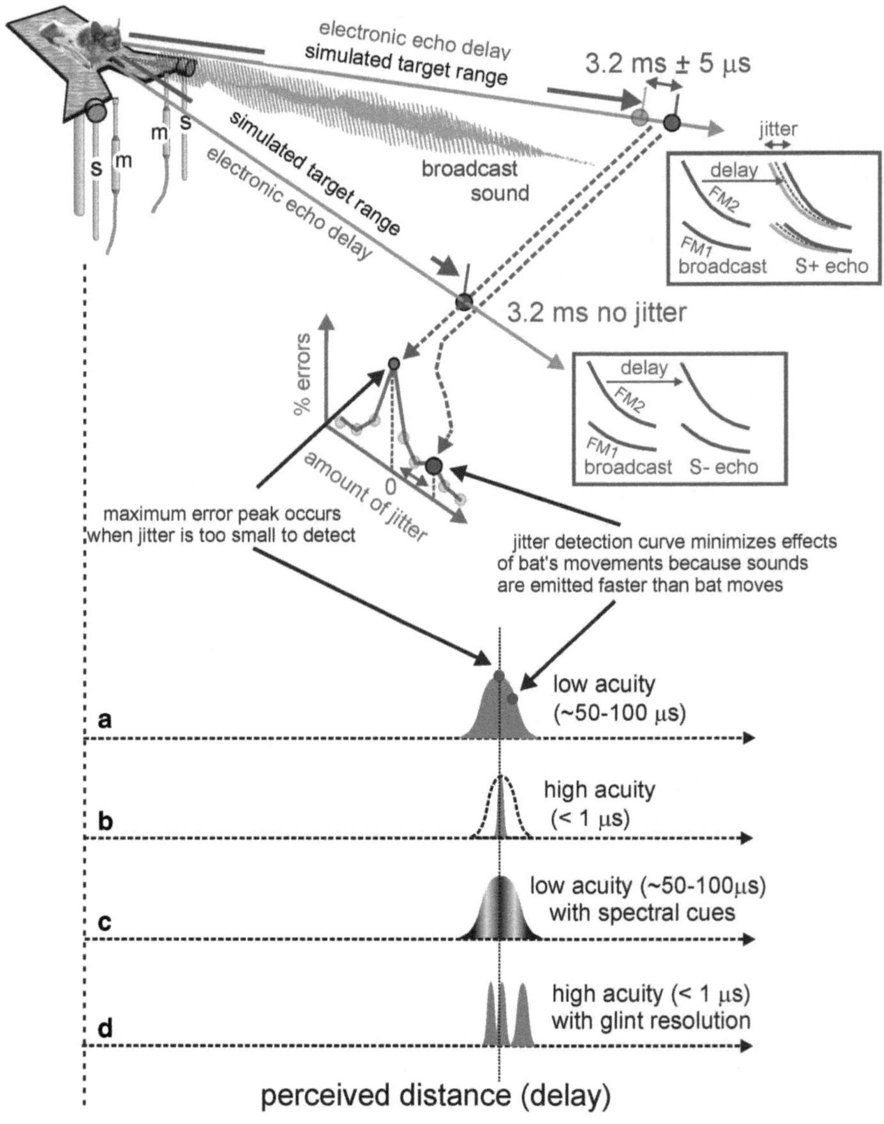

Fig. 6.6 The two-choice delay discrimination psychophysical test to determine the bat's echo delay accuracy, with diagram showing possible outcomes from the test (*m* microphones, *s* loudspeakers, *S+* rewarded stimuli, *S–* unrewarded stimuli). The letters *a–d* correspond to possible outcomes of procedures using single-glint and multiple-glint echoes. See text for details

microphone-loudspeaker channel set to the delay specified for S+, and the other echo from the channel set to the delay of S– (left or right, changed randomly). The shape of the error curve on the diagram in Fig. 6.6 for values of the S+ to S– delay difference traces the region where the bat perceives the two stimulus delays to be the same.

As an index of the delay-discrimination threshold, the half-width of the error curve half-way down from its peak (at 25 % errors) is taken to be the bat's echo delay resolution.

Figure 6.6a, b shows two extreme examples of alternate outcomes (hypothetical error curves) for the two-choice tests based on the shape of the cross correlation function of echoes (see Fig. 6.1). The examples assume single-glint echoes. In one extreme, the bat's intrinsic accuracy is hypothesized to be of the order of 100 μs (width of gray shaded curve in Fig. 6.6a). This corresponds to a target range accuracy of about a centimeter or two, and also to the width of the entire cross correlation function from Fig. 6.1, including the central peak and its side peaks. In the other extreme, the bat's intrinsic accuracy is hypothesized to be in the region of 1 μs or better (width of gray shaded curve in Fig. 6.6b), which corresponds to a target range accuracy of a fraction of a millimeter, and also to the width at the tip of the central peak in the cross correlation function from Fig. 6.1. This finding is satisfyingly similar to the precision of about 200 μs (3.4 cm) estimated from the bat's vocal sensorimotor control of broadcast duration and its reaching out to seize the target. One caveat, however, is that in a typical trial of a two-choice experiment the bat emits a series of broadcasts while scanning left and right to examine the two simulated objects. The size of these left-to-right head movements amounts to several centimeters, which changes the distance from the bat to the microphones and loudspeakers by up to 1–2 cm. This, in turn, changes the actual delay of the echoes reaching the bat's ears, by up to 50–100 μs. Therefore there remains a question as to whether the two-choice method has an intrinsic limitation that might conceal greater delay precision in the bat.

The next two examples in Fig. 6.6c, d are for S+ echoes simulating a multiple-glint target that returns two or three distinct reflections at small delay separations. The spectra are modified by interference nulls distributed at fixed frequency intervals across the spectrum, which are determined by the time separations of the glint reflections. Figure 6.6c shows a broad error peak, 50–100 μs wide, representing relatively low delay acuity coupled with an additional perceptual quality related to the spectral "coloration" from the nulls. In this case, there should not be an error peak at all because the bat can distinguish S+ from S– by the presence of this spectral coloration, even when the delay difference itself is small. Figure 6.6d shows a narrow error peak representing high range accuracy coupled with additional error peaks that mark delays where S– corresponds in delay to one or two additional glints in S+. The presence of the additional error peaks signify high delay resolution and represent the conversion of spectral nulls into time separation estimates by the bat. The curve in Fig. 6.6c implies that the bat perceives target glint structure in terms of the echo spectrum as a dimension orthogonal to echo delay (this dimension would prevent the error curve from having a peak), while the curve in Fig. 6.6d implies perception of target glint structure in terms of the distances to the glints within the target.

Experiments with detection of jitter in echo delay suggest the big brown bat's intrinsic delay precision might be as small as 10–20 ns (Simmons et al. 1990). This extraordinarily acute, psychophysically measured delay resolution implies very high delay precision for the bat. Nevertheless, however implausible it seems, a resolution of 10–20 ns is not an impossible result from an information-theoretic perspective; it is achievable given the bandwidth and signal-to-noise ratio of the bat's broadcasts (Sanderson and Simmons 2002). Because neural responses at various levels of the big brown bat's auditory system exhibit latency variability typically of hundreds of microseconds (Ferragamo et al. 1998; Valentine and Moss 1998; Sanderson and Simmons 2000), the bat's delay-perception mechanism clearly is distributed across populations of neurons and thus quite possibly does not have a conventional architecture (Simmons 2012).

6.4 Distortions of Perception for Target Range by Flying Bats

The occurrence of amplitude latency trading, or the physiological effect that causes a shift of response timing as a function of stimulus amplitude, introduces a bias in the bat's estimate of echo delay (Fig. 6.5a). In big brown bats, latencies of neuronal responses evoked by FM sounds or tone bursts in the auditory system, particularly the inferior colliculus, become longer by about 15 μs when stimulus amplitude is decreased by 1 dB (Simmons et al. 1990; Burkard and Moss 1994; Ma and Suga 2008). This affects the estimate's accuracy quite apart from the precision with which delay is perceived (variation around the mean perceived delay). Because the bat perceives echo delay in relation to the broadcasts—actually the neural responses evoked by echoes in relation to neural responses evoked by broadcasts (Simmons 2012; Simmons and Gaudette 2012)—the accuracy of the bat's perception of delay depends on the latencies of these responses being the same for both the broadcasts and the echoes. Both the broadcast and the echo undergo changes in amplitude during their propagation from the bat's larynx (Suthers 2004) to the inner ear (Veselka et al. 2010; Simmons and Gaudette 2012), and there is no easy way to estimate their equivalence as the proximal stimuli for perception of delay.

Concentrating just on changes in echo amplitude relative to broadcast amplitude during an interception maneuver would introduce a bias of underestimating the target's range, and this bias increases as the distance from the bat to the target increases. Such a bias toward a progressively increasing underestimation of the target's range could easily compromise the bat's ability to coordinate its actions during interception. There is evidence that the big brown bat largely avoids experiencing this bias, however. As target range shortens from at least 1–1.5 m down to about 0.2 m, the bat actively compensates for the expected 15–19 dB increase in echo strength due to declining target range by raising its echo detection thresholds (i.e., decreasing its echo detection sensitivity) by roughly the same amount (Kick and Simmons 1984; Simmons et al. 1992). This appears to be achieved as a consequence of the contraction of the bat's middle ear muscles synchronized to each vocalization. The relaxation of each contraction following the emission leads to progressively improving echo detection sensitivity along a track of 11–12 dB of improvement per doubling

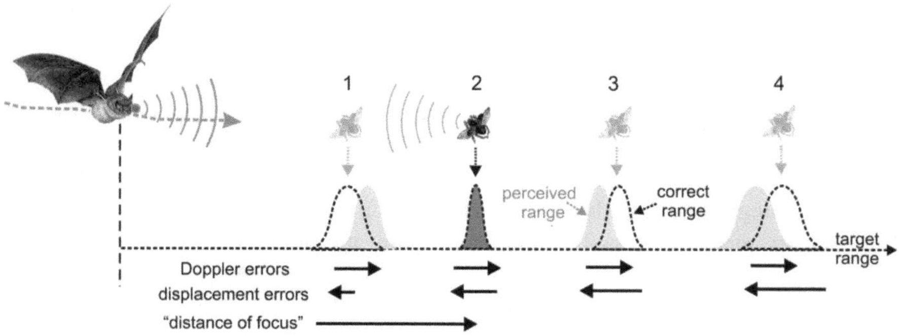

Fig. 6.7 Factors that affect the accuracy of target range determination (horizontal axis) by a flying bat. Four possible locations of the target are shown—at ranges 1, 2, 3, or 4 incremental "steps." The *gray* peaks show the bat's perceived image of the target at each range; *dashed black lines* over or near the *gray* peaks indicate the target's objective position at positions *1*, *2*, *3*, or *4*. The width of the *gray* peaks stands for the accuracy of the estimation process, which is blurred by the bat's continuous motion, while the difference in location between the *gray* peaks and the *black dashed lines* stands for systematic errors that occur as a result of this same forward motion. The magnitude and direction of both Doppler errors and displacement errors are demonstrated by the *black arrows*. See text for details

of echo delay from the moment of emission out to about 6–10 ms. The maximum attenuation achieved by the bat's middle ear muscles is about 30–35 dB, which implies a zone of target ranges extending over a factor of 6–7 times for which the actual strength of echoes reaching the inner ear is stabilized. The bat's peripheral auditory system intervenes between the target and the inner ear to regulate echo strength in a manner that cancels out the expected accumulation of bias in perception of target range so that the bat perceives more nearly the target's actual range during its approach to capture.

The foregoing discussion of bias in perception of target range caused by amplitude changes in echoes that lead to amplitude-latency trading raises a broader question about what "accuracy" really means for biosonar. There are other sources of range bias, and the "big picture" reveals that the bat's perception of target range floats on a sea of variability that belies the notion of real accuracy (Holderied et al. 2008). As considered in Sect. 6.3, this makes the distinction between accuracy and precision very important. Big brown bats are in flight in nearly all natural situations, so they are in constant motion while emitting sonar sounds and receiving echoes. Owing to the bat's forward progress, the echo's delay will be shorter than it should be for the bat's initial position (when the process of determining delay was activated). If the whole auditory process is taken as the bat's gauge for target range, the perceived range will be biased nearer than what was the true range at the moment when the broadcast was produced. This bias toward perceiving a too-short delay depends on the overall distance to the target. If the target is closer than 3 m, the bat will have traveled a shorter distance by the time the echo is received, and the size of the reduction in echo delay consequent on the bat's forward movement will be shorter, so the range bias will be smaller. Figure 6.7 illustrates two major sources of bias for perception of echo delay and target range that depend on the bat's flight. There are four targets depicted in this diagram, at increasingly longer ranges and larger shortening biases (1, 2, 3, 4).

However, there is another source of range bias, and it is in the opposite direction. Apart from displacement in flight, first, the broadcasts that impinge on the target and, second, the echoes that return to impinge on the bat's ears undergo Doppler shifts at a magnitude directly related to the bat's flight velocity. In terms of frequencies, higher approach velocities mean larger upward Doppler shifts. The big brown bat's sonar sounds are not single frequencies, however, but FM sweeps (Fig. 6.1), so the Doppler shift strictly is a compression of the FM waveform that shortens its duration while raising its frequencies. Because the FM signals always sweep downward, the upward Doppler shift in the FM sweeps results in a lengthening of the time that elapses between the occurrence of a given frequency in the broadcast and the occurrence of that same frequency in the echo. This causes the approaching bat to perceive Doppler-shifted echoes as biased toward a longer delay than is true for the target's actual range, which adds a Doppler error toward a longer range to its overall range bias (rightward arrows in Fig. 6.7). Unlike the shortening bias from displacement errors, which increase in size as target range increases, Doppler errors are a lengthening bias that is the same for all target ranges (Fig. 6.7). The total range bias—the result of the two effects—is the sum of the displacement and Doppler errors. For targets close to the bat, the Doppler error dominates the displacement error, and the target is perceived as being farther away that it really is when the broadcast is sent out (Fig. 6.7, distance step 1). As range increases, the displacement error increases, gradually overcoming the Doppler error so that the target is perceived as being increasingly nearer than its true range at the moment of the broadcast (Fig. 6.7, distance steps 3 and 4). At one particular range, the Doppler and displacements biases are equal but in opposite directions; they cancel each other out so the bias is zero (Fig. 6.7, distance step 2). This is the "distance of focus," where the target's range is correctly determined relative to the moment the broadcast is sent out (Holderied et al. 2008).

For representative bat sounds, the range bias varies in size from a lengthening of 1–2 cm for near targets to as much as 3–4 cm of shortening for far targets. Added to both of these biases is the effect of the target's direction in relation to the direction of the bat's flight. The size of the Doppler error and the displacement error decrease for targets located to the side of the bat's velocity vector by an amount proportional to the cosine of the angle of offset. The resultant bias thus also decreases. A plot of the overall range bias for targets located at different distances and in different directions shows a "force-field" of range-bias vectors that decrease in size as the offside angle of the target increases (Holderied et al. 2008). In the plane of range and offside angles (range and cross range), the distance of focus for correct estimation of range becomes a curve that extends to the bat's left and right. For perception of range, locations closer to the bat than the distance of focus are associated with a lengthening bias while locations farther away are associated with a shortening bias.

The bat appears to compensate for this bias by stabilizing the amplitude of echoes at different delays and may potentially mitigate the range bias associated with the bat's forward motion. Bats change the duration and sweep rate of their broadcasts as they fly nearer to prey in aerial interceptions (Fig. 6.2). As they do this, the distance of focus (equal Doppler and displacement errors; Fig. 6.7) comes nearer to the bat, as well, which raises the possibility that the bat takes advantage of the adaptability of its sounds to shift the distance of focus, too (Holderied et al. 2008). The question is whether the bat manipulates the distance of focus to keep either the target itself or some region near the target at the distance of focus. In that case, the adaptive changes in broadcasts, as shown in Fig. 6.2, may accomplish more in perceptual terms than merely matching the duration of broadcasts to echo delay so overlap does not occur, and to extend the sonar operating range ("field of view") to accommodate background clutter (Schnitzler et al. 2003). This is a difficult hypothesis to test, but it will require the effort because range bias has a potentially adverse effect on the perception of targets.

6.5 Perception of Target Shape: Echo Spectra and Glint Delays

The arrangement of objects along the distance axis is the target "scene," which is represented by the corresponding stream of echoes returning to the bat or dolphin. The target scene spreads out to the animal's left and right. Within the full-spectrum zone any changes in echo amplitude and spectrum can reasonably be attributed to the target itself. These include changes in overall echo amplitude from one moment to the next due to fluctuating target strength, and changes in the echo spectrum due to interference between multiple reflections from the target's glints. Additional modifications to returning echoes are dependent on the target's location in the echolocation beam. Echoes generated off of the main response axis of the echolocation beam undergo low-pass filtering because directionality is narrower at high frequencies than low frequencies, and the target's shape also affects the echo spectra. For small objects, such as flying insects (bats) or fish (dolphins), which consist of two or more prominent, closely spaced glints, such spectral patterning is the acoustic manifestation of target shape and size (Imaizumi et al. 2008; Au et al. 2009; Matsuo et al. 2009).

Virtually all materials have far greater acoustic impedance than the medium of air, so echoes consist of specular reflections from surfaces and points, which make target geometry the predominant object-related information carried by returning echoes. The situation is more complicated for dolphins, which may feed on species that have swim bladders as well as those that do not. The impedance of the tissues of fish are close to that of the sea water, leaving the swim bladder of prey species that contain them as the prominent source of signal backscatter and target geometry (Fig. 6.8; Au et al. 2007). However, dolphins and other species of echolocating odontocetes may also feed on flatfish, squid, and other animals that lack gas containing structures. In all cases, other ancillary structures (e.g., fins, mantles) and target geometry contribute to the fine echo structure and aspect dependence (Au et al. 2009).

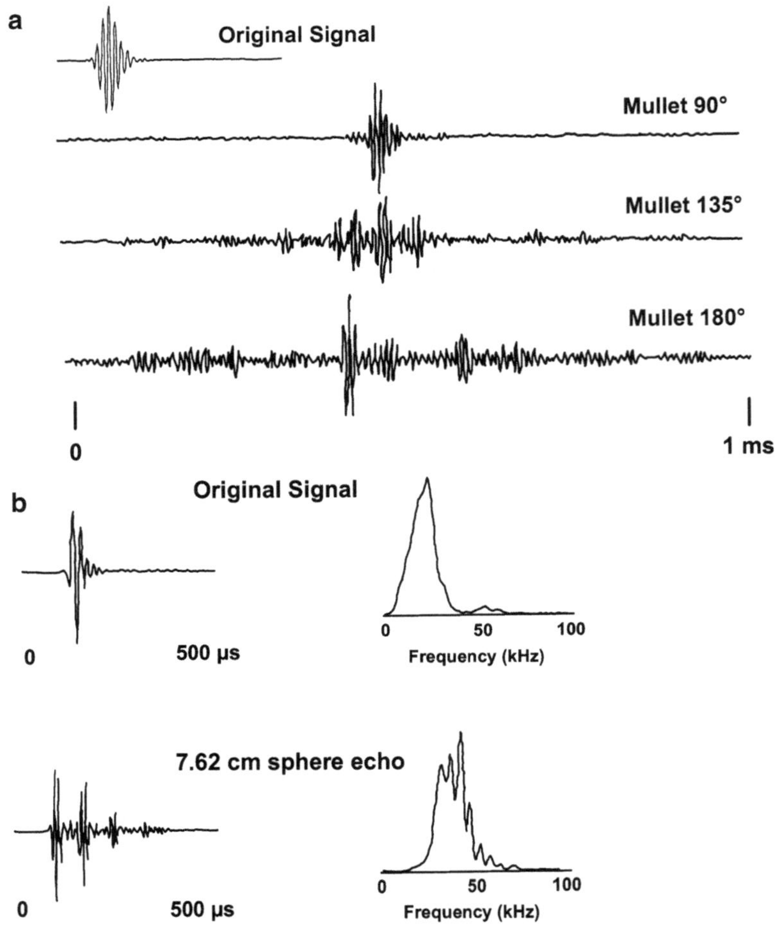

Fig. 6.8 Variation in target echoes for a simulated bottlenose dolphin click. (**a**) The echo from a mullet, a common prey item for the bottlenose dolphin, is highly variable owing to the air-filled swim bladder and fluctuates depending on target geometry. The angles listed correspond to the orientation of the incident signal with respect to the fish body (data from Au et al. 2009). (**b**) The echo from a solid stainless steel sphere yields three time separation reflections (waveform) which result in spectral notches in the frequency domain. If the echoes arrive within the integration time of the auditory system, the dolphin may use these spectral notches for target discrimination (data from Muller et al. 2007)

The spacing of glints has important implications to an echolocating animal's identification of the target from which it originates. If two echoes arrive closer together than the integration time of the auditory system, they merge together to create a single spectrogram with interference notches located at specific frequencies determined by the time separation of the reflections (Fig. 6.8). If the frequencies of these notches are known, the underlying time separation can be estimated. In psychophysical tests, big brown bats actually perceive the arrival times of closely spaced reflections, which they infer from the frequencies of the interference notches. This process amounts to deconvolution; in bats, it is effective for determining two-glint separations from 10 μs to about 300 μs, and even down to 2 μs. This degree of resolution is possible because the bat has nearly perfect knowledge of the transmitted signal (it hears the sound at the moment of emission) and can work backward from its internal replica of the broadcast to determine the pattern of reflections required to produce a given interference pattern in echoes. Psychophysical tests with dolphins suggest similar capabilities and potentially similar underlying processes (i.e., deconvolution based on knowledge of the transmitted signal). Bottlenose dolphins have a temporal integration time of approximately 265 μs (Vel'min and Dubrovskii 1976; Moore et al. 1984; Au et al. 1988), yet bottlenose dolphins can determine glint separations as small as 75 μs. The resolution of closely spaced glints is, however, dependent on the relationship between the glint interval and the amplitude of the respective glints (Helweg et al. 2003). In both systems, the limitations of the deconvolution process are crucial for understanding how the echolocating animals cope with clutter.

The wide bandwidth of the big brown bat (75–80 kHz) and bottlenose dolphin (85+ kHz) biosonar sounds allows them to form images that precisely depict the delay of echoes on a scale finer than 1–5 μs, which corresponds to range precision of less than a millimeter. The bat's delay precision depends on receiving the full broadcast bandwidth in echoes, but it still is very acute, far smaller than the different range biases that affect the accuracy of target ranging on a scale up to several centimeters. Why does the big brown bat have such high broadcast bandwidth and high delay precision? One reason appears to be incorporation of shape into the range images of targets (Simmons et al. 1995; Neuweiler 2000). Small targets such as flying insects are relatively simple objects in acoustic terms. A typical insect consists of two or three prominent body parts (the target's glints, e.g., head, wings, abdomen) that each reflect a discrete replica of the incident sonar sound (Simmons and Chen 1989; Moss and Zagaeski 1994). Big brown bats prey upon flying insects mostly with dimensions up to about 1–3 cm, so the largest delay separation between reflections will be roughly 60–180 μs. Across insect aspect angles, flight postures, and wing positions, the majority of delay separations between glint reflections distribute between 5–10 μs and 50–100 μs, with periodic transient excursions to longer separations at particular points in the insect's wing beat cycle. During approach, the bat's broadcasts are 2–10 ms long (Fig. 6.2), while the glint reflections are only a few tens of microseconds apart. Echoes returning from insects thus contain two or three reflections that arrive at such small delay differences the reflections overlap almost completely. They add together to reinforce or cancel at different frequencies.

Frequencies and frequency spacing of interference nulls in the echo spectrum depend on the time separation of the glint reflections, but typical values range from 50 kHz null separations for a 20 μs two-glint delay separation down to 10 kHz null spacings for a 100-μs delay separation. These nulls are recognized by neurons in the big brown bat's auditory system that are tuned to frequencies from 15 to 100 kHz with tuning widths from about 1–2 kHz to 10–12 kHz (Simmons 2012). When target shape is taken into account, the wide span of frequencies in the broadcast spectrum (Fig. 6.1) seems well adapted for registering the shape of targets from the pattern of

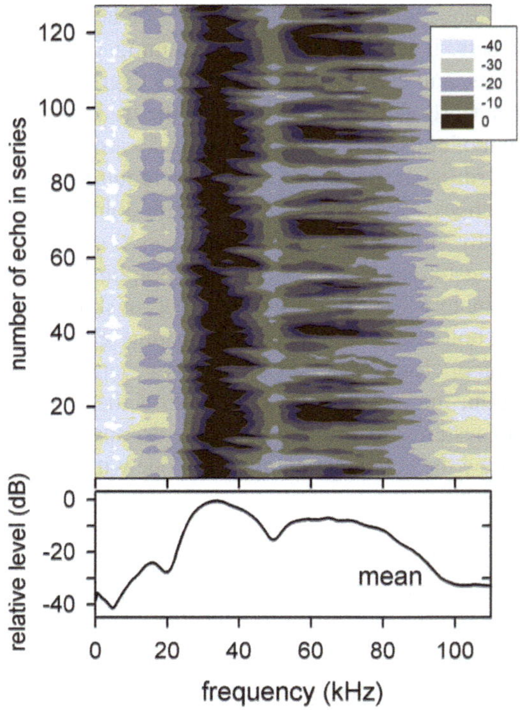

Fig. 6.9 Relative strengths of a series of 126 echoes from a tethered fluttering moth (data from Moss and Zagaeski 1994). (*Top*) The regular horizontal striping of moth echo spectra in the contour plot reflects the insect's periodic fluttering wingbeats across successive incident sounds. Contours are plotted with 0 dB as the maximum level that occurs across all spectra so that only spectral coloration is displayed. The first harmonic of the bat's sounds (FM1 approximately 23–50 kHz) arrives relatively intact in echoes, with minimal losses resulting from different acoustic effects. Higher harmonics especially are significantly affected by target orientations and wingbeat postures. (*Bottom*) Average spectra of three echoes

regularly spaced nulls in the spectrum caused by interference between reflections from the target's glints (Simmons 2012; Simmons and Gaudette 2012). If the target is located close to the bat, no further away than about 1 m, and on the axis of the broadcast beam (Masters et al. 1985; Ghose et al. 2006), only these nulls will affect the echo spectrum significantly because atmospheric absorption has not yet accumulated enough to lowpass filter the echoes by more than a few decibels (Fig. 6.3). Also, in most cases, the targets of interest to aerial-feeding bats have dimensions that are enough larger than the broadcast (i.e., incident) wavelengths to avoid Rayleigh scattering, which would lead to highpass filtering of echoes (Houston et al. 2004). (However, if the target is farther away than 1–2 m, or off the acoustic axis of the broadcast beam, then the echo spectrum manifests lowpass filtering caused by the frequency dependence of the directional beam and by the increased atmospheric absorption at higher frequencies. This effect is considered later in this section.)

The contour plot in Fig. 6.9 shows a vertical stack of 31 horizontal spectral-difference slices derived from a sequence of echoes recorded from a fluttering moth (data from Moss and Zagaeski 1994). In effect, these are slices of the moth's transfer function frozen as an acoustic snapshot at the moment the incident sound (a bat-like FM signal) impinged on the moth to capture its instantaneous wing posture. The contour plot thus displays variations in the transfer function over time. The mean spectra of three such echoes are shown at the bottom of Fig. 6.9. This transfer function contains a deep notch at 55–70 kHz as well as noticeable ripple spaced at intervals of about 15 kHz. The pattern of notches reveals interference between reflections from different parts of the moth, most likely between the two wings and another prominent body part such as the abdomen.

For a typical insect, the echo as a whole consists of one, two, or three discrete replicas of the incident sound arriving very close together in time according to the size of the target. Thus, a target with linear dimensions of less than 2 cm returns reflections at separations of 0 to about 100 μs depending on its orientation and the attitude of its wings in the wing-beat cycle. For example, in the case of two reflections arriving at a time separation of 50 μs, the spectrum has interference notches spaced at frequencies 20 kHz apart. Each of these reflections of course undergoes attenuation and lowpass filtering on the journey back to the bat's ears, but it is the pattern of the notches that distinguishes the target. Echoes from targets located closer than 1–2 m of range and within the 10° width of the full-spectrum beam arrive to convey spectral effects specific to the target's size and shape without significant lowpass filtering owing to the target's location. When it does occur, lowpass filtering due to distance or direction is qualitatively different from the pattern of interference notches related to shape. Consequently, the effects of the target's location in the beam readily can be segregated from the effects of the target's intrinsic reflectivity. Moreover, the bat actively participates in segregation of location from identity for any particular target of interest. By pointing its head and ears at the target, and so tracking the target's movements with the sonar version of gaze implied by its imaging beam, the bat nulls out fluctuations in positional lowpass filtering and keeps echoes as close as possible to the full original broadcast spectrum of FM1 and FM2, leaving only the effects of the target's geometry to be perceived.

6.6 Summary

The principal purpose of wideband sonar is the localization and classification of targets. Considerable variation in the echolocation system of bats and odontocetes exists between and within groups (i.e., different species) and prevents a unified model of biosonar, yet the fundamental problem of target localization through echo delay and target identification through echo spectra is a commonality. Within certain target distances, bats and dolphins resolve pulse-echo ambiguity by producing pulses or clicks at intervals that permit the return of desired target echoes from a biosonar emission before a subsequent emission. In both systems, this pattern dissolves during the terminal phase of prey capture when emissions occur at shorter time intervals than that which corresponds to target range. In dolphins, the production of click packets may also occur at target distances >100 m. This phenomenon may permit integration of echo information across multiple echoes when delays between echo returns from a distant target would otherwise limit the integration process. The accuracy and precision of echo delay perception are acute in both bats and dolphins owing in part to the broadband nature of the biosonar emission. However, at least in bats, variations on the conventional neural architecture due to the distribution of delay perception across populations of neurons may contribute to delay resolution. Target identification through echolocation requires the discrimination of target glints, a deconvolution process that is crucial to resolving target shape. Bats and dolphins demonstrate an ability to resolve spectral notches associated with glint separations on the order of tens of microseconds or less. The resolution of the deconvolution process permits differentiation of glints at spatial scales substantially smaller than the size of the target, regardless of the medium in which echolocation is used (i.e., air vs. water).

Though diverse echolocation strategies and mechanisms exist that match the diversity of habitats in which animals live, it can be concluded that bats and dolphins both have evolved sophisticated approaches to echo delay resolution and spectral processing that permit desired targets (e.g., prey) to be differentiated from clutter within the acoustic scene of their respective environments. However, a complete picture of the mechanisms associated with any natural biosonar system, particularly at the level of neural processing, remains elusive. Continued investigations into the mechanisms engaged by individual systems, both bat and dolphin, will be required before a more complete understanding is obtained of how biosonar "works."

References

Akamatsu, T., Wang, D., Nakamura, K., & Wang, K. (1998). Echolocation range of captive and free-ranging baiju (*Lipotes vexillifer*), finless porpoise (*Neophocaena phocaenoides*), and bottlenose dolphin (*Tursiops truncatus*). *Journal of the Acoustical Society of America*, 104(4), 2511–2516.

Altes, R. A., Dankiewicz, L. A., Moore, P. W., & Helweg, D. A. (2003). Multiecho processing by an echolocating dolphin. *Journal of the Acoustical Society of America*, 114(2), 1155–1166.

Au, W. L., & Snyder, K. J. (1980). Long-range target detection in open waters by an echolocating Atlantic bottlenosed dolphin (*Tursiops truncatus*). *Journal of the Acoustical Society of America*, 68(4), 1077–1084.

Au, W. W. L. (1993). *The sonar of dolphins*. New York: Springer-Verlag.

Au, W. W. L., & Turl, C. W. (1983). Target detection in reverberation by an echolocating Atlantic bottlenosed dolphin (*Tursiops truncatus*). *Journal of the Acoustical Society of America*, 73(5), 1676–1681.

Au, W. W. L., Floyd, R. W., Penner, R. H., & Murchison, A. E. (1974). Measurement of echolocation signals of the Atlantic bottlenose dolphin, *Tursiops truncatus* Montagu, in open waters. *Journal of the Acoustical Society of America*, 56(4), 1280–1290.

Au, W. W. L., Moore, P. W. B., & Pawloski, D. A. (1988). Detection of complex echoes in noise by an echolocating dolphin. *Journal of the Acoustical Society of America*, 83(2), 662–668.

Au, W. W. L., Benoit-Bird, K. J., & Kastelein, R. A. (2007). Modeling the detection range of fish by echolocating bottlenose dolphins and harbor porpoises. *Journal of the Acoustical Society of America*, 121(6), 3954–3962.

Au, W. W. L., Branstetter, B. K., Benoit-Bird, K. J., & Kastelein, R. A. (2009). Acoustic basis for fish prey discrimination by echolocating dolphins and porpoises. *Journal of the Acoustical Society of America*, 126(1), 460–467.

Aytekin, M., Mao, B., & Moss, C. F. (2010). Spatial perception and adaptive sonar behavior. *Journal of the Acoustical Society of America*, 128(6), 3788–3798.

Burkard, R., & Moss, C. F. (1994). The brain-stem auditory-evoked response in the big brown bat (*Eptesicus fuscus*) to clicks and frequency-modulated sweeps. *Journal of the Acoustical Society of America*, 96(2), 801–810.

Capus, C., Pailhas, Y., Brown, K., Lane, D. M., Moore, P. W., & Houser, D. (2007). Bio-inspired wideband sonar signals based on observations of the bottlenose dolphin (*Tursiops truncatus*). *Journal of the Acoustical Society of America*, 121(1), 594–604.

Ferragamo, M., Haresign, T., & Simmons, J. A. (1998). Frequency tuning, latencies, and responses to frequency-modulated sweeps in the inferior colliculus of the echolocating bat, Eptesicus fuscus. *Journal of Comparative Physiology A*, 182, 65–79.

Finneran, J. J. (2013). Dolphin "packet" use during long-range echolocation tasks. *Journal of the Acoustical Society of America*, 133(3), 1796–1810.

Ghose, K., Horiuchi, T. K., Krishnaprasad, P. S., & Moss, C. F. (2006). Echolocating bats use a nearly time-optimal strategy to intercept prey. *PLos Biology*, 4(5), e108.

Glaser, W. (1974). Zur hypothese des Optimalenfangs bei der Fledermausortung. *Journal of Comparative Physiology A*, 94(3), 227–248.

Helweg, D. A., Moore, P. W., Dankiewicz, L. A., Zafran, J. M., & Brill, R. L. (2003). Discrimination of complex synthetic echoes by an echolocating bottlenose dolphin. *Journal of the Acoustical Society of America*, 113(2), 1138–1144.

Hiryu, S., Bates, M. E., Simmons, J. A., & Riquimaroux, H. (2010). FM echolocating bats shift frequencies to avoid broadcast-echo ambiguity in clutter. *Proceedings of the National Academy of Sciences of the USA*, 107(15), 7048–7053.

Holderied, M. W., Baker, C. J., Vespe, M., & Jones, G. (2008). Understanding signal design during the pursuit of aerial insects by echolocating bats: Tools and applications. *Integrative and Comparative Biology*, 48, 74–84.

Houser, D. S., Helweg, D. A., & Moore, P. W. B. (1999). Classification of dolphin echolocation clicks by energy and frequency distributions. *Journal of the Acoustical Society of America*, 106(3), 1579–1585.

Houston, R. D., Boonman, A. M., & Jones, G. (2004). Do echolocation signal parameters restrict bats' choice of prey? In J. A. Thomas, C. F. Moss, & M. Vater (Eds.), *Echolocation in bats and dolphins* (pp. 339–345). Chicago: University of Chicago Press.

Imaizumi, T., Furusawa, M., Akamatsu, T., & Nishimori, Y. (2008). Measuring the target strength spectra of fish using dolphin-like short broadband sonar signals. *Journal of the Acoustical Society of America*, 124(6), 3440–3449.

Ivanov, M. P. (2004). Dolphin's ecolocation signals in a complicated acoustic environment. *Acoustical Physics*, 50(4), 469–479.

Jensen, F. H., Bejder, L., Wahlberg, M., & Madsen, P. T. (2009). Biosonar adjustments to target range of echolocating bottlenose dolphins (*Tursiops* sp.) in the wild. *Journal of Experimental Biology*, 212, 1078–1086.

Kick, S. A. (1982). Target-detection by the echolocating bat, *Eptesicus fuscus*. *Journal of Comparative Physiology A*, 145(4), 431–435.

Kick, S. A., & Simmons, J. A. (1984). Automatic gain control in the bat's sonar receiver and the neuroethology of echolocation. *Journal of Neuroscience*, 4, 2705–2737.

Ma, X., & Suga, N. (2008). Corticofugal modulation of the paradoxical latency shifts of inferior collicular neurons. *Journal of Neurophysiology*, 100, 1127–1134.

Masters, W. M., Moffat, A. J., & Simmons, J. A. (1985). Sonar tracking of horizontally moving targets by the big brown bat *Eptesicus fuscus*. *Science*, 228, 1331–1333.

Matsuo, I., Imaizumi, T., Akamatsu, T., Furusawa, M., & Nishimori, Y. (2009). Analysis of the temporal structure of fish echoes using the dolphin broadband sonar signal. *Journal of the Acoustical Society of America*, 126(1), 444–450.

Moore, P. W. B., Hall, R. W., Friedl, W. A., & Nachtigall, P. E. (1984). The critical interval in dolphin echolocation: What is it? [Letters to the Editor]. *Journal of the Acoustical Society of America*, 76, 314–317.
Moss, C. F., & Zagaeski, M. (1994). Acoustic information available to bats using frequency modulated sonar sounds for the perception of insect prey. *Journal of the Acoustical Society of America*, 95, 2745–2756.
Moss, C. F., & Schnitzler, H.-U. (1995). Behavioral studies of auditory information processing. In A. N. Popper & R. R. Fay (Eds.), *Hearing by bats* (pp. 87–145). New York: Springer-Verlag.
Moss, C. F., & Surlykke, A. (2010). Probing the natural scene by echolocation in bats. *Frontiers in Behavioral Neuroscience*, 4, 1–16.
Muller, M. W., Au, W. W. L., Nachtigall, P. E., Allen, J. R. III., & Breese, M. (2007). Phantom echo highlight amplitude and temporal difference resolutions of an echolocating dolphin, Tursiops truncatus. *Journal of the Acoustical Society of America*, 122, 2255–2262.
Muller, M. W., Allen, J. S., Au, W. W. L., & Nachtigall, P. E. (2008). Time-frequency analysis and modeling of the backscatter of categorized dolphin echolocation clicks for target discrimination. *Journal of the Acoustical Society of America*, 124(1), 657–666.
Murchison, A. E. (1980). Detection range and range resolution of echolocating bottlenose porpoise (*Tursiops truncatus*). In R. G. Busnel & J. F. Fish (Eds.), *Animal sonar systems* (pp. 43–70). New York: Plenum Press.
Neuweiler, G. (2000). *Biology of bats*. Oxford: Oxford University Press.
Penner, R. H. (1988). Attention and detection in dolphin echolocation. In P. E. Nachtigall & P. W. B. Moore (Eds.), *Animal sonar processes and performance* (pp. 707–713). New York: Plenum Press.
Petrites, A. E., Eng, O. S., Mowlds, D. S., Simmons, J. A., & DeLong, C. M. (2009). Interpulse interval modulation by echolocating big brown bats (*Eptesicus fuscus*) in different densities of obstacle clutter. *Journal of Comparative Physiology A*, 195(6), 603–617.
Saillant, P. A., Simmons, J. A., Bouffard, F. H., Lee, D. N., & Dear, S. P. (2007). Biosonar signals impinging on the target during interception by big brown bats, *Eptesicus fuscus*. *Journal of the Acoustical Society of America*, 121(5), 3001–3010.
Sanderson, M. I., & Simmons, J. A. (2000). Neural responses to overlapping FM sounds in the inferior colliculus of echolocating bats. *Journal of Neurophysiology*, 83, 1840–1855.
Sanderson, M. I., & Simmons, J. A. (2002). Selectivity for echo spectral interference and delay in the auditory cortex of the big brown bat, *Eptesicus fuscus*. *Journal of Neurophysiology*, 87, 2823–2834.
Schnitzler, H.-U., Moss, C. F., & Denzinger, A. (2003). From spatial orientation to food acquisition in echolocating bats. *Trends in Ecology and Evolution*, 18, 386–394.
Simmons, J. A. (2005). Big brown bats and June beetles: Multiple pursuit strategies in a seasonal acoustic predator-prey system. *Acoustic Research Letters Online*, 6, 238–242.
Simmons, J. A. (2012). Bats use a neuronally implemented computational acoustic model to form sonar images. *Current Opinion in Neurobiology*, 22, 311–319.
Simmons, J. A., & Stein, R. A. (1980). Acoustic imaging in bat sonar: Echolocation signals and the evolution of echolocation. *Journal of Comparative Physiology A*, 135(1), 61–84.
Simmons, J. A., & Grinnell, A. D. (1988). The performance of echolocation: Acoustic images perceived by echolocating bats. In P. E. Nachtigall & P. W. B. Moore (Eds.), *Animal sonar processes and performance* (pp. 353–385). New York: Plenum Press.
Simmons, J. A., & Chen, L. (1989). The acoustic basis for target discrimination by FM echolocating bats. *Journal of the Acoustical Society of America*, 86, 1333–1350.
Simmons, J. A., & Gaudette, J. E. (2012). Biosonar echo processing. *IET Radar, Sonar and Navigation*, 6, 556–565.
Simmons, J. A., Ferragamo, M., Moss, C. F., Stevenson, S. B., & Altes, R. A. (1990). Discrimination of jittered sonar echoes by the echolocating bat, *Eptesicus fuscus*: The shape of target images in echolocation. *Journal of Comparative Physiology A*, 167(5), 589–616.
Simmons, J. A., Moffat, A. J. M., & Masters, W. M. (1992). Sonar gain control and echo detection thresholds in the echolocating bat, *Eptesicus fuscus*. *Journal of the Acoustical Society of America*, 91, 1150–1163.

Simmons, J. A., Ferragamo, M. J., Saillant, P. A., Haresign, T., Wotton, J. M., Dear, S. P., & Lee, D. N. (1995). Auditory dimensions of acoustic images in echolocation. In A. N. Popper & R. R. Fay (Eds.), *Hearing in bats* (pp. 146–190). New York: Springer-Verlag.

Surlykke, A. (2004). The relationship of detection thresholds to the number of echoes in the big brown bat, Eptesicus fuscus. In J. A. Thomas, C. F. Moss & M. Vater (Eds.), *Echolocation in bats and dolphins* (pp. 268–272). Chicago: University of Chicago Press.

Suthers, R. A. (2004). Vocal mechanisms in birds and bats: A comparative view. *Anais da Academia Brasileira de Ciências*, 76, 247–252.

Valentine, D. E., & Moss, C. F. (1998). Sensorimotor integration in bat sonar. In T. H. Kunz & P. A. Racey (Eds.), B*ats: Phylogeny, morphology, echolocation and conservation biology* (pp. 220–230). Washington, DC: Smithsonian Institution Press.

Vel'min, V. A., & Dubrovskii, N. A. (1976). Critical interval of active hearing in dolphins. *Soviet Physics Acoustics–USSR*, 22, 351–352.

Veselka, N., McErlain, D. D., Holdsworth, D. W., Eger, J. L., Chhem, R. K., Mason, M. J., Brain, K. L., Faure, P. A., & Fenton, M. B. (2010). A bony connection signals laryngeal echolocation in bats. *Nature*, 463, 939–942.

Wahlberg, M., Jensen, F. H., Aguilar Soto, N., Beedholm, K., Bejder, L., Oliveira, C., Rasmussen, M., Simon, M., Villadsgaard, A., & Madsen, P. T. (2011). Source parameters of echolocation clicks from wild bottlenose dolphins (*Tursiops aduncus* and *Tursiops truncatus*). *Journal of the Acoustical Society of America*, 130(4), 2263–2274.

Chapter 7
On-Animal Methods for Studying Echolocation in Free-Ranging Animals

Mark Johnson

Keywords Bats • Echogram • Event detection • Foraging • Movement sensors • Sensor fusion • Sound recording • Sound-source parameters • Tags • Toothed whales • Visualization

7.1 Introduction

A great deal of what is known about echolocation has come from studying animals, both terrestrial and aquatic, in captivity. Captive studies have provided fundamental information about the capabilities of biosonar systems, revealing the remarkable sensitivity of echolocating animals in detecting and discriminating targets. However, to find out how animals use these capabilities as part of an integrated sensory–cognitive–locomotory system to navigate and obtain food requires the study of animals in the wild. Fundamentally different approaches are needed to study animals in the wild from those that can be used with captive and trained animals. Although the movements and sound production of constrained animals can be measured precisely, these parameters are often difficult to measure in the wild over the temporal and spatial scales of interest. Moreover, the complexity of, and lack of control over, natural environments tend to support observational rather than experimental approaches, radically changing the way in which studies are designed.

In recent years, a number of powerful new techniques have emerged for studying wild echolocators. These include remote-sensing methods such as synchronized arrays of cameras, microphones, or hydrophones, discussed elsewhere in this volume (Wahlberg & Surlykke, Chap. 4). This chapter examines the potential for, and limitations of, animal-attached devices such as radio-telemetry microphones and

M. Johnson (✉)
Sea Mammal Research Unit, University of St. Andrews, St. Andrews KY16 8LB, Scotland
e-mail: markjohnson@st-andrews.ac.uk

A. Surlykke et al. (eds.), *Biosonar*, Springer Handbook of Auditory Research 51,
DOI 10.1007/978-1-4614-9146-0_7,

acoustic recording tags for studying echolocation in the wild. Given the speed at which technology changes, the focus is on the issues involved in acquiring, processing, and interpreting data from echolocators rather than on current technological limits. In this vein, much of the discussion is centered on newer multisensor tags that provide dense data on the sounds and movements of echolocators, and for which the analytical challenges are most acute. To date, the size of these tags has restricted their use to aquatic echolocators but, as technology advances, multisensor tags will become small enough for use on flying animals.

Many important insights into echolocation have emerged from the synthesis of biology, physics, and engineering, and this active interdisciplinary atmosphere contributes to the excitement and complexity of echolocation as a study area. It can also mean that there is a substantial hurdle to overcome in learning sufficient of each discipline to assimilate, and add to, the literature. This chapter takes an engineering approach to data collection and analysis, complementing the more biologically centered chapters in this volume. Rather than review the achievements of tags in studying echolocation to date (see Madsen & Surlykke, Chap. 9), the discussion focuses on the key sensor modalities available in tags, and the signal processing methods used to interpret the data from these. Section 7.2 begins by reviewing the types of information that might be collected by a device attached to an echolocating animal. Practical limitations in terms of size, weight, and performance of tags are then discussed. Examples of currently used tags, from miniature radio microphones to multisensor archival tags, demonstrate the degree to which animal size and environment control how, and what, data can be collected. Despite the variety of tags and sensors, every signal acquisition scheme involves trade-offs in bandwidth, dynamic range, and power consumption that affect the utility of the data collected. These fundamental constraints in sound and sensor sampling are examined to establish the performance envelope of on-animal devices.

Tags that sample echolocation behavior produce an enormous quantity of data, the analysis of which can be daunting. Often data are collected simultaneously from multiple sensors, for example, sound, position, and orientation. The initial analysis task is to integrate these data sources so that data from different sensors can be compared on a common time base. Another common task is that of detecting events or signatures that relate to specific activities. A third challenge in data exploration is visualization. Section 7.3 discusses these processing steps and reviews some techniques for representing multisensor echolocation data visually that enable inferences about the capabilities and tactics of the animal. The quality of data collected by tags depends on a number of factors including the placement of the tag, the environmental conditions, and the behavior of the animal. As a result, some aspects of behavior can be measured less reliably than others from tag data and these limitations are discussed in Sect. 7.3. The chapter ends by looking at ways in which tagging technology, and the studies it will support, may improve over the coming years.

7.2 Animal-Borne Devices for Studying Echolocation

The temporal scales of interest in studying wild echolocators span at least 12 orders of magnitude from 10 s of microseconds (the oscillation period of most echolocation calls), to years, the interval over which the abilities of an echolocating animal translate into reproductive success. Spatial scales vary from a few millimeters (the distance between prey and predator in the final moments of prey capture) to at least kilometers (the distance covered by an animal in the course of a foraging bout). The breadth of these scales makes it extremely challenging to observe individuals with sufficient resolution and over meaningful temporal and spatial intervals.

Most studies of echolocation in the wild use sound recording arrays or camera systems, installed in, or driven through, areas frequented by the study species. These remote-sensing systems catch brief glimpses of individual behavior but sample a large number of individuals (albeit with little idea of how often an individual is resampled) providing an overview of how animals behave in a limited study zone (Surlykke and Kalko 2008). There are, however, many research questions in echolocation that require continuous information from individual animals that is difficult, if not impossible, to obtain by remote sensing. Examples are:

- How does an individual allocate time and energy between transport, foraging, rest, and socializing?
- What tactics do individual animals use to search for, select, and acquire prey?
- How does the complexity of the environment influence individual echolocation performance?
- How does echolocation function within a group of animals (e.g., interference vs. communication)?

On-animal measuring devices can potentially overcome many of the limitations of remote sensing in sampling individual behavior. A device attached to an echolocating animal can provide continuous information over hours or days, revealing individual foraging rates and tactics. Tags could measure the fine-scale movements of the animal and the environmental context (e.g., temperature, noise, light levels) at the precise location of the animal. It should also be possible to track a tagged animal more accurately than an untagged one, enabling localization of prey encounters. However, there are many practical difficulties in constructing tags for echolocators with the two echolocating taxa, bats and toothed whales, presenting very different challenges.

7.2.1 *Tags for Bats*

The over-arching constraint on tags for bats is weight. Insectivorous bats are agile foragers relying on rapid maneuvers to capture aerial prey. As tag weight increases, bats must fly faster to stay aloft, reducing their maneuverability (Aldridge and

Brigham 1988). Thus, though trained bats in indoor flight rooms can support tags weighing 10 % or more of their body weight, their flight style may be far from natural, impacting the quality of data collected. A rule of thumb quoted widely in telemetry studies of small flying animals is that tag weight should not exceed 5 % of body weight (Caccamise and Hedin 1985; Aldridge and Brigham 1988; Phillips et al. 2003). This implies a tag weight of less than 1 g for the widely studied big brown bat (*Eptisicus fuscus*). Despite this severe weight constraint, a variety of tags have been developed for bats including radio beacons for localizing animals (Meyer et al. 2005; Dechmann et al. 2009), temperature tags to study time spent foraging and roosting (Chruszcz and Barclay 2003), and acoustic telemetry tags (so-called telemikes) to study sounds made and heard by bats (Lancaster et al. 1992; Hiryu et al. 2005). All of these devices use radio telemetry to transmit data to a nearby receiver rather than storing data on the tag, a radio transmitter currently being lighter to construct than a digital recorder. In essence then, these tags comprise a sensor coupled to a transmitter, an antenna, and a battery (Fig. 7.1; see Lancaster et al. (1992) for a circuit diagram). For simplicity, data are transmitted in analog form and digital sampling occurs at the receiver.

Tags for bats are usually glued to the head or back of animals with the fur removed at the attachment site. Although tags may ultimately be shed when bats molt, for short-term attachments animals must be recaptured for tag recovery, which may cause discomfort for the bat. Even though the tag weight is a small proportion of body weight, tags can interfere with movements and may alter the balance of the animal when flying. The antenna wire in particular can interfere with wing movements impacting both the animal's comfort and the radio link performance. Interference can be reduced, at the expense of radio range, by using a very short antenna.

Several research groups have worked on developing sound telemetry tags for bats, but so far only few experimental data have been published. The first reported devices, developed by Henson, Lancaster and colleagues (Henson et al. 1987; Lancaster et al. 1992, 1995) weighed about 0.9 g and were used to study Doppler compensation and electromyographic potentials in 11-g mustached bats (*Pteronotus parnellii*). Hiryu and colleagues developed tags weighing less than 3 g to study large constant frequency (CF) and CF-frequency-modulated (CF-FM) bats (Hiryu et al. 2005). This group has since produced a smaller version weighing less than 0.6 g (Hiryu et al. 2007), which is suitable for a wide range of species including small FM bats. Sound telemetry tags have several common features, dictated by the need for lightweight circuits. Miniature condenser microphones such as the Knowles FG2239 are used with a 10–20 kHz high-pass filter to eliminate wind noise. The microphone output directly modulates the frequency of a radio transmitter in the very high frequency–ultra high frequency (VHF–UHF) bands (e.g., 100–400 MHz) and the signal is received by a customized commercial FM receiver. This modulation scheme is robust but not immune to amplitude variations that occur as the bat moves with respect to the receiving antenna (Lancaster et al. 1992). A wide audio frequency range (up to 100 kHz or more) is possible and multiple tags can be operated simultaneously at different carrier frequencies with minimal interference.

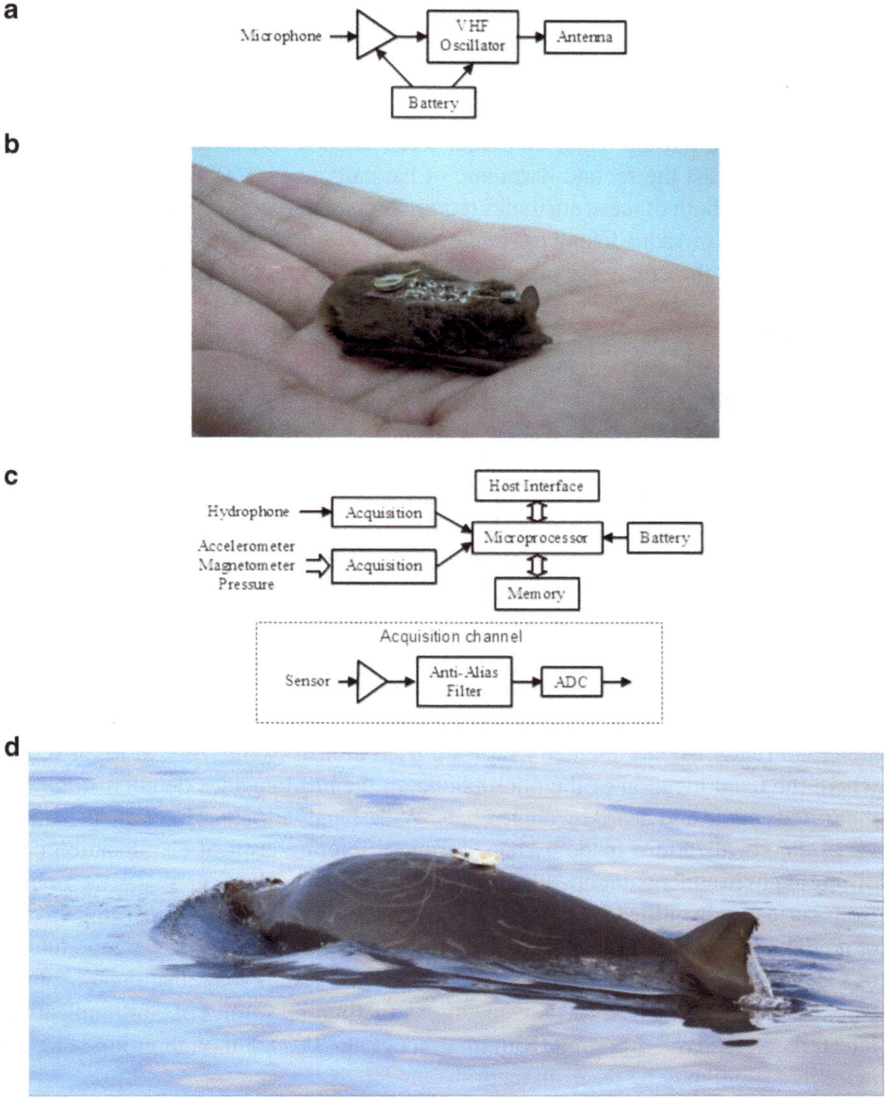

Fig. 7.1 Tags for studying echolocating animals. (**a**) Block diagram of a radio telemetry sound tag for bats (the *triangle symbol* represents an amplifier). (**b**) A tele-mike tag attached to a Japanese house bat (*Pipistrellus abramus*). (Photo courtesy of S. Hiryu.) (**c**) Simplified block diagram of a sound recording tag for odonotocetes adapted from Johnson and Tyack (2003). The *inset* shows the details of each sensor acquisition channel (*ADC* analog-to-digital convertor). (**d**) A DTAG sound and movement recording tag attached with suction cups to a Blainville's beaked whale (*Mesoplodon densirostris*). (Photo courtesy of Y. Bocalon/Univ. La Laguna)

Nonetheless, there are some fundamental performance tradeoffs in these devices, imposed by the severe weight and size constraints.

Ideally, a sound telemetry tag should have both a wide dynamic range (i.e., the difference between the highest and lowest signal levels that can be accommodated) to detect both outgoing sonar calls and weak echoes and a frequency range that extends to at least the second harmonic of bat calls (e.g., 150 kHz for some bats). Unfortunately, both of these attributes require high battery power and so are obtained at the expense of radio telemetry range and tag lifetime. Antenna size is another limiting factor: the efficiency of an antenna depends on its length in comparison to the wavelength of the radio frequency used (e.g., 2 m at 150 MHz). The short antennas suitable for use on bats are inefficient requiring relatively high transmit powers to achieve even modest telemetry ranges. Thus, telemetry range, lifetime, and audio performance all depend on battery size, which depends in turn on the permissible weight of the tag. As a result, a small wideband tag with a lifetime of several hours may have a reliable radio range of only some 20 m. Although this may be adequate for indoor flight rooms, it is critically limiting in outdoor studies. Free-flying bats instrumented with such tags would need to be followed closely with directional antennas to maintain radio contact while intervening foliage may cause dropouts especially with high carrier frequencies.

Study design and tag design must then be considered hand-in-hand to make the most of these performance limitations. For studies of free-flying bats, larger species with low vocalization frequencies are best suited for sound telemetry tags. For these species, heavier tags with radio ranges of several hundreds of meters may be possible. Currently, sound telemetry tags are used almost exclusively in flight rooms where the fixed location of the microphone on the head enables detailed studies of the sounds made and heard by tagged bats. Wide-bandwidth tags have enabled studies of the vocalizations made by bats during different phases of echolocation or in different environments independent of head movement (Hiryu et al. 2010). Echoes from surfaces ensonified by bats have been detected in some tags enabling studies of navigation through obstacles, and how call frequency or amplitude are modulated to match the environment (Hiryu et al. 2005, 2007). Tags may also detect sounds from other bats flying near the tagged animal, facilitating studies of communication, interference, or eavesdropping (Riquimaroux et al. 2007), although group behavior in a flight room may be far from natural.

There is considerable scope for advances in bat tags as smaller, more integrated electronic devices and batteries become available. Adding audio compression circuitry to extend the dynamic range may enable the detection of echoes from prey. Multichannel coding methods will allow low-rate sensor data from accelerometers or bioelectric sensors to be sent along with sound signals (Lancaster et al. 1995). Improved battery lifetime or transmission range may be possible by taking advantage of the low duty cycle of bat vocalizations to reduce transmission power between vocalizations. Diversity receiving techniques used in UHF wireless networks could enable greater radio range and more reliable connections. Looking further in the future, self-recording tags similar to devices used now on whales will eventually become practical for bats. A recording tag would avoid the outages, limited dynamic range, and trailing antennae that plague radio transmission at the expense of greater circuit complexity. The difficulties in producing such a tag are still substantial given that a single memory chip currently weighs about the same as an entire tele-mike, but the miniaturization of electronic components will predictably allow for a "great leap forward" in understanding natural bat echolocation by means of tags in the near future.

7.2.2 Tags for Toothed Whales

Aquatic echolocators vary in size from porpoises (about 1.5 m body length) to sperm whales (up to 18 m). Evidently, tags for these animals are not so severely constrained by size or weight as are bat tags but there are a number of other factors that complicate their design. Tags must add little to the hydrodynamic drag of animals and they must survive salt water immersion and hydrostatic pressure when the animal dives. As yacht racers know, a small flat surface on a streamlined keel can cause a major reduction in speed. Most cetaceans are highly streamlined to minimize drag and the cross-sectional area of tags must be kept small to reduce their impact on the swimming gait and energetic requirements of the host. A rule of thumb, extrapolated from Wilson et al. (1986), for a streamlined tag is that its frontal area is less than 2 % of the animal's frontal area. For a porpoise, this translates into a tag cross-sectional area of less than 10 cm^2. The length of the tag parallel to the flow has less impact on drag but tag length is ultimately limited by the need to attach to a flexing skin surface (Pavlov et al. 2007). Another constraint on tag design is that the electronic circuits inside are protected from salt water, which will corrode and destroy them. As some echolocating animals can dive to 2,000 m (Watkins et al. 1993; Tyack et al. 2006), where the pressure is 200 atm (i.e., 200 kg cm^{-2}), the tag must be sturdy to maintain its shape and seal the interior compartments against water ingress. Seawater also impacts tag design in another way: given the poor transmission of radio waves in the ocean, and the brief opportunities to transmit when the animal surfaces, high bandwidth tags for aquatic animals must store data internally and be recovered to access the data.

Although some smaller delphinids can be captured and temporarily restrained for tag attachment (Read and Westgate 1997), larger echolocating whales cannot be captured with safety. Tag attachment methods vary accordingly. Tags are attached to restrained animals with suction cups, harnesses, dorsal fin bolts, or sutures (Hooker and Baird 2001; Ezer et al. 2008). For free-moving animals, tags are attached with suction cups or skin-penetrating barbs, and are delivered by a pole, crossbow, or gun (Hooker and Baird 2001). Even if animals are captured to attach a tag, recapturing the animal to recover the tag may be impractical. Archival tags are typically designed to detach automatically when their memory is full and include floatation material so that they float at the surface after detachment, emitting a radio signal to aid recovery.

There are a number of species-dependent veterinary and ethical concerns with skin-penetrating attachments, and so the design and use of these must be considered very carefully. For short term attachments (up to a few days), suction cups are a practical alternative but they can slide or detach during fast maneuvers and social behaviors, making them less reliable. For longer attachments, there are few alternatives to implanted devices.

Tags with a variety of depth, temperature, and positioning sensors have been used since the 1960s to study aquatic animals and these continue to be relevant for studying the dive patterns and movements of echolocators over long time intervals (Ropert-Coudert and Wilson 2005). More recently, sound recording tags have been developed enabling detailed short-term studies of echolocation. Early devices, comprising a hard-drive or tape recorder in a water proof housing, were used briefly in the late 1990s (Fletcher et al. 1996; Burgess et al. 1998) but sound recording tags with semiconductor memory have been used almost exclusively since about 2000 (Madsen et al. 2002; Johnson and Tyack 2003; Burgess 2008). By eliminating the moving tape or hard-drive, these tags can be encapsulated in plastic to protect them from sea water. As most electronic components and some batteries are minimally affected by hydrostatic pressure, a heavy rigid pressure housing is not required, opening the way to miniaturization. It is presently feasible to construct a tag with a 150-kHz bandwidth and a recording capacity of two days in a volume of 50 cm^3 including battery and floatation material (Johnson et al. 2009).[1] With the phenomenal pace at which memory density is increasing, battery power is now the major factor limiting the recording duration of sound sampling tags.

Sound recording tags comprise a hydrophone, a digital sampling circuit, microprocessor, and an array of flash nonvolatile semiconductor memory (Fig. 7.1; Johnson and Tyack 2003). Included also are circuits to communicate with a host computer for data offloading and a VHF radio beacon to enable recovery of the tag. In addition to sound, a variety of sensors can be added to the tag. Widely used sensors include accelerometers, magnetometers (compasses), speed, pressure (i.e., depth), water temperature, and GPS. Sensor channels can be sampled at an appropriate rate by the microprocessor and folded into the recorded data stream. As all sampling rates are derived from a common clock, the sensor and audio data remain synchronized throughout deployments. The additional sensors may add little to the size of the device but can increase the power consumption and so shorten the tag life. A practical issue is then to decide which sensors are needed for a particular study and whether the resulting recording time will be acceptable.

An example of a sound-recording tag that has been used widely on echolocating whales is the so-called DTAG (Johnson and Tyack 2003). This is a multisensor device containing two sound channels (two hydrophones spaced 2.5 cm apart in the nose of the tag) as well as orientation and depth sensors. The latest version of this device has an audio bandwidth of 160 kHz, recording duration of more than 48 h, and a frontal area of about 15 cm^2. Another multisensor sound recording tag, the

[1] Flash memory density has more than doubled since 2009, enabling a wider recording bandwidth than described in Johnson et al. (2009).

Bio-Acoustic Probe (Burgess 2008), has been used less often in echolocating studies owing to its low sampling rate but a new version, the Acousonde, may be more suitable for use on odontocetes (see www.acousonde.com). Yet another device of this kind, the Bioacoustic Measurement Tool (Martin et al. 2005), is specifically designed for echolocation studies with trained dolphins. It has multiple acoustic channels, a range of sensors, and a high sampling rate but is too large for use on most wild animals. Video recording tags such as the Crittercam (Marshall et al. 2007) that have audio recording capability have also been used with echolocating animals, particularly sperm whales and narwhals (Dietz et al. 2007).

A different type of audio recording device has been developed for studying small high-frequency echolocators such as porpoises. Until recently, the high sampling rate needed to record vocalizations from these animals has meant short recording times given the amount of memory that can fit in a small tag. One solution to this problem is to detect vocalizations, and record the time of each detection, in the tag rather than make a continuous recording (Blomqvist and Amundin 2004; Akamatsu et al. 2005a). The A-tag (Akamatsu et al. 2005b) works on this principle recording the detection time, amplitude, and angle of arrival of each transient detected on a pair of hydrophones. This tag can also detect strong echoes from the sea surface and vocalizations from other nearby animals (Akamatsu et al. 2005c). A limitation with sound detecting tags is the possibility that weak vocalizations from the tagged whale will not be detected (Akamatsu et al. 2007). This may especially impact the reliability of detecting rapid clicks during buzzes which often seem to be produced at lower levels than regular clicks (Madsen et al. 2005). Increasing memory density is making in situ detection unnecessary for short-term tags but sound detecting tags remain an interesting option for long-term studies or if telemetry of data is required from an expendable tag.

Acoustic recording tags have opened up a rich array of studies on aquatic echolocators, as reviewed elsewhere in this volume (Madsen & Surlykke, Chap. 9). As component sizes continue to shrink, devices small enough to be used with all echolocating cetaceans will soon be feasible. More challenging limitations are the short recording time of these devices, and the complexity of the data they produce. Battery power and attachment tenacity are currently the two factors controlling recording life. There are few prospects for harvesting power on a tag without creating a local noise source and so longer recording times will require batteries with greater power density and more efficient electronics. Both of these will arrive in time but improvements are likely to be slow. There has been relatively little work on improving the longevity of suction cup attachments and substantial advances may be possible. Other surface attachment methods such as adhesives may also be possible, perhaps enabling week-scale attachments. However, the dense data collected by sound recording tags, even in the day-scale deployments currently achievable, are challenging to analyze. Some processing and graphical aids for exploring these complex data will be discussed in Sect. 7.3 but it is important to first consider in more detail the capabilities and limitations of the sensors used in tags.

7.2.3 Sound Acquisition

The high-frequency sounds produced by echolocating animals can be sensed in a tag using a condenser microphone (air) or a piezoelectric ceramic hydrophone (water). These sensors differ in their transduction mechanism but both produce a small voltage that is proportional to the instantaneous pressure on the detecting surface. Sensitivity is typically proportional to the size of the sensor so that large sensors have high sensitivity, allowing the detection of weak sounds over the noise floor of the following preamplifier circuit. Sensitivity must be balanced against the weight/size constraint for the tag, which often dictates smaller sensors. Small sensors are also desirable if an omnidirectional response is required at high frequencies: the 2.5 mm diameter microphones used on tele-mikes are equivalent to one half of a wavelength at 70 kHz and so become increasingly directional at and above this frequency. The wavelength of echolocation signals produced by marine mammals varies from about 10 cm (sperm whales) to 1 cm (porpoises), and hydrophones may need to be sized according to the study species to optimize performance (Madsen and Wahlberg 2007).

The low sensitivity and high impedance of microphones and hydrophones necessitate a preamplifier circuit before the signal can modulate a transmitter or be converted into digital form. At the high frequencies used by most echolocators, ambient noise is low and the electronic noise generated by circuits in the tag will likely set the minimum signal level that can be detected. This is especially true for the low power circuits used in a small, battery powered, tag which inevitably have higher noise than might be achieved otherwise. The preamplifier, being the circuit in the tag that handles the lowest signal levels, usually sets the overall noise floor of the device. The noise floor influences another key parameter of the sound recording system: the dynamic range (DR), defined as the difference in decibels (dB) between the highest sound level that the tag can record without distortion (called the clipping level) and the broadband noise floor of the instrument (Madsen and Wahlberg 2007). Ideally a sound recording tag for echolocators will be able to sample both the signal produced by the animal and the weak echoes returning from small scatterers such as prey. The difference in level between these two signals can be estimated using the sonar equation (Madsen & Surlykke, Chap. 9). For a small target, the returning echo level (EL) that would be received at a tag located close to the sound source is approximately:

$$\mathrm{EL} = \mathrm{SL} - 2\mathrm{TL} + \mathrm{TS} \tag{7.1}$$

where SL is the source level, TL is the one-way transmission loss to the target, TS is the target strength, and all quantities are in dB. The tag, being located a little behind the sound source, receives a weaker version of the outgoing signal due to the directionality of the source. The difference between the on-axis source level and the received level (RL) at the tag can be termed the front-to-back ratio (FBR), which is defined here as combining both the radiation pattern of the source and any attenuation incurred (with respect to the reference distance of the SL) in the propagation of sound from the source to the tag. Thus, the difference between the received level of the outgoing click and the echo level at the tag is:

$$\mathrm{RL} - \mathrm{EL} = \mathrm{SL} - \mathrm{FBR} - (\mathrm{SL} - 2\mathrm{TL} + \mathrm{TS}) = 2\mathrm{TL} - \mathrm{TS} - \mathrm{FBR} \tag{7.2}$$

The FBR describes directionality in a different way than the more commonly used directivity index and is not usually reported in beam pattern studies. Nonetheless, rough values can be deduced for the two echolocating taxa. For bats, FBR is likely to be relatively small owing to the low directivity of the sound source and the proximity of the microphone, while for odontocetes, the combination of a highly directional beam and a tag placement behind the skull leads to a large FBR. This parameter has an important influence on the dynamic range needed in a tag to record both the outgoing and echo signals. This dynamic range can be roughly estimated as follows:

1. A bat echolocating at 50 kHz on an insect at 2 m (reference distance is 10 cm).
   TL including absorption is about 30 dB (Lawrence and Simmons 1982).
   TS may be about −25 dB (Waters et al. 1995; Surlykke et al. 1999).
   FBR may be about 10 dB (e.g., 20 dB of directivity minus about 10 dB transmission loss reduction due to the tag being placed much closer to the mouth than the 10 cm reference distance).
   => RL − EL is about 75 dB (i.e., $2 \times 30 + 25 - 10$ dB).
2. A whale echolocating at 40 kHz on a fish at 10 m (reference distance is 1 m):
   TL including absorption is about 20 dB.
   TS may be about −40 dB (Madsen & Surlykke, Chap. 9).
   FBR may be about 40 dB (based on recordings from Blainville's beaked whales, Madsen & Johnson, unpublished data).
   => RL − EL is about 40 dB (i.e., $2 \times 20 + 40 - 40$ dB)

To be able to detect an echo above the noise floor of the recording system, a received signal-to-noise ratio of at least 10 dB is needed. At the other extreme, the gain of the system should be set so that the clipping level is at least 10 dB above the average RL of the outgoing calls to allow for fluctuations in the source level. Combining these, the recording system must have a dynamic range, in the frequency range of interest, of at least 95 dB for bats (i.e., $75 + 10 + 10$) or 60 dB for whales to sample both the outgoing and echo signals.

With careful design, a broadband dynamic range of about 90 dB is possible in a low-power sound recording tag, with the preamplifier noise and battery voltage controlling the lower and upper extremes of this range (Madsen and Wahlberg 2007). For a tele-mike, the DR is constrained also by the bandwidth of the transmitter and lower figures are likely (perhaps closer to 60 dB). The DR can be improved by filtering noise outside of the frequency range of interest. Many echolocators produce band-limited signals that occupy, instantaneously, only a portion of the tag's bandwidth. Filtering the received sound signal to match the spectrum of the outgoing signal improves the DR by about $10 \log_{10}(f/b)$ where f is the frequency range of the tag (over which the broadband DR is defined) and b is the signal bandwidth. Improvement will be minimal for the broadband transients produced by odontocetes but the benefit of narrowband processing may be substantial (e.g., 10 dB or more) for bat calls. Even so, the wide dynamic range needed to acquire both out-going calls and prey echoes from bats means that researchers must choose between these and set the gain of the tag accordingly. For odontocetes, on the other hand, it is possible to acquire both signals, at least on some species, with low-power recording equipment. However, to achieve this, the gain of the tag must be set carefully to optimize usage of the limited dynamic range (Johnson et al. 2013). If the gain is too low, good recordings of the outgoing signal will be obtained but echoes will be lost in the noise floor. At high gains, echoes may be detectable but the outgoing signal will overload the tag and will be clipped. Sounds that are clipped may still be used for some analyses (e.g., to time vocalizations) but cannot be used to determine sound level or spectrum. Low-frequency noise due to fluid flow over the sensor (i.e., wind or water noise) is another important cause of overloading in tags, and a high-pass filter (sometimes referred to as a whitening filter) is usually included in the preamplifier to reduce low-frequency sensitivity.

Another consideration affecting the detectability of echoes is whether or not there is an unobstructed sight line between the echoic organism and the tag. If the returning echo is shielded by the body of the animal, the transmission loss will be greatly increased and echoes may be impossible to detect. This seems to be the case with dorsally attached tags on sperm whales and pilot whales, perhaps due to their bulbous head and broad maxillae. Echoes from the sea floor are detectable in tags on these species (Teloni et al. 2008) but echoes from organisms are rarely observed. Tags on the two species of beaked whales tagged to date (Cuvier's and Blainville's) have been more productive in this regard, regularly yielding echoes from organisms in the water including targeted prey (Johnson et al. 2004). This may be a result of the sloping forehead and possible upward-tilted acoustic beam of these species (Zimmer et al. 2005a). In bats, echoes from obstacles are detectable using tele-mikes (Hiryu et al. 2007) but echoes from prey have not been reported presumably because of the high dynamic range required to detect these weak signals.

Ultimately, all audio signals are sampled with an analog-to-digital converter (ADC) and stored in a digital format. In archival tags, sampling occurs on the device itself while for tele-mikes, sampling occurs at the output of the radio receiver. In either case, the sampling rate must be at least twice the highest frequency of interest although a factor of three simplifies the anti-alias filter design and improves performance (Orfanidis 2010). The number of bits, b, needed to represent each sample depends on the dynamic range according to $b >= \mathrm{DR}/6$ (Orfanidis 2010). Given the practical limit of about 90 dB in a battery-powered device, each sound sample can be represented by 16 bits (2 bytes) with negligible performance degradation. For an archival tag, the recording duration, d, is then given by (Johnson et al. 2009):

$$d(\text{hours}) \approx 150g / nf_s \tag{7.3}$$

where g is the memory size in gigabytes (GB), n is the number of sensors (microphone or hydrophone) channels, and f_s is the sampling rate in kHz. For example, a 32 GB tag sampling at 200 kHz will be capable of recording continuously for 24 h.

There is a clear benefit in compressing sound data to increase recording time. The DTAG uses lossless audio compression (Johnson et al. 2013) achieving an average compression ratio of 3–5, enabling at least a tripling of the recording time. Lossless compression permits exact recovery of the original signal. Higher compression ratios are possible with lossy compression algorithms in which an approximation of the original signal is recovered after decompression. Popular compression algorithms such as MP3, based on psychoacoustic models of human hearing, can give compression ratios of 8 or more. However, these algorithms introduce significant (but, by design, inaudible) artefacts in the recording which can greatly reduce the value of the data for echolocation analysis (Liu et al. 2008).

7.2.4 Nonacoustic Sensors

Given the tight weight constraint, there is currently limited scope for additional sensors in tags for bats. However a great variety of sensors have been built into tags for marine animals for which tag size is less constrained (Cooke et al. 2004). Examples are video and still cameras (Davis et al. 1999), bioelectric sensors (Andrews et al. 1997), stomach and skin surface temperature sensors (Ancel et al. 1997; Westgate et al. 2007), and jaw motion sensors (Ropert-Coudert et al. 2004). Arguably sensors for position, orientation, and movement have received the most attention, at least in foraging studies of marine animals (Wilson and Wilson 1988; Johnson and Tyack 2003; Mitani et al. 2003; Johnson et al. 2009). The ability to say where an animal is and how it moves while producing or hearing a sound is of such fundamental value that all currently used sound recording tags include, or are deployed with, some form of positional or movement sensor.

Six parameters are needed to represent position and orientation in three dimensions, namely, latitude, longitude, depth, pitch, roll, and heading. An additional six parameters are required to characterize movement (i.e., speed or acceleration) in the linear (x, y, and z) and angular (pitch, roll, and yaw) dimensions. Owing to space and power limitations, only a subset of these parameters is measured in tags. Depth and orientation are straightforward to measure with high resolution using miniature pressure, acceleration, and magnetic field sensors (Johnson and Tyack 2003; Mitani et al. 2003), and these are the commonest sensors in sound recording tags. The position of marine animals at the sea surface (and bats out of the forest canopy) can be obtained using an on-animal GPS sensor (MacLean 2009). However, satellite signals attenuate rapidly in sea water and no global method exists for tracking animals below the surface. For most marine echolocators, this means that GPS positions are unavailable when the animal is foraging. Two alternative methods are used widely to position animals during dives: passive acoustic tracking and dead reckoning. In passive acoustic tracking, the position of a vocalizing animal is calculated from the arrival time of its sounds at an array of receivers with known positions placed around the animal (see Wahlberg & Surlykke, Chap. 9; Spiesberger and Fristrup 1990). For animals tagged with an acoustic recorder or tele-mike, improved accuracy and robustness of acoustic tracking is achieved by combining tag and array data (Zimmer et al. 2003; Johnson et al. 2009; Ward Shaffer et al. 2013). Acoustic tracking can give accurate, geo-referenced positions of diving animals every time they vocalize but has the major drawback of needing several receiving stations within acoustic range of the animal.

In dead reckoning, the track of a tagged animal is estimated solely from sensors in the tag (Wilson and Wilson 1988; Wilson et al. 2007). The track is formed by integrating an estimate of the three-dimensional velocity vector of the animal with respect to time, starting from a known position. Although velocity is the integral of acceleration, it cannot be estimated directly from accelerometers in the tag because these sensors measure both the animal's specific acceleration and its orientation with respect to the gravity vector (Grewal et al. 2007). The usual approach is to assume that the animal moves in a direction given by its orientation (i.e., its pitch and heading angles). This direction, combined with a forward speed estimate, gives the velocity vector (Wilson et al. 2007). There are a number of sources of error in this approach. Although forward speed can be estimated with moderate accuracy (e.g., Blackwell et al. 1999; Shepard et al. 2008), differences between the true direction of motion and the orientation of an animal can be significant and, when integrated to form the dead-reckoned track, result in positional errors which grow with time easily reaching hundreds of meters during a long foraging dive. Thus, dead reckoning offers good time resolution and moderate short-term accuracy but gives only a rough idea of the absolute position of the animal. This type of information may nonetheless be useful in examining stereotypical movements during encounters with prey.

Another aspect of animal movement is gaining prominence in foraging studies. Rapid movements during locomotion, maneuvers, and foraging strikes can be detected by wide-bandwidth accelerometers attached to animals (Cooke et al. 2004; Wilson et al. 2006), providing information about the magnitude and timing of these events. Acceleration signals cannot be precisely separated from orientation changes preventing a direct estimate of force and work done. However, relative measures of energy investment can be derived by differentiating or filtering the acceleration signals to characterize the jerk (i.e., the acceleration rate; Johnson et al. 2004; Simon et al. 2012) or the overall dynamic body acceleration (Gleiss et al. 2011). To acquire acceleration data for dynamic analysis it is critical that the tag is attached firmly to the animal's body. For suction cup attachments, this can be achieved by using three or more cups in a planar array.

The presence of high-frequency information in acceleration signals also provides a reminder of the need to filter sensor signals correctly prior to digitization. All sensor signals are ultimately sampled digitally and a separate anti-alias filter is required on each sensor channel before analog-to-digital conversion (Fig. 7.1; Orfanidis 2010). Failure to filter signals before conversion reduces their dynamic range and can lead to ambiguity when interpreting the signals (Shiomi et al. 2010).

For example, an accelerometer sampled at 2 Hz attached to a flying or swimming animal stroking at 1.5 Hz will register a stroke rate of 0.5 Hz without anti-alias filtering. With adequate filtering, the same sensor will correctly report no locomotion activity within its frequency range.

7.2.5 Impact of Tags

Tagging studies inherently involve a greater degree of impact on wild animals than do most observational studies and ethical issues must be considered carefully both in the design of tags and in the studies that use them. Relevant questions are whether the tag is suitable for the animal; whether similar data already exist, making a new study unnecessary; or whether it is possible to measure similar data using a less invasive method. There are probably very few situations in which an animal does not respond to a tag or the tagging procedure, no matter how innocuous it may seem. A key issue is then whether the change in behavior elicited by the tag is small enough that the resulting data can be used to answer the scientific question at hand. Tag data from an animal that responds strongly to a tag can be difficult to interpret and may be unusable. For long-duration tags it is important to assess how much the tag affects foraging performance or energy consumption with potential consequences on fitness (Wilson and McMahon 2006).

It is difficult to assess the impact of tagging and tags on a wild animal because of the lack of a suitable control. Though foraging or locomotion can be quantified in a tagged animal, similar measures are often impossible to obtain over comparable intervals in untagged animals. One approach is to vary the size of the tag while maintaining the same sensor suite to see how tag size impacts animals (Wilson et al. 1986). To date, this type of study has been attempted on very few species and, in lieu of better data, anecdotes are often used to justify tag size or tagging procedure. It is tempting to conclude that affect is minimal because an animal does not seem to be bothered by a tag or because it appeared to return to normal behavior soon after tagging. But visual assessment of impact is highly subjective and it may not be possible to observe many critical behaviors, for example, those that take place below the sea surface. Both from an ethical viewpoint and to maximize scientific quality, better information is needed about how tags impact animals and therefore how they should be used and improved.

7.3 Exploring and Visualizing On-Animal Echolocation Data

Sound and movement sampling tags offer enormous potential for studying the ethology, ecology, and sensory physiology of echolocating animals in the wild but this capacity is unlocked only by creative analysis of the complex data recorded by these devices. Most field experiments involve the collection of data from multiple sources, each sampled at different rates. For example, a tagged animal might be simultaneously recorded by a far-field microphone/hydrophone array while also being the subject of focal behavioral sequencing. The tag itself may contain sound, acceleration, and depth sensors all recorded at different sampling rates. Combining these heterogeneous data gives a far more complete picture of the animal's behavior than can be achieved with any one sensor but requires careful processing to align the sensor channels and extract data from them. Data analysis software such as R, Scilab, Octave, or Python (all available free on the Internet), or commercial products such as Matlab and Igor Pro are needed to perform these functions. These software products provide a range of built-in data processing and graphical display tools as well as scripting languages for developing new algorithms. Irrespective of the program used, the analysis of tag data in echolocation studies generally involves four steps: sensor fusion (i.e., combining data from difference sensors), event detection, visualization, and quantification, and these are discussed in the following sections.

7.3.1 Sensor Fusion

Sensor fusion is the process of combining data from multiple sensors to estimate quantities that are not directly measured by a single sensor. For example, the angle-of-arrival of a sound can be deduced by combining data from two hydrophones (Akamatsu et al. 2005a) while compass heading can be estimated by combining data from triaxial accelerometers and magnetometers in a tag (Johnson and Tyack 2003; Mitani et al. 2003). Resampling data from one sensor to synchronize with events found in a second sensor is another fusion task. An example of this is estimating the orientation of an animal at the moment that it produces each echolocation call. The data analysis programs mentioned above provide a variety of tools for processing and combining data streams greatly simplifying sensor fusion. However, if incorrectly implemented, sensor fusion can produce errors and artifacts that can mask important behaviors or can even be misinterpreted as behaviors themselves (e.g., Shiomi et al. 2010).

The first task in combining sensor data is to resample signals to a common time base so that sample *k* from sensor 1 represents precisely the same time as sample *k* from sensor 2. As sensors may be sampled at different rates and at different time instants, this involves both resampling to a common sampling rate and synchronizing the sampling moments. Although this could be achieved simply by selecting the temporally closest samples from each signal, the result will have more error and greater risk of aliasing artifacts than if surrounding samples are also considered. Better performance will be obtained using the signal processing operations of decimation (to resample data) and interpolation (to time-align data) (Orfanidis 2010). These are essentially filtering operations which produce new signals at the desired sampling rate and time offset. To implement these processing steps, precise information is required about the relative sampling rates and time delays in each sensor channel. In multisensor tags, sensor sampling rates are usually derived from a common clock and so it is straightforward to determine the required decimation factors.

It is important, though, to remember that almost all filters, whether digital or analog, introduce time delays that depend on the type of filter and its spectral characteristic (Orfanidis 2010). Thus, each sensor channel can have a different time delay due to differences in the way the signals are processed, and this must be taken into account when aligning sensor data. The significance of time delays depends on the rate of change, and therefore the bandwidth, of the signals: for wide-band signals (with respect to the sampling rate), even small timing errors between signals can cause unexpected errors.

Sensor fusion is more complicated when combining data from separate recording devices such as a recording tag and a far-field hydrophone (e.g., Zimmer et al. 2003). The sampling clocks of these devices will drift with respect to each other, requiring time-varying time-alignment. A small drift of 20 ppm, typical of the clocks used in sound recorders, adds up to 0.3 s after 4 h, which would be an unacceptable error for acoustic tracking if left uncorrected. This problem is inevitable with recording tags but is avoided in tele-mikes by recording the signal from the tag and from far-field microphones on the same multichannel recorder. In general, to minimize timing errors, calibration signals should be generated periodically to measure the time delays between sensors. This can be done before and after an experiment or, preferably, throughout the experiment provided that the calibration signals themselves will not impact the study animal. Thus, sensor fusion should not be an afterthought but should be considered when planning experiments and designing measurement devices.

Once resampled and synchronized, the corresponding samples from different sensors can be combined directly to form new signals. In doing so, it is essential to also evaluate how errors in the individual sensors translate into errors in the combined signal. In simple situations, the standard deviation of the derived signal can be calculated directly from the error distributions of the component sensors (e.g., Wahlberg et al. 2001). For more complex systems, the Monte Carlo method can be used to estimate errors. In this method, random errors in each sensor are simulated and passed through the sensor fusion algorithm to estimate the distribution of errors in the measurement. Evaluation of errors may reveal that the derived signal is more dependent on errors in some sensors than in others. A number of techniques are available such as weighted least squares and Kalman filtering (Bar-Shalom et al. 2001) for optimally combining sensor signals so as to minimize errors. These methods may also help to interpolate over short outages in sensor data, for example, as occur frequently in acoustic tracking.

7.3.2 Event Detection

Echolocation is inherently a discrete-time process comprising series of repeating events (echolocation calls, prey captures, foraging forays, etc.), making event detection a critical part of data analysis. The ability to detect events consistently in data, and ascribe a biological interpretation to them, opens the way to powerful visualizations and statistical analyses. The multiple sensor channels in acoustic recording tags offer a plethora of possibilities for identifying events but do not necessarily make detection trivial. Individual echolocation calls, for example, should be easily detected in tele-mike or acoustic tag recordings under normal conditions but can be difficult to detect if the ambient noise is high or the source level is low, for example, during buzzes (Madsen et al. 2005). The location of the tag on the animal can be another confounding factor: weak sounds such as buzzes and respirations may be hard to detect in recordings from posterior positions on marine animals but these locations will yield clear locomotory signals from accelerometers. Thus, in some cases, the challenge is to detect an event in the presence of noise or interference. In other cases, the events themselves may not be clearly defined or may have variable signatures. For example, the start time of buzzes can be defined precisely in species that change call type in buzzes (e.g., beaked whales, Johnson et al. 2006) but is hard to judge in species (e.g., sperm whales or bats) for which the transition from regular clicking to buzzing tends to be gradual. Similarly, changes in movement patterns associated with the location or capture of prey may be hard to detect consistently unless they are highly stereotyped. For example, a rolling prey capture tactic (Akamatsu et al. 2010) is readily detected in accelerometer signals if the animal is horizontal but is unobservable with an accelerometer if the animal is vertical. Thus, similar physical events can create quite different sensor signals depending on the larger context. The art of event analysis is then to choose cues that have clear biological significance and that can be detected reliably in the study species with the available sensors, while making sure that the ensuing analyses are robust to occasional mistaken detections.

A key step in event detection is to process signals so as to accentuate the event of interest. Filtering can be used to improve the signal-to-noise ratio of signals and is especially effective for enhancing signals that change rapidly, such as at a sudden behavioral transition. This can be achieved using a high-pass filter with a cut-off frequency located between the maximum change rate of normal signals and the elevated change rate of the signal of interest. The signal analysis programs listed previously provide tools to design and apply such a filter. More complex filters are needed if the spectrum of the signal of interest overlaps the spectra of other signals in the recording. Matched filters are the extreme case of a filter designed to enhance a specific signal but these work well only if the signal of interest is known and invariant (Kay 1998) which rarely occurs with biological signals. To detect more variable signals, additional classification rules are required that constrain the detector to find only signals with certain characteristics. Even so, supervision is often required to minimize errors. In a supervised detector, each detection is presented to the researcher for approval along with unexpected gaps between detections which may contain a missed event.

7.3.3 Visualization

Visualization of multidimensional data such as obtained from sound recording tags is a challenge requiring both creativity and programming skills. Creativity is critical: using the same visualizations as another researcher may simply show the same things that they have already reported. Some options for data visualization

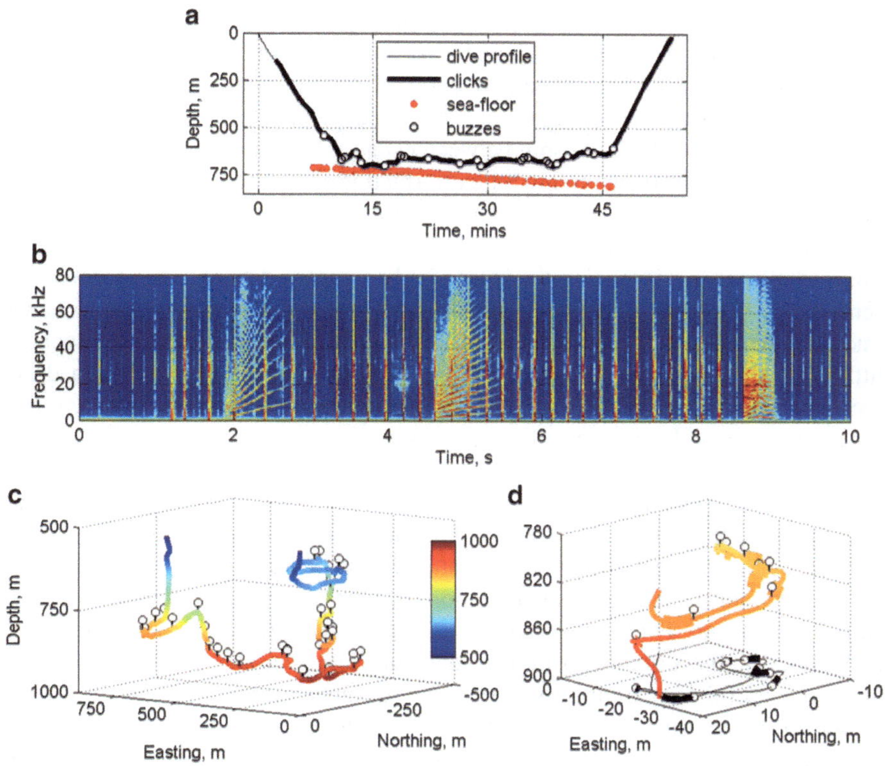

Fig. 7.2 Visualization methods for data from multisensor acoustic recording tags. (**a**) Dive profile of a sperm whale (*Physeter macrocephalus*) foraging dive annotated with information collected from a sound recording tag. The wide track indicates when the whale is producing echolocation clicks; the *open circles* indicate buzzes, which are presumed prey capture attempts; the *red dots* below the profile indicate the sea floor depth determined from timing echoes received on the tag. (**b**) Spectrogram of 10 s of sound recorded by a tag on a short-finned pilot whale (*Globicephala macrorhynchus*) showing a mixture of echolocation and social sounds produced by the tagged whale and possibly others nearby. *Red* and *blue* colors indicate strong and weak signals, respectively. It can be difficult to distinguish which sounds are produced by the tagged whale when whales are in tight groups. (**c**) Three-dimensional ribbon plot reconstructing the track of a tagged Blainville's beaked whale (*M. densirostris*) during 32 min of a foraging dive. *Open circles* indicate buzzes and the *color* (as well as the vertical scale) indicates depth. (**d**) Detailed view of 2 min of the track showing tortuous movements during a sequence of buzzes possibly to exploit a prey patch. Wider track segments indicate when a buzz is being produced. The horizontal projection of the track is shown at the base of the figure in *black*. (**c**, **d**) adapted from Johnson et al. (2009)

are reviewed below, followed by a more detailed examination of one method, the echogram, which is especially pertinent for exploring sound recordings made by tags on echolocating animals.

Familiar plots useful for exploring echolocation data from tags are dive/altitude profiles, spectrograms, and ribbon plots (Fig. 7.2). The first two of these show position (in a single dimension) or sound production as a function of time and are used to explore the temporal sequence of events. Ribbon plots show the three-dimensional track taken by a tagged animal, for example, estimated by dead-reckoning. Additional data can be displayed along the track line by modulating the color, width, and shading of the trackline (Ware et al. 2006). Ribbon plots are useful for exploring how animals move while foraging or navigating by echolocation but are poor at showing the temporal relationship between events. Dive profiles and spectrograms have the opposite problem, giving little information about the spatial relationship of events. Given the number of dimensions in play, satisfactory ways to combine both temporal and spatial information are hard to conceive and such displays will be a future challenge in visualizing the increasingly complex data obtained from multi-sensor tags. Another challenge, especially relevant when plotting dead-reckoned tracks, is that of displaying uncertainty. With most sensors, error variances change little with time and can be conveyed, for example, by line thickness. In dead-reckoned tracks, relative movements over short intervals may be accurately represented throughout the track but absolute position errors grow with time and can end up exceeding the scale of the ribbon plot itself.

Yet another challenge lies in visualizing both large and small-scale features on the same plot. Inevitably, fine details are lost as the plot covers a larger temporal or spatial extent, making it difficult to find similarities or trends across events. One way to overcome this is to display small sections of time- or spatially indexed data, synchronized to events. For example, short track segments can be displayed synchronized to buzzes with the absolute position at each buzz set to zero (Fig. 7.3). This kind of display is useful for exploring stereotypicity in movements, for example, in prey capture or maneuvers. It takes advantage of the high resolution of dead-reckoned tracks without conveying unreliable information about the distance covered by the animal over longer intervals.

Another type of synchronized plot, familiar from boat depth-sounder displays, is the echogram. In this plot, short segments of sound data, synchronized to outgoing echolocation calls, are displayed as stacked colored or shaded bars (Fig. 7.4). Each sound segment is filtered to accentuate the signals of interest and the envelope (or another measure of instantaneous signal level) of the filtered signal is plotted as a colored bar, usually with a log (i.e., decibel) scaling. Colored bars from a series of calls are stacked to form an echogram with the axes being call number (the x-axis in Fig. 7.4) and time elapsed since each call (the y-axis). If the time between consecutive calls is short, echoes from the same obstacle or organism will show up in an echogram as sloping lines spanning several calls. Drawing on the ability of the human eye to recognize patterns in a cluttered image, weak echoic targets can be identified in echograms even though the echoes may be too weak to be detected individually. The resolution and contrast in echograms can be improved by adjusting the filter used in preparing the sound segments and by changing the mapping between signal level and color. Matched filters may be helpful if the outgoing calls are well characterized and are not distorted spectrally by reflection but may offer little advantage over bandpass filtering for reverberant or complex targets (Kay 1998).

Echograms can be constructed from any sound recording, not just those made by a tag, provided that individual echolocation calls can be reliably detected.

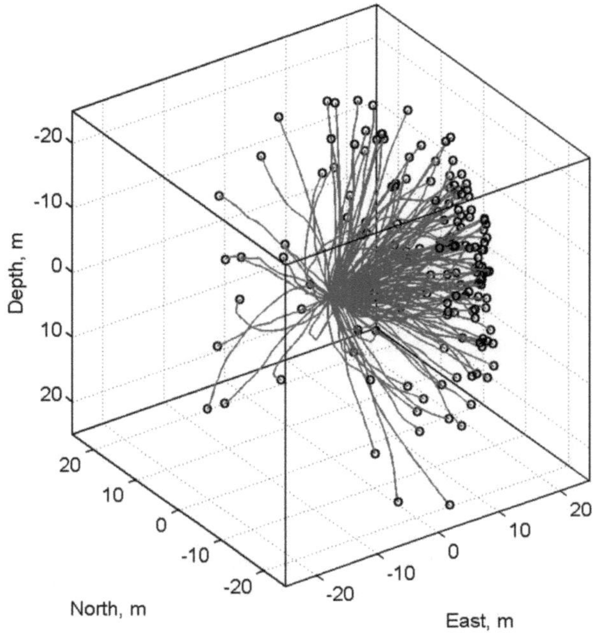

Fig. 7.3 A synchronized movement plot showing the three-dimensional tracks of a sperm whale (*P. macrocephalus*) during 150 prey approach and capture attempts. Each line shows the path taken by the whale during the 15 s before the end of a buzz. The position of the whale at the end of each buzz is set to (0, 0, 0). The start of each track is indicated by a *black circle*. There appears to be no vertical preference in approach direction but the predominance of easterly approaches suggests that the whale encounters prey while swimming westward and is able to capture prey with relatively little maneuvering, for example, compared to the whale in Fig. 7.2c

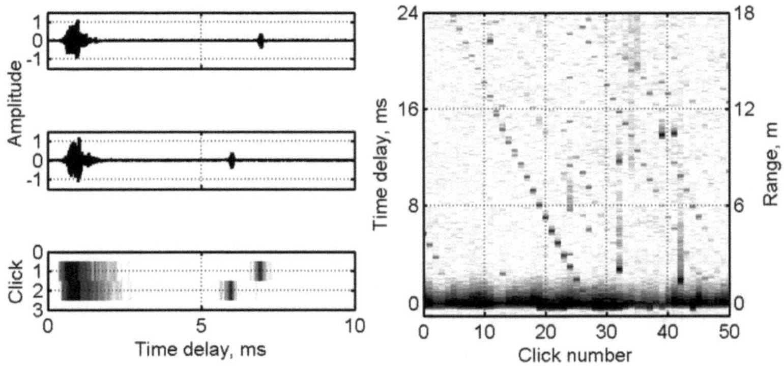

Fig. 7.4 Constructing an echogram from a sound recording made by a tag on a Blainville's beaked whale (*M. densirostris*). *Left*: 10-ms segments of audio synchronized to two consecutive outgoing clicks show an echo at 7 and then 6 ms. The audio segments can be represented more compactly by *colored* or *shaded bars* (*lower left*). *Right*: Bars from 50 consecutive clicks are collected in a stack plot to create an echogram. The left-hand axis shows the time elapsed since the outgoing calls. The elapsed-time axis can be converted to range (right-hand axis) by multiplying by one half of the sound speed. The simple click detection algorithm used here gives some synchronization errors in the outgoing clicks leading to a ragged edge on the 0-time delay line. These errors will lead to estimation errors in the closing speed and should be corrected by a click alignment step. Some strong single echoes (e.g., near click number 40 and 10 m range) are actually clicks from other animals foraging near the tagged animal and represent a form of interference

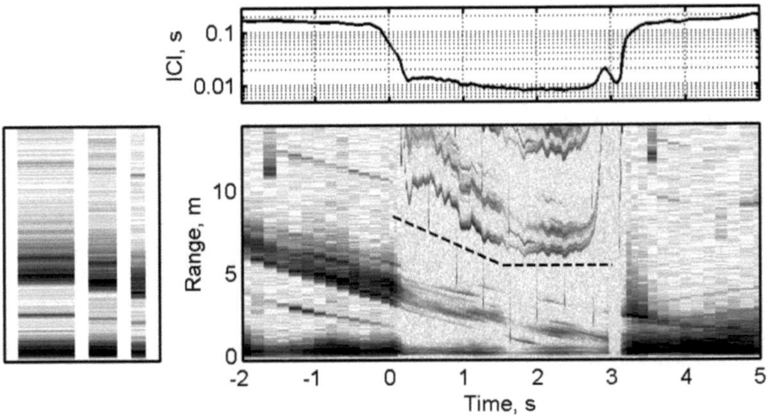

Fig. 7.5 Adjusting the bar width to make a time-range echogram for a buzz. *Above*: Interclick interval (ICI) can vary over two orders of magnitude during buzzes, making it difficult to interpret conventional echograms. By modulating the width of the bars used to represent each sound segment according to the ICI (*left*), a time-range echogram can be constructed (*lower right*). Closing speed can be measured directly from the slope of target lines on this echogram. Short ICIs during buzzes mean that subsequent clicks, and the echoes from these, can also appear in echograms. In the echogram here, the subsequent click trace is the *ragged line* above the *dashed line*. This echogram was produced from a buzz made by *M. densirostris* and may represent foraging on a small shoal of prey, some of which seem to swim away from the whale after the buzz

However, echograms made from far-field sound recordings are difficult to interpret as they represent effectively a bistatic sonar in which the source and receiver positions are constantly changing. Echograms constructed from on-animal sound recordings are more valuable because the source and receiver are closely spaced and fixed with respect to each other. The time elapsed from the outgoing call to a received echo in an on-animal echogram is very close to the two-way travel time to the target as experienced by the echolocating animal itself and can be converted to distance by multiplying by one half of the speed of sound. Echograms formed from on-animal sound recordings can be used to identify echoes from obstacles and prey, and therefore to study navigation, prey selection, and capture from the perspective of the animal itself.

For foraging studies, a modified version of the echogram can be easier to interpret (Fig. 7.5). In a standard echogram, each call is given equal space on the call axis even though the intercall interval (ICI) may vary widely. The ICI during buzzes, for example, can vary over two orders of magnitude in less than a second (Madsen et al. 2013). Displaying each call in a buzz as a constant width bar leads to a warped representation of time, making it hard to assimilate the true time scale of predator–prey interactions. An alternative approach is to use a linear-time x-axis in the echogram and position the colored bars at the time corresponding to each outgoing call (Fig. 7.5). To make the echogram image continuous, the width of each bar is made equal to the ICI between the current and the following call. Produced this way, the echogram has linear time–time axes (or time–distance if the echo-delay axis is

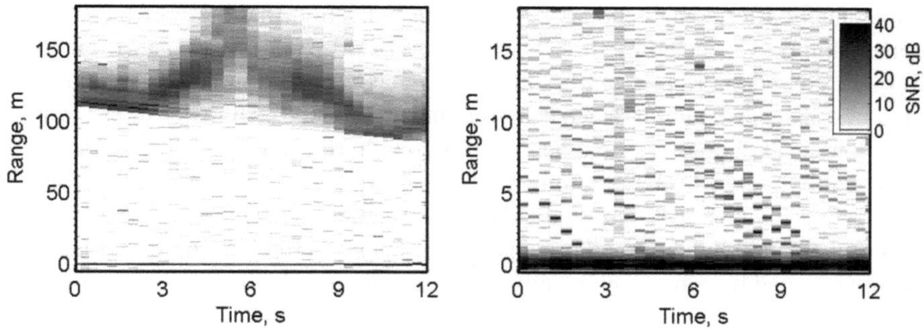

Fig. 7.6 Echograms showing reverberant sea floor echoes (*left*) and clutter from numerous nearby organisms (*right*). Note the different range scales in these plots. Speckle in the left-hand image is due to clicks from other whales foraging near the tagged whale. The received level of echoes depends on the tag location on the animal, among other factors, and so *color* (or *shade*) scaling in echograms based on the absolute level of echoes is usually not informative. The echograms here are scaled by the signal-to-noise ratio (SNR) calculated by comparing the magnitude of each pixel to the lower quartile pixel magnitude (an estimate of the background noise if echoes are fairly sparse). Such SNR scaling may be useful when comparing echograms from different locations or animals. (Adapted from Arranz et al. 2011)

converted into target range as in Fig. 7.5) and so portrays relative movements accurately. The closing speed and the relative timing of events such as prey escapes or predator strikes can be measured directly from this plot.

Echograms contain a lot of information and take some practice to interpret. Echoes from individual targets appear as sloping lines with the slope (in a time–distance echogram) being approximately equal to the net closing speed of the tagged animal on the target. For marine animals, the closing speed is almost always positive, consistent with a forward-moving animal searching ahead of itself with a narrow sonar beam. Echoes can come from organisms in the water column as well as from the sea surface and the sea floor, and it is important to be able to distinguish between these. Echoes from the sea surface and sea floor can be detected at ranges of hundreds of meters depending on the orientation of the animal (Arranz et al. 2011). Because these surfaces are distributed targets which can scatter sound back toward the source from a range of ensonification angles, bottom and surface echoes often have a long duration (Fig. 7.6). Sound can also reverberate within the rock, sand, and mud layers of the sea floor, re-radiating to form a slow-decaying echo. This decay is useful to distinguish bottom echoes from sea surface echoes which are not typically reverberant. When bottom echoes can be clearly identified, the range to the earliest arriving bottom echo is likely a good estimate of the altitude of the whale above the bottom, which can help identify the foraging layer exploited by the whale (Arranz et al. 2011).

Echoes from organisms are typically much shorter in duration than echoes from the sea floor or surface, and are usually detectable only at short ranges (e.g., <30 m) from the whale. These organisms may be anything from plankton to potential prey or even competitors for those prey. The occurrence of buzzes and fast movements immediately after some echo sequences is an indication that these echoes at least come from prey selected by the whale (Johnson et al. 2004). However, for reasons to be discussed shortly, it can be difficult to make reliable inferences about what kind of organism is giving rise to echoes (Jones et al. 2008). High densities of indistinct low-level echoes are obtained at some depth ranges (Fig. 7.6) and these are presumably caused by large numbers of small organisms, as, for example, in the deep-scattering layer (Arranz et al. 2011). These cluttered echoic scenes may mask echoes from prey but the small organisms may also be a food resource for the larger nekton targeted by echolocators and so form a biological landmark to help interpret the movements of tagged animals (Madsen et al. 2005).

Echograms are a powerful way to display sound data from echolocating animals but they are only as clear as the sound recording itself. Three forms of interference can make echograms difficult to interpret. Calls from other animals, especially conspecifics foraging nearby, show up as speckle in echograms (e.g., Fig. 7.4). Wideband ambient noise decreases the signal-to-noise ratio of echoes and so reduces the contrast of echograms. Finally, subsequent calls produced by the tagged animal itself can be mistaken for echo traces. This is especially an issue during buzzes when the ICI is short and variable (Fig. 7.5). There are also numerous sources of information that echograms do not display. Echo level is displayed but not the spectra of echoes or their angles of arrival. Likewise, the motion of the tagged animal is shown relative to the organisms it ensonifies but not in absolute terms and this leads to ambiguity in identifying whether relative motion is a result of predator of prey movements. New kinds of combined plots are needed to display both the motion of the predator derived from movement sensors in the tag, and the movement of its prey via echography.

7.3.4 *Quantifying Tag Data from Echolocating Animals*

Tag-borne sensors can provide unprecedented detail about the behavior of wild echolocators but the quality of these data, and therefore the inferences that arise from them, depend on the species tagged, its environment, and even the location of the tag on the animal (Johnson et al. 2009). While some physical and behavioral parameters have been successfully measured with tags, others remain challenging.

7.3.4.1 Sound Source Parameters

The sounds produced by echolocating animals are difficult to record consistently in the field using remote microphones or hydrophones because of the constantly changing orientation and distance of the animal with respect to the receiver. Unfortunately, tags do not solve this problem directly. Although tags have a fixed spatial relation with the sound source, they are inevitably attached behind the source putting them in a poor position to acquire the outgoing signal. They may also be in the near field of the sound source and there may be no unobstructed air/water path

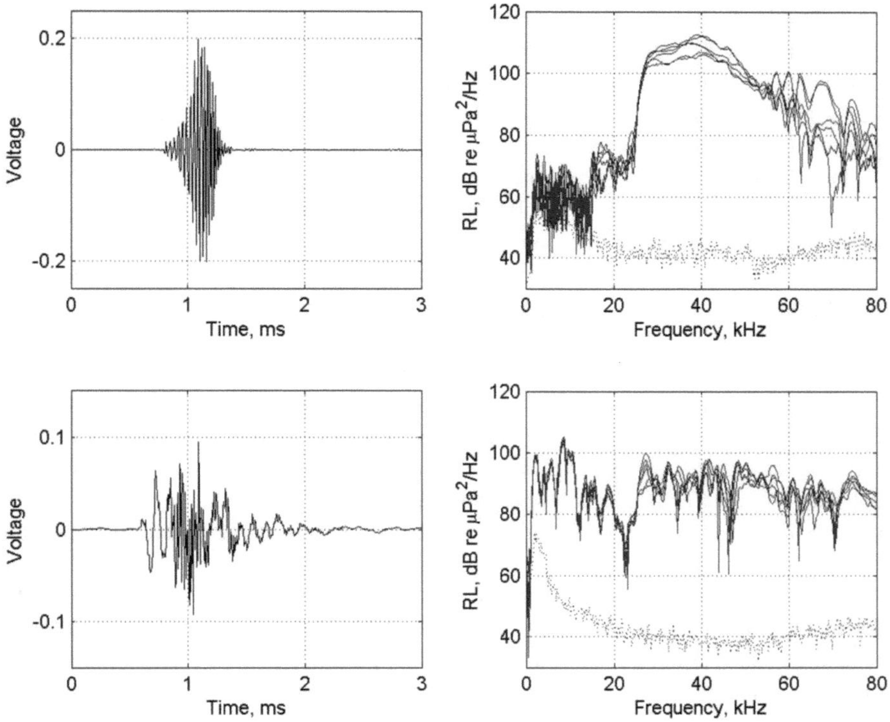

Fig. 7.7 Echolocation sounds recorded from *M. densirostris* in the far-field and on the dorsal surface of a whale. *Upper*: Far-field waveform and power spectrum of clicks produced by an untagged whale near the tagged whale. *Lower*: Waveform and power spectrum of clicks produced by the tagged whale. The waveform of a single click and the spectrum of six successive clicks are shown in each case. The *dotted lines* in the *right hand panels* show the average ambient and system noise level taken 20 ms before each click. Note the longer duration and low frequency spectral content of the on-animal recorded clicks. (From Johnson et al. 2009)

between the source and receiver. Though lower frequency components can refract around the animal's body and reach the tag through the surrounding media, higher frequency components have to pass through the body to arrive at the tag, resulting in a frequency dependent transmission path. As a result, the signal recorded by a tag can be quite different from the on-axis signal and many useful parameters of the outgoing signal cannot be precisely determined from on-animal recordings (Johnson et al. 2009). These include the on-axis source level, duration, spectrum, and harmonic content of echolocation sounds (Fig. 7.7). Clicks recorded by tags on odontocetes are many times longer in duration than the on-axis signal, have more low-frequency energy, and may contain deep spectral peaks or notches that are absent in the on-axis signal. Signal characteristics can also vary markedly as the animal turns its head or flexes its body, changing the position of the tag relative to the sound source. These kinds of spectral distortions may be more of an issue for marine echolocators than for bats because the obstacles are larger relative to the wavelength of the sounds. Microphones on bats can be placed within three to six wavelengths of the sound source (Hiryu et al. 2007, 2008) while tags on whales are usually at least 20 wavelengths behind the sound source. To reach a tag on a whale's dorsal surface, high-frequency sound has to pass through several tissue layers including bone, each of which will reflect and refract sound, giving rise to the complex spectra found in tag recordings.

Despite these issues, quantitative measurements can be made from on-animal recordings. The start times (but not durations) of calls recorded by tags should be reliable enabling acoustic tracking, echo ranging, and studies of call rate adaptation. Though the absolute level of on-axis calls cannot be determined, it may be possible to estimate relative levels across a sequence of calls by arguing that a change in the on-axis level should be reflected in a change in the level recorded by the tag. This approach has been used to investigate the relative level of regular and buzz clicks in odontocetes (Madsen et al. 2005), and automatic gain control in bats (Hiryu et al. 2007). By the same argument, the harmonic levels recorded by a tag may not match those in outgoing tonal calls from bats but changes in harmonic level from call-to-call maybe reliable. The instantaneous frequency of tag-recorded tonal calls is also reliable enabling studies of Doppler compensation and call adaptation (Lancaster et al. 1992; Hiryu et al. 2008).

For the wideband echolocation transients produced by odontocetes, the waveforms recorded by tags bear little resemblance to the on-axis signals, and far-field recordings are essential to establish source parameters (Fig. 7.7). Calls from nearby conspecifics are often recorded by tags on animals swimming in a group providing a source of quality far-field recordings (Zimmer et al. 2005a; Johnson et al. 2006). In tags with multiple hydrophones, sequences of calls from distinct animals can be distinguished by their angles-of-arrival (Akamatsu et al. 2005a; Johnson et al. 2006). The received levels of calls in these sequences will change with the orientation of the calling animal and so the strongest call in each sequence provides an approximation of the on-axis signal (Møhl et al. 2000). Sea-surface echoes of calls from tagged animals when ascending provide another opportunity to acquire on-axis recordings (Akamatsu et al. 2005c). Tag recordings can also be used to identify calls from the tagged animal in far-field recordings opening the possibility of estimating the beam pattern by combining tag-based orientation sensors with far-field received levels (Zimmer et al. 2005b; Ward Shaffer et al. 2013).

7.3.4.2 Echo Parameters

Echoes from organisms and obstacles recorded by tags on echolocating animals are a rich source of information about the animal's environment and how it explores and interacts with this. Most attractively, this information is collected in the same way that the animal itself samples it. However, interpreting and quantifying echoes can be challenging. There are two primary sources of information in echoes. The echo signal (i.e., its wave shape or spectrum) contains information about the type and size of the organism being ensonified while the time of arrival of echoes and how this changes from call-to-call reflects the relative movements of predator and prey.

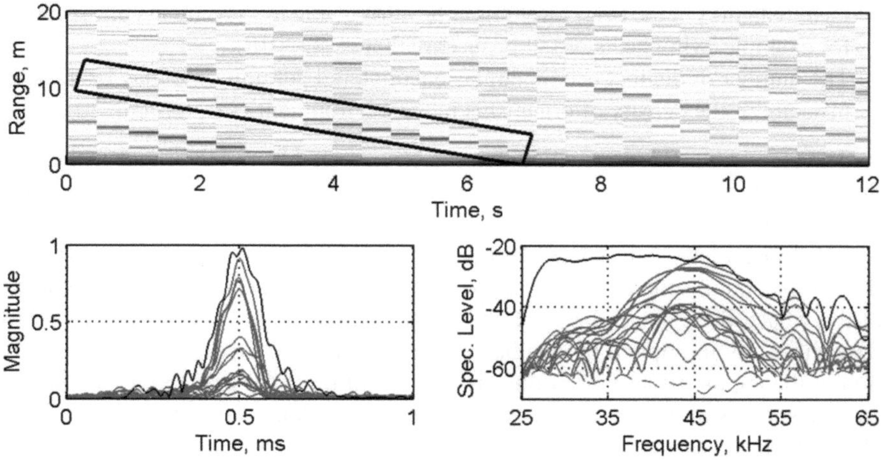

Fig. 7.8 Call-to-call variation in echoes received from the same organism can be substantial. *Upper*: Time-range echogram formed from 12 s of sound recorded by a tag on a *M. densirostris*. Echo sequences from several organisms are evident. The *box* encloses a sequence of echoes received from the same organism on 15 consecutive clicks. *Lower left*: Envelopes of a far-field *M. densirostris* click (*black*) and of the 15 echoes (*gray*). Echo duration is fairly constant and similar to the outgoing pulse length but received level is highly variable. *Lower right*: Spectra of the outgoing click (*black*) and the echoes (*gray*). The background noise level is shown by the *dashed line* at the base of the figure. All of the echoes show a high-pass characteristic, suggesting a small target size, but otherwise the spectra vary widely

The echo signal received by a tag is dictated by both the outgoing signal and the way that this is reflected by the ensonified organism, both of which depend on several factors. The target strength and frequency response of an organism are size and taxa dependent (Simmonds et al. 1996) but are also highly dependent on its aspect (Love 1977), that is, its orientation with respect to the sound source, which is largely unknown and can change from call to call. The level and spectrum of the call produced by the echolocator can also vary with time (Au and Benoit-Bird 2003; Madsen et al. 2005), changing the way that the organism is ensonified. For marine echolocators with narrow beam patterns, the level and spectrum of the outgoing call change radically a few degrees off the acoustic axis (Au 1993; Zimmer et al. 2005c) and so the sound arriving at an organism depends also on its position relative to the beam center. Thus, although the echo signal encodes valuable information about the target, this is combined with major sources of variability that must be carefully accounted for (Jones et al. 2008) and the interpretation of echo spectra remains a challenging problem (Fig. 7.8). However, some other biologically relevant echo parameters may be more straightforward to measure and interpret, for example:

- The duration of echoes may be a useful indicator of prey size (Burwen et al. 2003), especially if durations can be collected from the same target over a range of aspects.

- Fast dynamics in the echo level might be attributable to movements of the target, for example, wing beats or tail beats (Mueller et al. 2010).
- The number of distinct echoic targets encountered per unit time may be a useful measure of the density of organisms in the environment (Arranz et al. 2011).

The second type of information that can be extracted from echoes lies in the time difference of arrival between the outgoing call and the echo. Errors in measuring this echo-delay may be of order tens of microseconds giving rise to target ranging errors of just a few centimeters. Another source of error can arise from the separation of the sound source and the tag on odontocetes. If the target is directly ahead of the whale, the distance between the sound source and target is simply given by the echo-delay multiplied by one half of the sound speed, irrespective of the location of the tag on the animal. If the target does not lie in the axis formed by the tag and the sound source, the relationship between echo-delay and range is more complicated. However, the relative error arising from using the usual formula is small except for very close targets. Thus the distance between the tagged animal and echoic organisms or surfaces such as the sea floor can be measured accurately. The altitude of an odontocete above the sea floor, in combination with the echo density, provides a high-resolution summary of the foraging environment of the tagged animal that would be difficult to obtain at the same scale with a ship-based hydroacoustic survey (Arranz et al. 2011). The distance to echoic organisms ensonified in consecutive calls can be compared to parameters of the outgoing calls and to the movements of the tagged animal to study call adaptation (Houser et al. 2005; Madsen et al. 2005; Johnson et al. 2006; Hiryu et al. 2007) and capture tactics (Johnson et al. 2008). The number of calls ensonifying the same object may also be a useful measure of interest on the part of the predator, that is, does the tagged animal maneuver or turn its head to keep the target in the beam (Ghose and Moss 2003, 2006; Wisniewska et al. 2012)?

The echo-delay defines the distance but not the location of an echoic target. To pinpoint the location of a target relative to the tagged animal, the bearing of each echo is also required. Three microphones/hydrophones are needed to measure bearing unambiguously in three dimensions, increasing the power consumption and reducing the recording time of a tag. Two-channel tags have been used to deduce the angle-of-arrival of outgoing click sounds to track head movements (Akamatsu et al. 2005a; Ward Shaffer et al. 2013) but combined range-and-bearing tracking of echoes has not been attempted. When this does become practical, it will be possible, in conjunction with movement sensors on the tagged animal, to quantify separately how the predator and prey move during capture attempts.

A simpler but less accurate way to estimate prey motion is via the closing speed, that is, the slope of an echo trace in a time–distance echogram (Fig. 7.5). The closing speed can be estimated from the change in target range over a sequence of calls divided by the time elapsed between the first and last call in the sequence. This is the speed at which the predator and prey are approaching each other (or separating if the speed is negative) and represents the net result of predator and prey motions. Because these motions occur in three-dimensional space but are reduced to the single dimension of closing speed over the paths taken by each animal, it is difficult to infer much about prey speed from the closing speed alone. For example, a closing speed of 2 m s^{-1} could result from the predator closing at this speed on a stationary prey, or from the predator swimming at 3 m/s towards a prey escaping ahead at 1 m s^{-1}, etc. Knowing something about the movements of the tagged animal helps to delimit the possibilities. For example, if the tag includes a speed sensor and the speed reading is greater than the closing speed, then the prey may be moving away from the predator but it could also be stationary and off to one side of the predator. If the predator does not change course much during an approach, it may be reasonable to assume that the prey is directly ahead of the predator. In this case, the difference between the predator's speed and the closing speed gives an underestimate of the prey speed. For example, if the tagged animal is moving at 2 m s^{-1} and the closing speed is 1 m s^{-1}, the prey must be moving at, at least, 1 m s^{-1} in a bearing that takes it away from the predator. An underestimate of prey speed may be useful to determine the minimum capabilities of prey and their reactions to approaching predators.

7.4 Summary and Future Directions

Although still relatively new, sound sampling tags are proving to be powerful tools for studying echolocation in the wild. Compared to remote camera, microphone, or hydrophone systems that receive only fleeting glimpses of individuals, a tag follows an animal for periods of hours or even days as it navigates and forages in its natural environment. Sensors can be added to sample movements, physiological parameters, and the environment of the animal. Tags thus seem ideal in many respects for studying how free-moving animals use biosonar to find and capture prey, and to navigate. Some of this potential is already being realized in studies on odontocetes in which, given the large body size, complex multisensor recording tags can be used. These have enabled fine-scale studies of prey search, selection, and capture behaviors that are expanding our knowledge of species traditionally thought of as hard to study. The severe weight constraint on tags for bats has proven to be more difficult to overcome and short-range telemetry devices are still the only option for on-animal sound sensing. These have been used to study signal adaptation in captive animals but telemetry range is too limited for most outdoor use. As electronic component sizes shrink, it will soon be possible to build recording tags small enough to use on some bat species, bypassing the limitations of telemetry.

Although tags provide great detail about the sounds and movements made by some echolocating animals, there are numerous challenges still to be overcome in analyzing these data. The quality of tag data, and the range of inferences that can be drawn, depend on many factors including the species, season, environmental context, gender and age, as well as the location of the tag on the animal and the level of disturbance induced by tagging. Analysis methods capable of separating and interpreting these different factors are still largely lacking: most tag-based studies involve small numbers of replicates with little power to explore variability in behavior. Standardized ways to exchange and compare data are needed to combine these small data sets into a larger ecological picture. Reliable attachment of tags to some species also remains a major issue, especially as longer recording durations become practical. Some marine species react strongly to tags whereas other species seem to be impractically difficult to locate and approach at sea. New field techniques and study sites will be required to work with these challenging species.

On-animal sound recordings can be difficult to interpret, not only because of a lack of replicates, but also because of limitations associated with the location of the tag on the animal and the density of the resulting data. Sounds recorded by a tag placed behind the head are very different from on-axis signals impacting the use of tags in studies of source properties and signal adaptation. The problem can be overcome by combining tags with remote arrays of microphones/hydrophones in integrated studies. However, this adds to the already difficult task of analyzing and visualizing the high-resolution multisensor data from tags. Currently available methods for visualizing sound and movement data offer good temporal or spatial resolution but seldom both. On-animal echograms are an especially appropriate and powerful way to visualize tag sound recordings but, as with other complex visualizations, these displays are challenging to interpret and offer ambiguous information that must be treated carefully. As ever denser and more varied data are collected by tags on animals, new creative ways to explore multidimensional datasets will be essential to unlock the full potential of these devices in echolocation studies.

Acknowledgements Thanks to T. Horiuchi, P. Madsen, N. Aguilar, T. Hurst, and A. Surlykke for helpful discussions. Funds from the National Oceanographic Partnership Program, a Woods Hole Oceanographic Institution senior technical award, the Marine Alliance for Science and Technology, Scotland, and a Marie Curie Career Integration Grant, supported the preparation of this chapter.

References

Akamatsu, T., Matsuda, A., Suzuki, S., Wang, D., Wang, K., Suzuki, M., Muramoto, H., Sugiyama, N., & Oota, K. (2005a). New stereo acoustic data logger for free-ranging dolphins and porpoises. *Marine Technology Society Journal*, 39(2), 3–9.

Akamatsu, T., Wang, D., Wang, K., & Naito, Y. (2005b). Biosonar behaviour of free-ranging porpoises. *Proceedings of the Royal Society of London B: Biological Sciences*, 272(1565), 797–801.

Akamatsu, T., Wang, D., & Wang, K. (2005c). Off-axis sonar beam pattern of free-ranging finless porpoises measured by a stereo pulse event data logger. *Journal of the Acoustical Society of America*, 117(5), 3325–3330.

Akamatsu, T., Teilmann, J., Miller, L. A., Tougaard, J., Dietz, R., Wang, D., Wang, K., Siebert, U., & Naito, Y. (2007). Comparison of echolocation behaviour between coastal and riverine porpoises. *Deep-Sea Research Part II: Topical Studies in Oceanography*, 54(3–4), 290–297.

Akamatsu, T., Wang, D., Wang, K., Li, S., & Dong, S. (2010). Scanning sonar of rolling porpoises during prey capture dives. *Journal of Experimental Biology*, 213, 146–152.

Aldridge, H. D. J. N., & Brigham, R. M. (1988). Load carrying and maneuverability in an insectivorous bat: a test of the 5 % "rule" of radio telemetry. *Journal of Mammalogy*, 69(2), 379–382.

Ancel, A., Horning, M., & Kooyman, G. L. (1997). Prey ingestion revealed by oesophagus and stomach temperature recordings in cormorants. *Journal of Experimental Biology*, 200(Pt 1), 149–154.

Andrews, R. D., Jones, D. R., Williams, J. D., Thorson, P. H., Oliver, G. W., Costa, D. P., & Le Boeuf, B. J. (1997). Heart rates of northern elephant seals diving at sea and resting on the beach. *Journal of Experimental Biology*, 200(Pt 15), 2083–2095.

Arranz, P., Aguilar de Soto, N., Madsen, P. T., Brito, A., Bordes, F., & Johnson, M. P. (2011). Following a foraging fish-finder: diel habitat use of Blainville's beaked whales revealed by echolocation. *PLoS ONE*, 6(12), e28353.

Au, W. W. L. (1993). *The sonar of dolphins*. New York: Springer-Verlag.

Au, W. W. L., & Benoit-Bird, K. J. (2003). Automatic gain control in the echolocation system of dolphins. *Nature*, 423(6942), 861–863.

Bar-Shalom, Y., Li, X. R., & Kirubarajan, T. (2001). *Estimation with applications to tracking and navigation: Theory algorithms and software*. New York: John Wiley & Sons.

Blackwell, S. B., Haverl, C. A., Le Boeuf, B. J., & Costa, D. P. (1999). A method for calibrating swim-speed recorders. *Marine Mammal Science*, 15, 894–905.

Blomqvist, C., & Amundin, M. (2004). An acoustic tag for recording directional pulsed ultrasounds aimed at free-swimming bottlenose dolphins (*Tursiops truncatus*) by conspecifics. *Aquatic Mammals*, 30(3), 345–356.

Burgess, W. C. (2008). *A miniature acoustic recording tag: Applications to assess marine wildlife response to sound* (No. GS0105A-0801). Santa Barbara, CA: Greeneridge Sciences, Inc.

Burgess, W. C., Tyack, P. L., Le Boeuf, B. J., & Costa, D. P. (1998). A programmable acoustic recording tag and first results from free-ranging northern elephant seals. *Deep-Sea Research Part II: Topical Studies in Oceanography*, 45(7), 1327–1351.

Burwen, D. L., Fleischman, S. J., Miller, J. D., & Jensen, M. E. (2003). Time-based signal characteristics as predictors of fish size and species for a side-looking hydroacoustic application in a river. *ICES Journal of Marine Science: Journal du Conseil*, 60(3), 662–668.

Caccamise, D. F., & Hedin, R. S. (1985). An aerodynamic basis for selecting transmitter loads in birds. *Wilson Bulletin*, 97(3), 306–318.

Chruszcz, B. J., & Barclay, R. M. (2003). Prolonged foraging bouts of a solitary gleaning/hawking bat, *Myotis evotis*. *Canadian Journal of Zoology*, 81(5), 823–826.

Cooke, S. J., Hinch, S. G., Wikelski, M., Andrews, R. D., Kuchel, L. J., Wolcott, T. G., & Butler, P. J. (2004). Biotelemetry: A mechanistic approach to ecology. *Trends in Ecology & Evolution*, 19(6), 334–343.

Davis, R. W., Fuiman, L. A., Williams, T. M., Collier, S. O., Hagey, W. P., Kanatous, S. B, Kohin, S., & Horning, M. (1999). Hunting behavior of a marine mammal beneath the Antarctic fast ice. *Science*, 283(5404), 993–996.

Dechmann, D. K., Heucke, S. L, Giuggioli, L., Safi, K., Voigt, C. C., & Wikelski, M. (2009). Experimental evidence for group hunting via eavesdropping in echolocating bats. *Proceedings of the Royal Society of London B: Biological Sciences*, 276(1668), 2721–2728.

Dietz, R., Shapiro, A. D., Bakhtiari, M., Orr, J., Tyack, P. L., Richard, P., Eskesen, I. G., & Marshall, G. (2007). Upside-down swimming behaviour of free-ranging narwhals. *BMC Ecology*, 7(1), 14.

Ezer, T., Hobbs, R., & Oey, L.-Y. (2008). On the movement of beluga whales in Cook Inlet, Alaska: Simulations of tidal and environmental impacts using a hydrodynamic inundation model. *Oceanography*, 21(4), 186–195.

Fletcher, S., Le Boeuf, B. J., Costa, D. P., Tyack, P. L., & Blackwell, S. B. (1996). Onboard acoustic recording from diving northern elephant seals. *Journal of the Acoustical Society of America*, 100(4 Pt 1), 2531–2539.

Ghose, K., & Moss, C. F. (2003). The sonar beam pattern of a flying bat as it tracks tethered insects. *Journal of the Acoustical Society of America*, 114(2), 1120–1131.

Ghose, K., & Moss, C. F. (2006). Steering by hearing: A bat's acoustic gaze is linked to its flight motor output by a delayed, adaptive linear law. *Journal of Neuroscience*, 26(6), 1704–1710.

Gleiss, A. C., Wilson, R. P., & Shepard, E. L. C. (2011). Making overall dynamic body acceleration work: On the theory of acceleration as a proxy for energy expenditure. *Methods in Ecology and Evolution*, 2(1), 23–33.

Grewal, M. S., Weill, L. R., & Andrews, A. P. (2007). *Global positioning systems, inertial navigation, and integration*. New York: John Wiley & Sons.
Henson, O. W., Jr., Bishop, A. L., Keating, A., Kobler, J. B., Henson, M. M., Wilson, B., & Hansen, R. (1987). Biosonar imaging of insects by *Pteronotus p. parnellii*, the mustached bat. *National Geographic Research*, 3, 82–101.
Hiryu, S., Katsura, K., Lin, L. K., Riquimaroux, H., & Watanabe, Y. (2005). Doppler-shift compensation in the Taiwanese leaf-nosed bat (*Hipposideros terasensis*) recorded with a telemetry microphone system during flight. *Journal of the Acoustical Society of America*, 118(6), 3927–3933.
Hiryu, S., Hagino, T., Riquimaroux, H., & Watanabe, Y. (2007). Echo-intensity compensation in echolocating bats (*Pipistrellus abramus*) during flight measured by a telemetry microphone. *Journal of the Acoustical Society of America*, 121(3), 1749–1757.
Hiryu, S., Shiori, Y., Hosokawa, T., Riquimaroux, H., & Watanabe, Y. (2008). On-board telemetry of emitted sounds from free-flying bats: Compensation for velocity and distance stabilizes echo frequency and amplitude. *Journal of Comparative Physiology A*, 194(9), 841–851.
Hiryu, S., Bates, M. E., Simmons, J. A., & Riquimaroux, H. (2010). FM echolocating bats shift frequencies to avoid broadcast-echo ambiguity in clutter. *Proceedings of the National Academy of Sciences of the USA, 107(15), 7048–7053.*
Hooker, S. K., & Baird, R. W. (2001). Diving and ranging behaviour of odontocetes: A methodological review and critique. *Mammal Review*, 31(1), 81–105.
Houser, D., Martin, S. W., Bauer, E. J., Phillips, M., Herrin, T., Cross, M., Vidal, A., & Moore, P. W. (2005). Echolocation characteristics of free-swimming bottlenose dolphins during object detection and identification. *Journal of the Acoustical Society of America*, 117(1), 2308–2317.
Johnson, M. P., & Tyack, P. L. (2003). A digital acoustic recording tag for measuring the response of wild marine mammals to sound. *IEEE Journal of Oceanic Engineering*, 28(1), 3–12.
Johnson, M. P., Madsen, P. T., Zimmer, W. M. X., Aguilar de Soto, N., & Tyack, P. L. (2004). Beaked whales echolocate on prey. *Proceedings of the Royal Society of London B: Biological Sciences*, 271(Suppl 6), S383–S386.
Johnson, M. P., Madsen, P. T., Zimmer, W. M. X., Aguilar de Soto, N., & Tyack, P. L. (2006). Foraging Blainville's beaked whales (*Mesoplodon densirostris*) produce distinct click types matched to different phases of echolocation. *Journal of Experimental Biology*, 209(24), 5038–5050.
Johnson, M. P., Hickmott, L. S., Aguilar de Soto, N., & Madsen, P. T. (2008). Echolocation behaviour adapted to prey in foraging Blainville's beaked whale (*Mesoplodon densirostris*). *Proceedings of the Royal Society of London B: Biological Sciences*, 275(1631), 133–139.
Johnson, M. P., Aguilar de Soto, N., & Madsen, P. T. (2009). Studying the behaviour and sensory ecology of marine mammals using acoustic recording tags: A review. *Marine Ecology Progress Series*, 395, 55–73.
Johnson, M. P., Partan, J., & Hurst, T. (2013). Low complexity lossless compression of underwater sound recordings. *Journal of the Acoustical Society of America*, 133(3), 1387–1398.
Jones, B. A., Stanton, T. K., Lavery, A. C., Johnson, M. P, Madsen, P. T., & Tyack, P. L. (2008). Classification of broadband echoes from prey of a foraging Blainville's beaked whale. *Journal of the Acoustical Society of America*, 123(3), 1753–1762.
Kay, S. (1998). *Fundamentals of statistical signal processing*, Vol. 2: *Detection theory*. Upper Saddle River, NJ: Prentice Hall.
Lancaster, W. C., Keating, A. W., & Henson, O. W. (1992). Ultrasonic vocalizations of flying bats monitored by radiotelemetry. *Journal of Experimental Biology*, 173(1), 43–58.
Lancaster, W. C., Henson, O. W., & Keating, A. W. (1995). Respiratory muscle activity in relation to vocalization in flying bats. *Journal of Experimental Biology*, 198(1), 175–191.
Lawrence, B. D., & Simmons, J. A. (1982). Measurements of atmospheric attenuation at ultrasonic frequencies and the significance for echolocation by bats. *Journal of the Acoustical Society of America*, 71(3), 585–590.

Liu, C. M., Hsu, H. W., & Lee, W. C. (2008). Compression artifacts in perceptual audio coding. ***IEEE Transactions on Audio, Speech, and Language Processing*, 16(4), 681–695.**

Love, R. H. (1977). Target strength of an individual fish at any aspect. *Journal of the Acoustical Society of America*, 62(6), 1397–1403.

MacLean, G. (2009). Weak GPS signal detection in animal tracking. *Journal of Navigation*, 62(1), 1–21.

Madsen, P. T., & Wahlberg, M. (2007). Recording and quantification of ultrasonic echolocation clicks from free-ranging toothed whalese. *Deep-Sea Research Part I: Oceanographic Research Papers*, 54(8), 1421–1444.

Madsen, P. T., Payne, R., Kristiansen, N. U., Wahlberg, M., Kerr, I., & Møhl, B. (2002). Sperm whale sound production studied with ultrasound-time/depth-recording tags. *Journal of Experimental Biology*, 205(13), 1899–1906.

Madsen, P. T., Johnson, M. P., Aguilar de Soto, N., Zimmer, W. M. X., & Tyack, P. L. (2005). Biosonar performance of foraging beaked whales (*Mesoplodon densirostris*). *Journal of Experimental Biology*, 208(2), 181–194.

Madsen, P. T., Aguilar de Soto, N., Arranz, P., & Johnson, M. P. (2013). Echolocation in Blainville's beaked whales (*Mesoplodon densirostris*). *Journal of Comparative Physiology*, 199(6), 451–469.

Marshall, G., Bakhtiari, M., Shepard, M., Tweedy III, J., Rasch, D., Abernathy, K., Joliff, B., Carrier, J., & Heithaus, M. R. (2007). An advanced solid-state animal-borne video and environmental data-logging device (Crittercam) for marine research. *Marine Technology Society Journal*, 41(2), 31–38.

Martin, S. W., Phillips, M., Bauer, E. J., Moore, P. W., & Houser, D. S. (2005). Instrumenting free-swimming dolphins echolocating in open water. *Journal of the Acoustical Society of America*, 117(4 Pt 1), 2301–2307.

Meyer, C. F. J., Weinberg, M., & Kalko, E. K. V. (2005). Home-range size and spacing patterns of *Macrophyllum macrophyllum* (Phyllostomidae) foraging over water . *Journal of Mammalogy*. 86(3), 587–598.

Mitani, Y., Sato, K., Ito, S., Cameron, M. F., Siniff, D. B., & Naito, Y. (2003). A method for reconstructing three-dimensional dive profiles of marine mammals using geomagnetic intensity data: Results from two lactating Weddell seals. *Polar Biology*, 26(5), 311–317.

Møhl, B., Wahlberg, M., Madsen, P. T., Miller, L. A., & Surlykke, A. (2000). Sperm whale clicks: Directionality and source level revisited. *Journal of the Acoustical Society of America*, 107(1), 638–648.

Mueller, A., Burwen, D. L., Boswell, K. M., & Mulligan, T. (2010). Tail-beat patterns in dual-frequency identification sonar echograms and their potential use for species identification and bioenergetics studies. *Transactions of the American Fisheries Society*, 139(3), 900–910.

Orfanidis, S. J. (2010). *Introduction to signal processing*. Rutgers University. (available free at: www.ece.rutgers.edu/~orfanidi/intro2sp).

Pavlov, V. V., Wilson, R. P., & Lucke, K. (2007). A new approach to tag design in dolphin telemetry: Computer simulations to minimise deleterious effects. *Deep-Sea Research Part II: Tropical Studies in Oceanography*, 54(3), 404–414.

Phillips, R. A., Xavier, J. C., & Croxall, J. P. (2003). Effects of satellite transmitters on albatrosses and petrels. *The Auk*, 120(4), 1082–1090.

Read, A. J., & **Westgate, A. J. (1997). Monitoring the movements of harbour porpoises (*Phocoena phocoena*) with satellite telemetry. *Marine Biology*, 130, 315–322.**

Riquimaroux, H., Osawa, Y., & **Watanabe, Y. (2007). Strategy for vocalization taken by FM bats during group flight measured by a wireless microphone system. *Journal of the Acoustical Society of America*, 121(5), 3037 (abstract only).**

Ropert-Coudert, Y., & Wilson, R. P. (2005). Trends and perspectives in animal-attached remote sensing. *Frontiers in Ecology and the Environment*, 3(8), 437–444.

Ropert-Coudert, Y., Kato, A., Liebsch, N., Wilson, R. P., Muller, G., & Baubet, E. (2004). Monitoring jaw movements: A cue to feeding activity. *Game and Wildlife Science*, 21(1), 1–20.

Shepard, E. L. C., Wilson, R. P., Liebsch, N., Quintana, F., Laich, A. G., & Lucke, K. (2008). Flexible paddle sheds new light on speed: A novel method for the remote measurement of swim speed in aquatic animals. *Endangered Species Research*, 4(1–2), 157–164.

Shiomi, K., Narazaki, T., Sato, K., Shimatani, K., Arai, N., Ponganis, P. J., & Miyazaki, N. (2010). Data-processing artefacts in 3-D dive path reconstruction from geomagnetic and acceleration data. *Aquatic Biology*, 8, 299–304.

Simmonds, E. J., Armstrong, F., & Copland, P. J. (1996). Species identification using wideband backscatter with neural network and discriminant analysis. *ICES Journal of Marine Science: Journal du Conseil*, 53(2), 189–195.

Simon, M., Johnson, M. P., & Madsen, P. T. (2012). Keeping momentum with a mouthful of water: Behavior and kinematics of humpback whale lunge feeding, *Journal of Experimental Biology*, 215, 3786–3798.

Spiesberger, J. L., & Fristrup, K. M. (1990). Passive localization of calling animals and sensing of their acoustic environment using acoustic tomography. *American Naturalist*, 135, 107–153.

Surlykke, A., Filskov, M., Fullard, J. H., & Forrest, E. (1999). Auditory relationships to size in noctuid moths: Bigger is better. *Naturwissenschaften*, 86, 238–241.

Surlykke, A., & Kalko, E. K. V. (2008). Echolocating bats cry out loud to detect their prey. *PLoS ONE* 3(4), e2036.

Teloni, V., Johnson, M., Miller, P. J. O., & Madsen, P. T. (2008). Shallow food for deep divers: Dynamic foraging behavior of male sperm whales in a high latitude habitat. *Journal of Experimental Marine Biology & Ecology*, 354, 119–131.

Tyack, P. L., Johnson, M. P., Aguilar de Soto, N., Sturlese, A., & Madsen, P. T. (2006). Extreme diving of beaked whales. *Journal of Experimental Biology*, 209(21), 4238–4253.

Wahlberg, M., Møhl, B., & Madsen, P. T. (2001). Estimating source position accuracy of a large-aperture hydrophone array for bioacoustics. *Journal of the Acoustical Society of America*, 109(1), 397–406.

Ward Shaffer, J., Moretti, D., Jarvis, S., Tyack, P. L., & Johnson, M. P. (2013). Effective beam pattern of the Blainville's beaked whale (*Mesoplodon densirostris*) and implications for passive acoustic monitoring. *Journal of the Acoustical Society of America*, 133(3), 1770–1784.

Ware, C., Arsenault, R., Plumlee, M., & Wiley, D. (2006). Visualizing the underwater behavior of humpback whales. *IEEE Computer Graphics and Applications*, 26(4), 14–18.

Waters, D. A., Rydell, J., & Jones, G. (1995). Echolocation call design and limits on prey size: A case study using the aerial-hawking bat *Nyctalus leisleri. Behavioural Ecology & Sociobiology* 37, 321–328.

Watkins, W. A., Daher, M. A., Fristrup, K. M., & Howald, T. J. (1993). Sperm whales tagged with transponders and tracked underwater by sonar. *Marine Mammal Science*, 9(1), 55–67.

Westgate, A. J., McLellan, W. A., Wells, R. S., Scott, M. D., Meagher, E. M., & Pabst, D. A. (2007). A new device to remotely measure heat flux and skin temperature from free-swimming dolphins. *Journal of Experimental Marine Biology and Ecology*, 346(1), 45–59.

Wilson, R. P., & Wilson, M. P. (1988). Dead reckoning: a new technique for determining penguin movements at sea. *Meeresforschung*, 32, 155–158.

Wilson, R. P., & McMahon, C. R. (2006). Measuring devices on wild animals: What constitutes acceptable practice? *Frontiers in Ecology and the Environment*, 4(3), 147–154.

Wilson, R. P., Grant, W. S., & Duffy, D. C. (1986). Recording devices on free-ranging marine animals: Does measurement affect foraging performance? *Ecology*, 67(4), 1091–1093.

Wilson, R. P., White, C. R., Quintana, F., Halsey, L. G., Liebsch, N., Martin, G. R., & Butler, P. J. (2006). Moving towards acceleration for estimates of activity-specific metabolic rate in free-living animals: The case of the cormorant. *Journal of Animal Ecology*, 75(5), 1081–1090.

Wilson, R. P., Liebsch, N., Davies, I. M., Quintana, F., Weimerskirch, H., Storch, S., Lucke, K., Siebert, U., Zankl, S., Müller, G., Zimmer, I., Scolaro, A., Campagna, C., Plötz, J., Bornemann, H., Teilmann, J., & McMahon, C. R. (2007). All at sea with animal tracks: Methodological and analytical solutions for the resolution of movement. *Deep-Sea Research Part II: Topical Studies in Oceanography*, 54(3), 193–210.

Wisniewska, D. M., Johnson, M. P., Beedholm, K., Wahlberg, M., & Madsen, P. T. (2012). Acoustic gaze adjustment during active target selection in echolocating porpoises. *Journal of Experimental Biology, 215(24), 4358–4373.*

Zimmer, W. M. X., Johnson, M. P., D'Amico, A., & Tyack, P. L. (2003). Combining data from a multisensor tag and passive sonar to determine the diving behavior of a sperm whale (*Physeter macrocephalus*). *IEEE Journal of Oceanic Engineering*, 28(1), 13–28.

Zimmer, W. M. X., Johnson, M. P., Madsen, P. T., & Tyack, P. L. (2005a). Echolocation clicks of free-ranging Cuvier's beaked whales (*Ziphius cavirostris*). *Journal of the Acoustical Society of America*, 117(6), 3919–3927.

Zimmer, W. M. X., Tyack, P. L., Johnson, M. P., & Madsen, P. T. (2005b). Three-dimensional beam pattern of regular sperm whale clicks confirms bent-horn hypothesis. *Journal of the Acoustical Society of America*, 117(3 Pt 1), 1473–1485.

Zimmer, W. M. X., Madsen, P. T., Teloni, V., Johnson, M. P., & Tyack, P. L. (2005c). Off-axis effects on the multipulse structure of sperm whale usual clicks with implications for sound production. *Journal of the Acoustical Society of America*, 118(5), 3337–3345.

Chapter 8
Analysis of Natural Scenes by Echolocation in Bats and Dolphins

Cynthia F. Moss, Chen Chiu, and Patrick W. Moore

Keywords Biosonar • Auditory perception • Spatial hearing • Complex environment • Psychoacoustics • Adaptive motor behaviors • Sonar beam aim • Temporal patterning • Head and ear movements • Acoustic glints • Integration window • Active sensing • Obstacle avoidance • Prey capture • Figure-ground segregation

8.1 Introduction

How do animals select out and organize auditory events from an acoustically complex environment? Research efforts aimed at addressing this question were pioneered by Albert Bregman, who carried out foundational experiments on the perceptual organization of sound in humans. Bregman's book, *Auditory Scene Analysis* (Bregman 1990), presents a comprehensive overview of 25 years of human research that applied Gestalt principles to studies of human hearing.

Experimental work has revealed organizational principles in human hearing that may inform our understanding of auditory scene analysis by echolocation in bats and dolphins. Using pure tones, harmonic complexes, speech, and a variety of other acoustic stimuli, Bregman showed that human listeners perceptually organize sound stimuli into auditory streams. A classic example of Bregman's experiments involves the presentation of pure tones that alternate between high and low frequencies. When the frequency separation between the tones is comparatively low (e.g., less than an octave) and the intervals between successive tones is comparatively long, a

C.F. Moss (✉) • C. Chiu
Department of Psychological and Brain Sciences, Johns Hopkins University, Baltimore, MD 21218, USA
e-mail: cynthia.moss@jhu.edu; chiuchen@gmail.com

P.W. Moore
National Marine Mammal Foundation, San Diego, CA 92106, USA
e-mail: Patrick.Moore@nmmpfoundation.org

A. Surlykke et al. (eds.), *Biosonar*, Springer Handbook of Auditory Research 51,
DOI 10.1007/978-1-4614-9146-0_8, © Springer-Verlag New York 2014

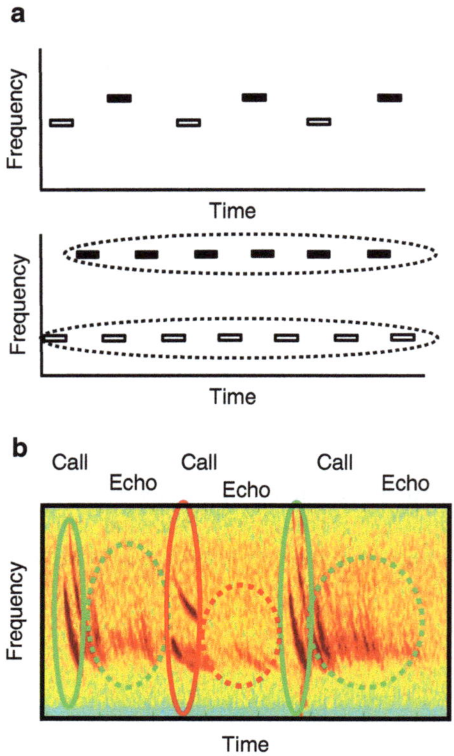

Fig. 8.1 Frequency and temporal separation diagram in human and bat studies. (**a**) Schematic illustration of the time and frequency parameters that can influence auditory stream segregation in psychophysical studies of auditory scene analysis studies in humans (see Bregman 1990). *Upper panel*: Human listeners tend to report hearing out individual tones presented in a sequence when the sounds alternate in frequency with comparatively long interstimulus intervals. *Lower panel*: Human listeners tend to perceive two separate auditory streams (indicated by *dashed lines* encircling tone sequences) when presented with tones alternating in frequency with larger spectral separation and comparatively short interstimulus intervals. (**b**) Spectrograms of echolocation calls and echoes that may give rise to perceptual segregation of auditory streams in echolocating bats. Frequency (kHz) is plotted against time (ms). The echolocation signals of two different bats are displayed, one *circled* in *red* and the other *circled* in *green. Solid lines* encircle the calls and *dashed lines* encircle the echoes. The timing and frequency of echo returns may contribute to the bat's perceptual analysis of auditory scenes

human listener typically reports hearing out individual tones in the pattern (Fig. 8.1a, upper panel). However, when the tone frequency separation increases and the time interval between them decreases, human listeners instead report hearing two streams of sounds, one high in pitch and the other low (Fig. 8.1a, lower panel). The spectro-temporal separation of tones required for a listener to hear out high- and low-frequency streams depends on details of the stimulus parameters and, to some extent, on the individual listener. Moreover, the perception of auditory streams tends to build up over time, indicating that auditory stream segregation depends on cognitive-perceptual processes. Such processes are likely to operate in a broad range of animal systems as well (Bee and Micheyl 2008 Fig. 8.1b); however, phenomenological reports, the dependent measure in many human auditory scene analysis studies, are not amenable to animal research. Further, the perceptual organization of simplified stimuli, such as tone sequences, holds little biological relevance to animals that rely on natural sounds for species-specific communication, territorial defense, foraging, and navigation. An additional challenge to researchers who wish to understand the perceptual organization of sound in echolocating animals, such as bats and dolphins, is the animal's active control over the timing and spectral content of their sonar signals, which immediately impact the acoustic information that comprises their experience of an auditory scene.

8.2 Characterizing Auditory Scenes of Echolocating Animals

The sensory world of an animal is acoustically complex and dynamic. From a barrage of auditory stimuli, echolocating animals face the challenge of detecting, sorting, grouping, and tracking biologically relevant signals to communicate with conspecifics, seek food, engage in courtship, avoid predators, and navigate in space. Sections 8.2.1 and 8.2.2 present an overview of the acoustic information that comprises the natural scene of bats and dolphins in their habitats in air and under water.

8.2.1 Bats

Echolocating bats live and forage in a variety of environments, including dense vegetation, open space, edges of forests, and close to water surfaces (Schnitzler et al. 2003). At night, vision is limited, and echolocation allows animals to orient and forage successfully using sound. Echolocating bats produce high-frequency sonar signals and listen to the retuning echoes to determine the three-dimensional location and features of objects (Griffin 1958; Moss and Schnitzler 1995). Echo returns from the bat's sonar signals come not only from targets of interest (e.g., food), but also from obstacles, such as trees, buildings, and other animals (Fig. 8.2a). Tracking echoes from isolated objects in open space is a comparatively simple task for the bat, but not one that it regularly encounters. Even for open space foragers, clusters of insects present the acoustic challenge of many overlapping echoes, from which a bat must select and pursue a single prey item at a time (Griffin et al. 1960). In addition, there may be other bats seeking food in the vicinity, also creating an acoustically complex mix of conspecifics' sonar sounds and echoes from moving prey and obstacles. For a bat hunting insects or fish over water, it must listen to echoes that reveal the object on the surface or water disturbances created by moving prey (Schnitzler et al. 1994; Kalko et al. 1998). In dense vegetation the bat's auditory scene is far more complex: Echo returns from closely spaced shrubs, trees, branches, and food items create a cascade of echoes, arriving at the bat's ears from

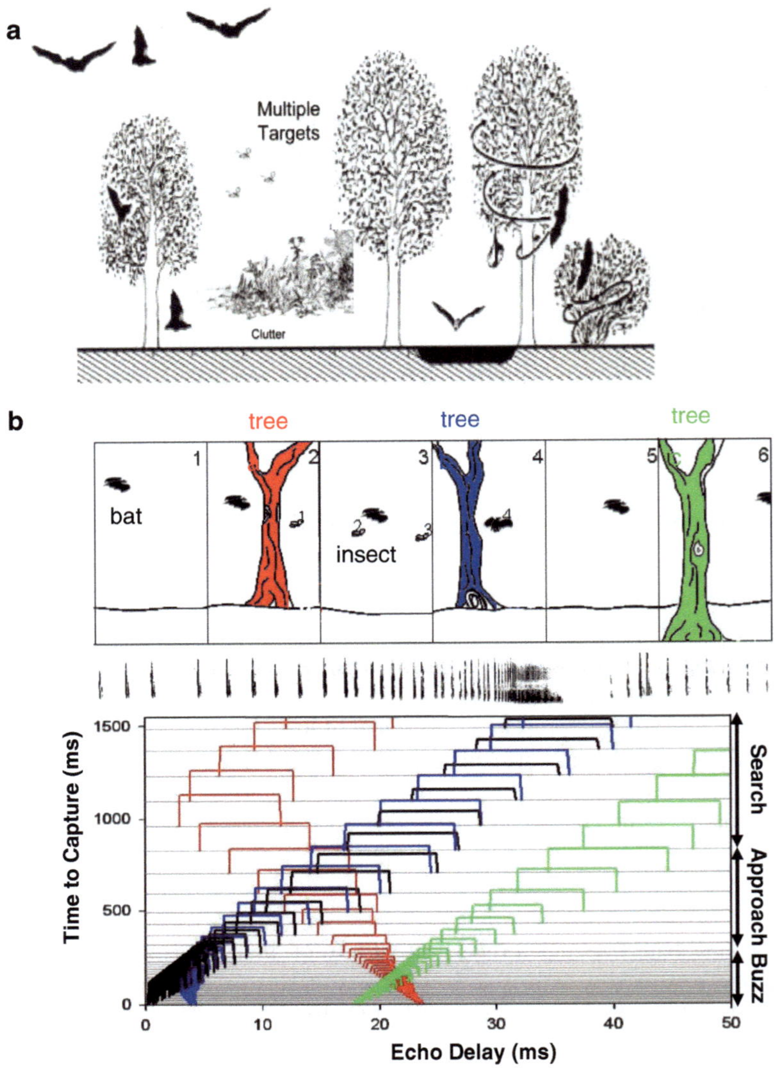

Fig. 8.2 A bat producing echolocation calls and getting echoes from various prey items and objects in the environment. (**a**) Complex acoustic environment for the echolocating bat. (Modified from Neuweiler, 1989.) A bat generates echolocation pulses and listens to returning echoes to track prey and avoid obstacles. It is relatively simple task when there is only one bat and one target in open space. However, a bat often encounters several echolocating conspecifics/heterospecifics, pursues multiple targets and forages in cluttered environment. (**b**) A schematic illustrates the timing of echo returns from different objects and prey in the environment. (Adapted from Moss, C. F., & Surlykke, A. Auditory scene analysis by echolocation in bats. *Journal of the Acoustical Society of America* [2001] 110, 2207–2226; reprinted with permission from Acoustical Society of America.) The *upper panel* shows a cartoon of a bat pursuing insect prey in the vicinity of trees. The *numbers* mark selected instances when the bat produces an echolocation call and the insect's position when the bat sonar signal ensonifies the insect. The *middle panel* shows echolocation call spectrograms generated by the bat. The echo delay from different objects in the environment is displayed in the *lower panel* for different phases of a bat's prey capture sequence. The *left* y-axis shows the time before prey capture and the *right* y-axis shows the prey capture phases (search, approach, buzz). The *x*-axis shows echo delay. The *color* corresponds to the echoes reflected from different objects in the environment, tree a (*red*), tree b (*blue*) and tree c (*green*), and insect target (*black*)

different directions and distances (Moss and Surlykke 2010; Fig. 8.2b). Stationary and large obstacles produce relatively strong echoes, but may be separable from fluttering insect echoes, which contain rapid amplitude and spectral variation produced by moving wings (Schnitzler and Flieger 1983; von der Emde and Menne 1989; von der Emde and Schnitzler 1990; Fig. 8.3a). If fruit hangs stationary amidst vegetation clutter, the bat must discriminate echo features from the fruit and nearby branches and leaves. Bats that take insects from substrate may use two streams of acoustic information, one from active echolocation and the other from passively listening to prey-generated signals (Barber et al. 2003).

To understand how an echolocating animal analyzes its acoustic environment, we begin with a review of the acoustic information carried by echoes returning from various objects, such as insect prey, fruit, flowers, and vegetation, and then design experiments to explore their discrimination and classification of these objects. Schnitzler et al. (1983) and Moss and Zagaeski (1994) recorded sonar echoes from fluttering insect prey, with the goal of characterizing the acoustic information that may be used by echolocating bats to detect and possibly discriminate prey (Fig. 8.3b). Acoustic "glints" in echoes from long constant frequency (CF) bat signals arise from beating wings of flying insects. The glints are characterized by spectral broadening and amplitude peaks that occur in each wingbeat cycle (Schnitzler et al. 1983), and may occur several times in a single echo, depending on the duration of the sonar signal and the insect's wingbeat frequency. By contrast, echoes reflecting from brief frequency modulated (FM) calls provide the bat with an acoustic "snapshot," a brief segment of an insect's wingbeat cycle, because the duration of the FM bat's signal is shorter than the wingbeat period of even the fastest fluttering insects (Moss and Zagaeski 1994). This means that FM bats must integrate echoes over time if they are to represent the changing profile of dynamic targets.

Yovel et al. (2011) reviewed studies of sonar echoes from objects in the echolocating bat's natural environment and proposes how this animal may classify complex sonar stimuli, such as vegetation. Researchers broadcast FM or CF signals, similar to bat echolocation calls, directed at objects from different angles and recorded the echoes. Yovel and colleagues described several models, which can be used to classify echoes from different objects (Yovel et al. 2009; Fig. 8.3c). By using statistical models, such as discriminant function analysis (Stilz 2004), or machine learning classifier (Yovel et al. 2008), it is possible to classify correctly most echoes reflected from different plants. Although using statistical methods can correctly classify the vegetation from sonar echoes, behavioral experiments must be carried out to explicitly study the animal's perception and classification of natural objects.

8.2.2 *Dolphins*

Dolphins are widely distributed throughout the oceans of the world, and individual dolphin species can be found in a variety of environments, but they generally live close to plentiful sources of food (Benoit-Bird and Au 2009). Dolphin groups (pods)

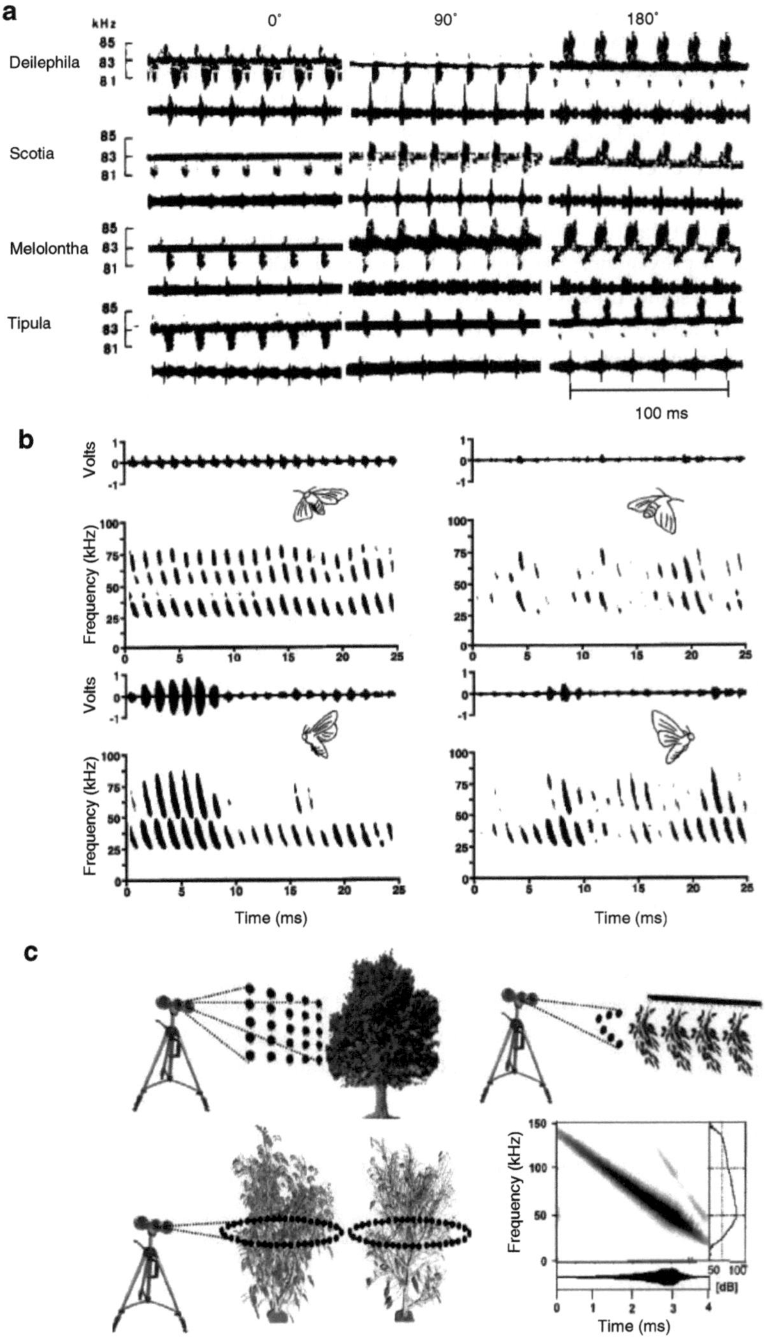

Fig. 8.3 Echo recordings from various objects. (**a**) Echo recordings from four different insect species (von der Emde and Schnitzler 1990). The *upper* traces show spectrograms and the *lower* traces show oscillograms. The signal used in this study was a constant 83-kHz tone, which imitated the CF part of *Rhinolophus ferrumequinum*'s echolocation calls. The tested insect faced three different

have been reported in shallow coastal and riverine environments, where observations tend to be easily made, and animals are accessible by boats or other water craft. For the most part dolphins are a noisy group of animals; they emit whistles, buzzes, clicks, squeals, and a host of other sounds. Dolphins produce broadband biosonar clicks, with energy in the frequency range of about 20–120 kHz, and the sound energy propagates forward from the animal's head, transmitted from the nasal area of the forehead. Dolphin sonar transmission characteristics are described in other chapters of this volume (e.g., Fenton, Jensen, Kalko, & Tyack, Chap. 2; Simmons & Houser, Chap. 6) (Fig. 8.4).

The ocean abounds with flotsam and jetsam, and a dolphin must be able to use its biosonar to determine which echo returns are natural and which are not. Echoes from prey, surrounding obstacles, clutter, the ocean bottom, and the reflective undersurface of the water-air boundary return to the animal in the form of an acoustic cauldron; a mix from which the animal must perceive and extract information relevant to its survival in the natural environment. Research has documented over the past 25 years that dolphins can process complex echo information to determine size, shape, material composition, and other properties of objects (Nachtigall and Moore 1988; Au 1993; Harley et al. 2003), which suggests that dolphin biosonar supports natural auditory scene perception in these animals.

For open water foraging in echolocating dolphins, the ability to detect and track prey targets is of primary importance for capturing fish either alone or in cooperative feeding bouts. Dolphin biosonar is assumed to be a relatively short-range, high-resolution active sensing system, i.e. hundreds of meters. In open water, free from reverberation and the interference of cluttering objects, dolphin detection range has been estimated using the noise-limited transient form of the sonar equation (Urick 1983; Au 1993). The detection range of engineered sonar is a function of several variables (Urick 1983):

$$\mathrm{DT} = \mathrm{SL} - 2\mathrm{TL} + \mathrm{TS} - (\mathrm{NL} - \mathrm{DI}),$$

Fig. 8.3 (continued) directions, 0°, 90°, and 180°, with 0° being head-on. All insects fluttered at 50 Hz. These four insect species are Deilephila: *Deilephila elpenor*, sphingid moth, Lcpidoptera; Scotia: *Scotia exelamationes*, noctuid moth, Lepidoptera; Melolontha: *Melolontha melolontha*, scarabid beetle, Coleoptera; Tipula: *Tipula oleracea*, cranefly, Tipulidae, Diptera. (**b**) Echo recordings from a fluttering army worm moth facing four different directions. (Adapted from Moss, C. F., & Zagaeski, M. Acoustic information available to bats using frequency modulated sonar sounds for the perception of insect prey. *Journal of the Acoustical Society of America* [1994] 95, 2745–2756; reprinted with permission from Acoustical Society of America.) The *upper* trace of each panel shows the oscillogram and the *lower* trace of each panel shows the spectrogram. Each panel represents one direction that the moth was facing. The moth drawing in each panel indicates the angle of ensonification. (**c**) Echo recordings from various plants. (Adapted from Yovel et al. 2009). The *upper left panel* is from a field recording. The *upper right panel* is from a plastic model plant (a single elevation angle and five horizontal angles). The *bottom left panel* shows ensonification of a Ficus plant with decreasing leaf density from 36 angles around the plant. The *bottom right panel* shows the time signal and spectrogram and spectrum of the emitted signal

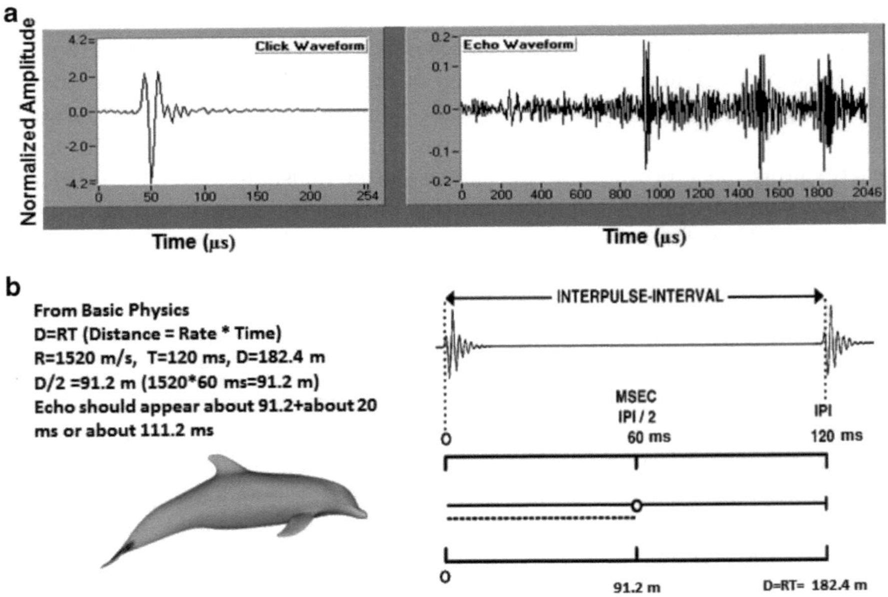

Fig. 8.4 The dolphin echolocation click train consists of a series of emitted clicks (signals) that usually have an interclick interval exceeding the two-way travel time to the target of interest. The target echo will appear midway between the clicks in a train (plus a few milliseconds). A method of estimating the dolphin attention range is to split the time between two emitted clicks. (**a**) Echolocation emitted click and target echo (spheroid). The echo is about 2.5 times the duration of the click (about 85 μs). The two-way travel time has been removed for clarity. (**b**) This panel shows the concept of two-way travel time and the method by which target distance can be estimated from interclick interval (ICI). At a range of about 91 m, the ICI is about 1,400 times the click duration and about 60 times the echo duration

where DT is the detection threshold, SL the source level, TL is the transmission loss, TS is the target strength, NL is the background noise level, and DI the receiver directivity index. As it relates to dolphin biosonar DT, SL, and DI are biologically determined variables and must be estimated based on animal performance in psychoacoustic experiments (see Fig. 8.5; Au 1993, pp. 143–151 for overview; Au et al. 2007).

Dolphin biosonar has been shown to be highly adaptive, and the animal has control over various aspects of the signal, such as source level, peak frequency, bandwidth, and beam geometry. A variety of environmental conditions, as well as task difficulty, animal age, and experience also influence the animal's echolocation signals. The biosonar beam is not fixed in either range or cross section and can vary considerably (up to about 32°; Fig. 8.6) on a click-to-click basis (Moore et al. 2008). Even the spectral energy distribution within the beam may vary dynamically between echolocation clicks (Starkhammar et al. 2010).

Dolphin identification of objects is based on the echo spectral returns produced by biosonar clicks and is due to material; size; shape; and whether the object is solid

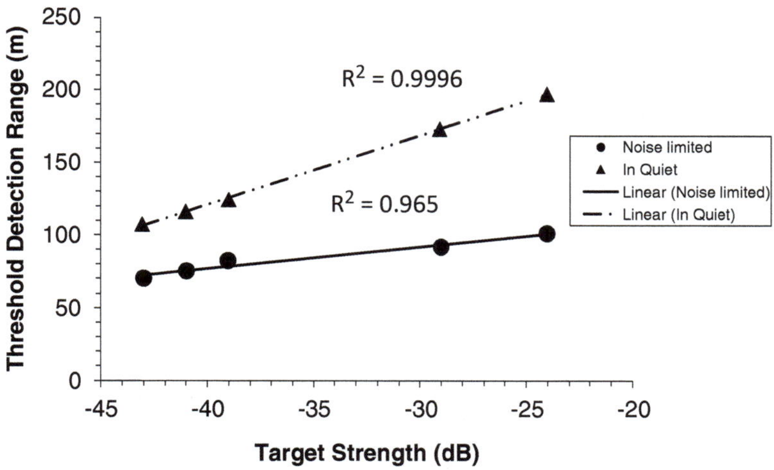

Fig. 8.5 Given the noise-limited form of the sonar equation: DT = SL − 2TL + TS − (NL − DI), the detection threshold range of a target can be estimated based on target strength. In this figure a generalized detection model that predicts the threshold detection range of a fish by an echolocating dolphin in quiet and in noise created by snapping shrimp

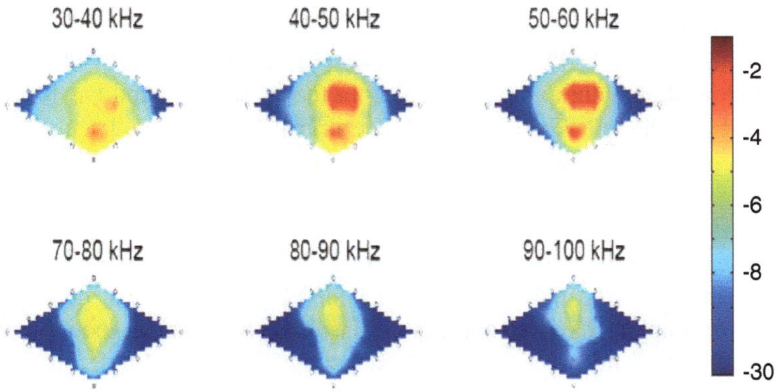

Fig. 8.6 In this single dolphin click, energy in the 30–40, 40–50, and 50–60 kHz bands show clustering in two different spatial regions based on the spectral magnitude of the band-limited frequency distribution of energy (*color bar* shows normalized SL). (Data from Moore et al. 2008, with permission)

or hollow, which can be discriminated by a dolphin using biosonar (Nachtigall and Moore 1988). Several experiments have investigated the ability of the dolphin to integrate, identify, and resolve various spectral cues within a target echo and the ability of dolphins to perceive and detect multiple echo returns (Vel'min and Dubrovskiy 1976; Moore et al. 1984; Au and Banks 1998). A temporal integration window in dolphin echo *detection* has been suggested and is termed the *critical interval* (Vel'min and Dubrovskiy 1976). Within this 265-μs interval, all the echo

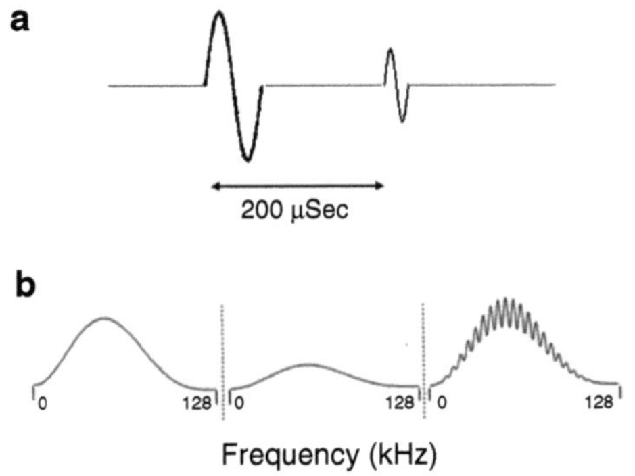

Fig. 8.7 The stimuli used in a dolphin temporal order discrimination experiment used unequal amplitude click-pairs, human listeners can discriminate these stimuli when the clicks are separated by only a few milliseconds. A spectral analysis of these stimuli show that there is no effect on the power spectrum by reversing the temporal order of the pairs and the discrimination was assumed to be from phase sensitivity. For the dolphin, however, the cue for discrimination was asserted to be the ripple in the power spectrum. (**a**) The click-pair stimuli used by Johnson et al. (1989) with a large-amplitude click 200 μs before a small-amplitude click. (**b**) The resulting power spectra of the large, small, and the ripple effect of combing the two clicks

energy appears to be summed and individual echo highlights outside this interval are not. However, Johnson and colleagues (1989) demonstrated that pulses presented within this interval could be *discriminated* when either a low-amplitude pulse, followed by a high-amplitude pulse, or the reverse; Au and Pawloski (1989) speculated that the relevant cue for this discrimination may have been spectral rippling in echoes (Fig. 8.7).

Helweg et al. (2003) examined complex multi-highlight echoes and found that a dolphin's *discrimination* performance was high when multiple complex echo highlights occurred both within a single integration window and when these highlights were distributed across many integration intervals. These results, taken together, indicate that dolphins have the ability to isolate and process brief acoustic events of lower amplitude echo highlights, while rejecting higher amplitude highlight features, a process that is adaptive for discrimination in reverberant environments, which prevail in the animal's natural ocean niche. Previous dolphin echolocation research purported the notion that each emitted echolocation click is triggered by the proceeding clicks echo return (e.g. emit a click, wait for the echo, emit the next click) a few other, less known observations, indicated that at long detection ranges dolphins emitted groups of closely timed packets of clicks. Ongoing investigations by Finneran (2013) and associates studying long range dolphin echo detection and discrimination indicate that the dolphin can change its click emission strategy. As the target range is extended beyond 75–100 m. the animal may switch to packet emissions or may simply increase the repetition rate of the overall click train; overlapping emitted clicks and returning echoes. These results further complicate the issue of how, exactly, does the animal integrate echo information and present exciting avenues of new research into the methods and processes of how dolphins perceive the auditory scene via echolocation.

8.3 Studies of Auditory Scene Analysis in Echolocating Animals

Over the last several decades, echolocation research in bats and dolphins has detailed the acoustic cues used to localize and discriminate sonar targets (see Moss and Schnitzler 1995); however, there remains an incomplete understanding of the larger problem of auditory scene analysis, namely how echo features from the natural environment are perceptually organized in the animals sonar receiver. Auditory scene analysis in echolocating animals may involve the combination of passive listening (e.g., communication calls and sonar signals of conspecifics and other natural sounds in the habitats) and active sonar. Here we emphasize the perceptual organization of sounds generated through echolocation.

8.3.1 *Bats*

Some psychophysical studies have revealed components of echo perception in bats that contribute to auditory scene analysis. The greater horseshoe bat, *Rhinolophus ferrumequinum*, for example, produces long CF signals, preceded and followed by FM sweeps. This bat species exhibits Doppler shift compensation as it flies (Schnitzler 1968), adjusting the frequency of its sonar emissions with its flight velocity, to ensure that Doppler shifted echoes fall in the region of its highest hearing sensitivity and frequency selectivity (Long and Schnitzler 1975). Doppler shift compensation behavior allows the bat to hear amplitude and frequency modulations in CF echoes introduced by fluttering insect prey, and the greater horseshoe bat can discriminate small changes in wingbeat rate (von der Emde and Menne 1989). Moreover, this bat species can *recognize* fluttering insects from novel echoes, suggesting that it represents complex acoustic patterns as an auditory object (von der Emde and Schnitzler 1990; Fig. 8.8).

Simmons et al. (1990) conducted a series of psychophysical experiments on the FM bat, *Eptesicus fuscus*, which suggest that this species converts spectral information in sonar echoes from complex targets to represent the underlying spatial separation of closely spaced reflecting surfaces. In an echo playback experiment, bats were trained to discriminate two-glint echoes, separated by 100 μs, and a single-glint echo. The two-glint playback simulated the reflection from two surfaces, separated by ~1.75 cm and contained spectral notches at 10-kHz intervals, created by the 100 μs offset of its component echoes; the single-glint echo simulated the return from a point target. The delay and attenuation of the two-glint echoes remained fixed across trials, but the delay and attenuation of the single-glint echo changed. Bats showed an increase in errors when the delay of the single-glint target coincided with the arrival time of either the first or second echo of the two-glint target, and these errors also depended on the amplitude of the single-glint target, because time-intensity trading influenced the bat's perception of the single-glint's target range. The results of this study suggest that the bat converts the spectral information carried by the interference pattern of the overlapping echoes of the two-glint target into the underlying delay or distance separation of two reflecting surfaces.

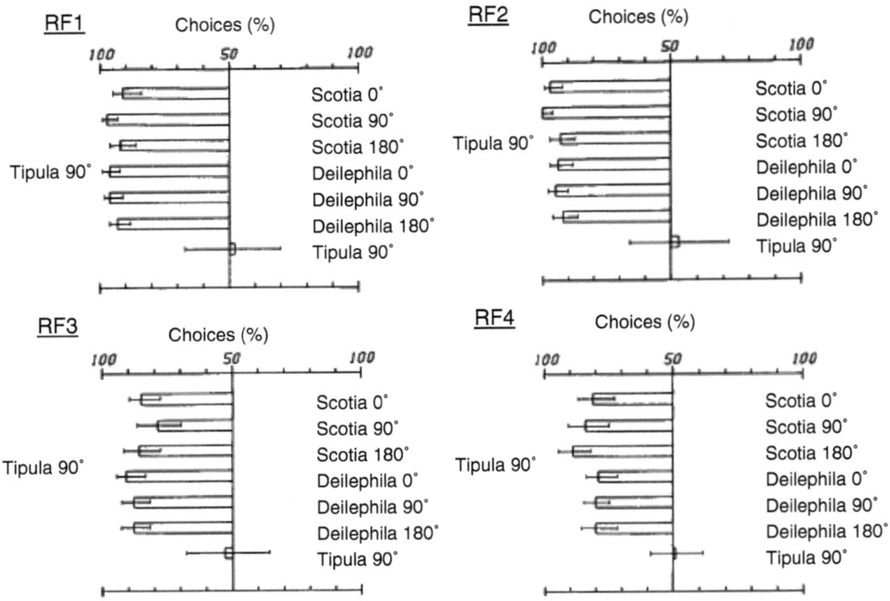

Fig. 8.8 Illustration of playback experiment showing the choices of four greater horseshoe bats (*Rhinolophus ferrumequinum*, RF1–RF4) between different insect echoes (von der Emde and Schnitzler 1990). The bat chose between the echo of the insect, *Tipula*, turned with its side toward the bat (*Tipula* 90°) and the echo of another insect species facing in one of three different directions (0°, 90°, and 180°). Each bar in the horizontal histograms shows the percentage of trials the bat chose a certain echo playback. All four individual bats showed a 90 % preference for the echo to which they were trained (*Tipula*, 90°)

When an echolocating bat forages in an acoustically complex environment, each sonar vocalization results in a cascade of echoes arriving from different directions and distances. Further, the bat's position changes between successive echolocation calls and echoes. If the bat were to respond to single echoes, it would surely fail to intercept moving insects, particularly in the presence of obstacles, such as vegetation and other flying bats. Success requires that the bat integrate the features of echoes over time and use this information to plan appropriate motor behaviors for prey interception and obstacle avoidance. Moss and Surlykke (2001) studied the echolocating big brown bat's ability to integrate delay information across echoes. In a two-alternative forced choice echo playback experiment, bats were required to discriminate between orderly sequences of echoes with decreasing or increasing delay and random sequences of echoes containing the same delay values (Fig. 8.9a, b). It was not

Fig. 8.9 (continued) electronically delayed, attenuated, low-pass filtered, and broadcast back to the bat through a loudspeaker (s). (**b**) There are two sets of delay pattern, one is sequential (S) echoes and the other is random (R) echo playbacks. (**c**) Performance of two bats (G-6 and M-6) trained in this experiment to discriminate sequential and random echo playback signals. The *dashed line* in each panel indicates 75 % of correct and this is the criterion to determine if the bat made a successful discrimination. Each block means trials recorded from different days

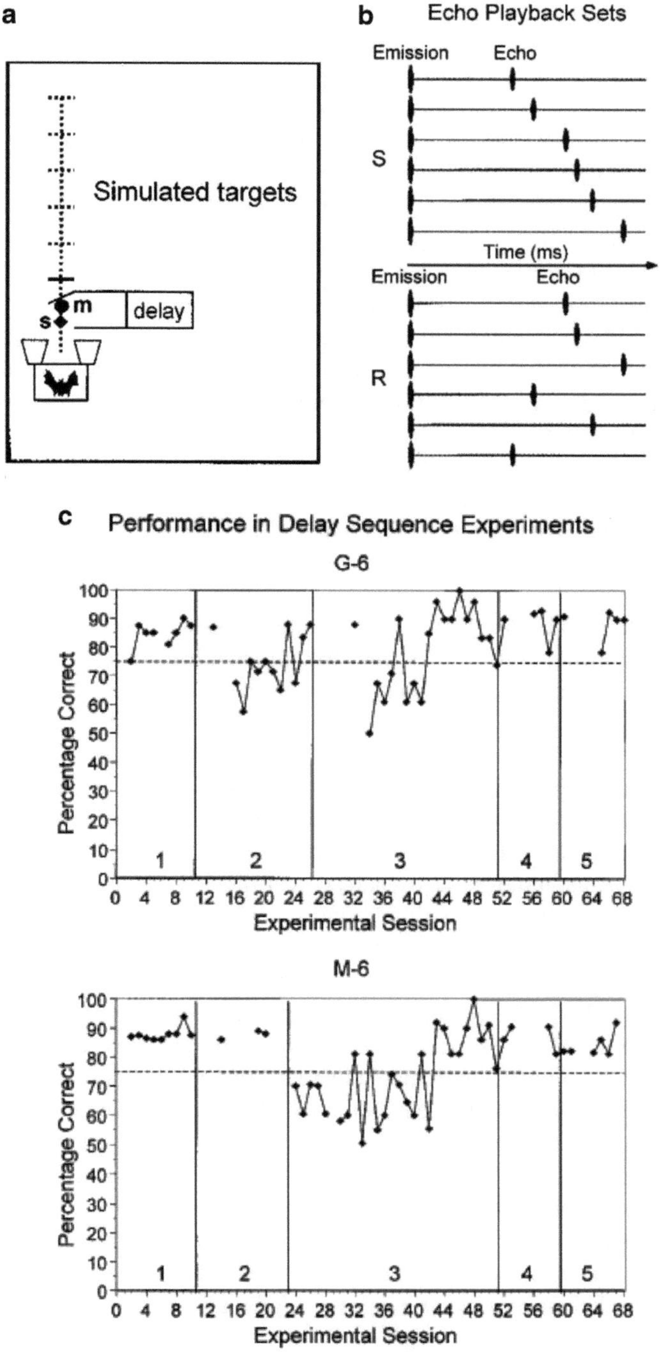

Fig. 8.9 Experimental setup and performance of two big brown bats (*Eptesicus fuscus*) in an echo integration experiment. (From Moss, C. F., & Surlykke, A. Auditory scene analysis by echolocation in bats. *Journal of the Acoustical Society of America* [2001] 110, 2207–2226; reprinted with permission from the Acoustical Society of America.) (**a**) The setup of the echo integration experiment. The bat was trained to sit on the Y-shaped platform and produce echolocation calls into a 1/8-in. microphone (m). The echolocation signals were amplified, band-pass filtered, digitized,

surprising that big brown bats could successfully perform a task that required integration of echo delay information over time (Fig. 8.9c), but it was important to establish that the bat's sonar receiver supports this basic operation required for auditory scene analysis.

A recent study uncovered an important feature of FM-bat sonar that would support auditory scene analysis in complex, cluttered environments, where echoes from vegetation could interfere with the perception of on-axis sonar prey. Bats actively lock the sonar beam pattern axis on selected targets (Ghose and Moss 2003; Surlykke et al. 2009), which results in off-axis echo returns from other objects. Bates et al. (2011) noted that off-axis echoes contain less energy at higher frequencies than on-axis echoes, due to the directional characteristics of the sonar beam, the head-related transfer function, and spherical spreading losses. This means that temporal registration of the fundamental and higher harmonics would be disrupted in the bat's sonar receiver, due to time-intensity trading. In other words, the second harmonic of the echo would be represented in the bat's sonar receiver at a longer delay than the fundamental, because weaker sounds evoke activity in the auditory system at longer latencies than stronger sounds (Simmons et al. 1990). Although the fundamental and harmonics arrive at the same time at the bat's ear, the auditory system would register a temporal offset of the weaker higher harmonics. Bates et al. (2011) measured the bat's range discrimination performance when the timing of the fundamental and second harmonic was electronically offset, and they found that the bat's perception of distance was degraded. They interpret this finding to suggest that off-axis echoes from clutter objects would be defocused, which would minimize their masking effect on selected on-axis target echoes. In effect, on-axis targets are represented as sharp, and off-axis targets are represented as blurred.

Advances in technology have enabled real-time playbacks of modified sonar calls to simulate echoes from simple and complex targets at different distances. Although research using these methods have advanced our understanding of the resolution of biological sonar systems, they are not suitable to the study of auditory scene analysis. The microphones and loudspeakers that comprise these playback systems are echo reflectors that interfere with the animal's perception of the simulated echoes of a complex environment by reducing the perceptual salience of phantom target echoes. Indeed, this realization led Cynthia Moss and colleagues to shift research efforts on the bat's perception of complex auditory scenes from phantom target playback studies to quantitative analyses of adaptive sonar behaviors. The bat's adaptive sonar behavior provides indirect information about its perception, because changes in sonar vocalizations indicate the information an animal has processed from echoes and what information it seeks. Therefore, studies of the bat's adaptive sonar behavior provide a window to the animal's perception of complex echo environments.

Bat echolocation engages adaptive sonar behaviors that contribute directly to accurate localization and tracking of objects. The features of sonar calls produced by a bat to probe its environment directly impact the information available to its acoustic imaging system. In turn, the bat's perception of the auditory scene influences the features of subsequent sonar vocalizations. The bat's adjustments in echolocation call parameters, such as duration, interval, direction, and spectrum, provide insight to the acoustic information used to solve the perceptual problem of sorting and tracking echoes arriving from different directions and distances.

The sonar beam patterns of echolocating bats are directional and vary with sound frequency and across species (Hartley and Suthers 1989). Nasal emitters produce complex beam patterns that are shaped by the noseleaf (Schnitzler and Grinnell 1977; Vanderelst et al. 2010). The sonar beam patterns of oral emitters typically show less complex spatial profiles but can contain more than one lobe. Laboratory studies have documented that oral echolocators accurately adjust the directional aim of sonar calls to sequentially inspect closely spaced objects (Moss et al. 2011). In one study that required the big brown bat to engage in both obstacle avoidance and prey capture, the animal sequentially pointed its sonar beam axis at the edges of a net opening to find its way through an obstacle and gain access to a food reward (Surlykke et al. 2009). This FM bat also adjusted the duration of its echolocation calls to avoid overlap with sonar vocalizations and echoes from the objects it was inspecting. When the big brown bat shifted its sonar gaze to more distant objects, it tolerated overlap between calls and echoes from nearby obstacles (Surlykke et al. 2009).

In a target discrimination study, free-flying big brown bats exhibited similar adjustments in call direction and duration to inspect small tethered objects sequentially with different textures (Falk et al. 2011). Recordings of sonar returns from the textured objects showed echo-to-echo variation in spectrum, with different patterns of change for each of the textured stimuli. The larger the differences in echo-to-echo spectral profile between stimuli, the higher the bat's target discrimination performance. This finding suggests that the bat listened to changes in echo profiles over time to perform the texture discrimination task. Further research is needed to better understand the bat's perception of target texture through echolocation.

When bats fly in groups, they face the challenge of sorting their own calls and echoes from those of conspecifics, and they must adopt strategies to avoid sonar jamming (Ulanovsky et al. 2004; Gillam et al. 2007; Fig. 8.10a, b). Laboratory experiments showed that individual big brown bats adjusted the start and end frequencies of their FM sweeps, along with FM bandwidth, when they were foraging with another bat in a large flight room (Surlykke and Moss 2000; Chiu et al. 2009). The magnitude of the vocal adjustment depended on the baseline similarity of the bats' calls (when they flew alone) and their spatial separation in the room: Bats with similar baseline call design made adjustments in spectral call characteristics, and bats with different baseline call design maintained spectral differences without adjusting their sonar signals. Bats made the largest adjustments in the spectral characteristics of their calls when they flew less than 1 m apart (Chiu et al. 2009). It is also noteworthy that bats in this competitive foraging situation sometimes went silent (ceased vocalizing for at least 200 ms), and this behavior appears at least in part to be driven by jamming avoidance. Bats showed the most silent behavior when their baseline call design was similar to that of its competitor and when the two bats flew less than 1 m apart (Chiu et al. 2008), a result that parallels the spectral adjustment data (Moss et al. 2011). It is possible that the silent bat listened passively to

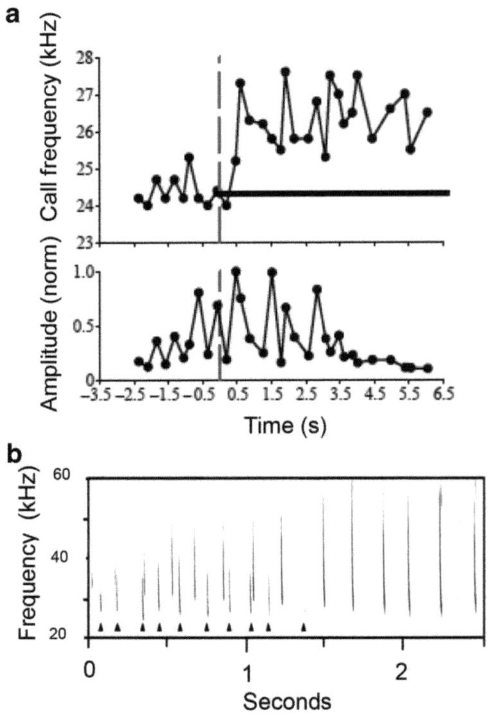

Fig. 8.10 Echolocating bats adjust their call frequency to avoid signal jamming with conspecifics. (**a**) The bat (*Tadarida brasiliensis*) adjusted its call frequency in response to the playback signals similar to its call frequency. (From Gillam et al. (2007), with permission from *Proceedings of the Royal Society of London*.) The *dashed line* indicates the time the playback signals (a frequency of 24.3 kHz) were switched on. This is an exemplary trial from the recordings. (**b**) A recording from two big brown bats (*Eptesicus fuscus*) flying together. (Adapted from Surlykke, A., & Moss, C. F. Echolocation behavior of the big brown bat, *Eptesicus fuscus*, in the field and the laboratory. *Journal of the Acoustical Society of America* [2000] 108(5), 2419–2429; reprinted with permission from Acoustical Society of America.) Before 1.5 s, one bat was calling at relatively low frequency (indicated by *triangle*) and the other bat was calling at relatively high frequency. After 1.5 s, the low-frequency one stopped calling and the high-frequency one lowered its call frequency

the calls produced by the vocalizing bat, and these signals contributed to this animal's perception of the auditory scene.

Field experiments in multiple bats flight conditions also indicated that bats change the temporal or spectral features of their echolocation calls in order to avoid signal interference with conspecifics. Past studies have demonstrated that when flying with conspecifics, several bat species (big brown bats, *Eptesicus fuscus*, and free-tailed bats, *Tadarida brasiliensis* and *T. teniotis*) shift their call frequencies either upward or downward to avoid possible call interference and overlap by neighboring individuals (Ulanovsky et al. 2004; Gillam et al. 2007; Bates et al. 2008). A playback experiment showed that *T. brasiliensis* raised the end frequency of the FM sweep in response to playback jamming signals, whose frequencies were equal to the average end frequencies of this species' sonar calls (Gillam et al. 2007; Fig. 8.10a). Both laboratory and field studies have identified the strategy that the bat increases differences between its own sonar call design and those of conspecifics when flying in groups. This strategy presumably helps the bat segregate its own echolocation pulses from others in proximity.

Studies of the echolocating bat's adaptive vocal behaviors provide a window to the acoustic information an animal has processed and the information it is seeking from its environment. Quantitative analysis of adaptive sonar behavior may therefore contribute to our understanding of auditory scene analysis by echolocation. It is important, however, to caution the reader that inferences made from adaptive sonar behavior are not direct measurements of perception, and other research methods, such as psychophysical tasks, can generate complementary data that serve to deepen our understanding of natural scene perception.

8.3.2 Dolphins

The dynamic quality of the dolphin echolocation signal, coupled with the animal's ability to perceive and discriminate changes in the acoustic information contained within the echolocation frequency range, make these signals well suited for auditory scene analysis. Although there is almost no direct information on the ability of the dolphin echolocation system to perform auditory scene analysis as described by Bregman (1990) for human listeners, one observation of auditory stream segregation in dolphins has been reported (Moore and Finneran 2011). We know that dolphins change signal level and frequency to overcome masking stimuli, and they shift signal frequency with age presumably due to hearing loss (Moore et al. 2004; Kloepper et al. 2010). We know little or nothing about why or how multiple animals adjust their signals when echolocating during foraging because it is extremely difficult to monitor and record individual animals accurately in the open ocean.

When dolphins echolocate in either a detection or discrimination experiment they emit click trains that comprise hundreds of dynamically changing signals. In the past it has been nearly impossible to determine exactly what causes the changes in individual emitted clicks in the train. However, using new high-speed phantom echo generators (PEGs; Finneran et al. 2010) experimenters can track each click in the train and precisely time when a target appears (detection) or when a target changes (discrimination). Armed with this precision, an examination of how the animal changes its signals in response to the changing target stimuli is possible. These experiments are now underway and hopefully will result in a much better understanding of the signal emission strategies of echolocating dolphins. In one phantom target discrimination study, following the methods of Moss and Surlykke (2001) with an echolocating dolphin, a series of phantom echoes, representing a sphere target, was programmed to systematically approach, recede, or appear randomly along the range axis (Moore and Finneran 2011). Each emitted echolocation signal triggered a target echo return that was presented to the dolphin's lower jaw via a jaw-phone (suction attached transducer at the acoustic window). The results clearly indicated discrimination between the systematically approaching/receding echo streams versus the random stream, demonstrating the both the ability to integrate information over time and short term memory for acoustic events. This result parallels that reported by Moss and Surlykke (2001) for the echolocating bat. Other indirect evidence of stream segregation for auditory scene analysis in dolphins is suggested by their ability to perform echolocation delayed-match-to-sample (DMTS) tasks. Roitblat et al. (1990) reported an experiment of dolphin echolocation DMTS that required a blindfolded dolphin to correctly choose, from a selection of three objects, a previously presented object. The ability to perform this task at 90 % or better clearly demonstrates the animal's ability to engage short-term memory for complex target returns. These studies provide evidence that the dolphin biosonar system possesses, at the minimum, the rudimentary requirements for auditory scene analysis.

8.3.2.1 The Littoral Ocean (Noisy, Reverberant, and Cluttered)

Most dolphins live in and along the coastal regions in shallow waters, bays, estuaries, and riverine environments, which are very noisy, notoriously non-Gaussian and non-stationary (Urick 1975). Dolphins have evolved biosonar that is adapted for this noisy, reverberant, and highly cluttered environment.

Dolphin biosonar has been shown to be a highly refined acoustic sense that these animals use for detection and discrimination of targets (see Simmons & Houser, Chap. 6). For dolphins hunting prey in very shallow water, or prey that is buried beneath the ocean bottom, reverberation plus noise impose limits on the perception of returning echoes. To be successful in the detection of targets, dolphins must be able to overcome competing returns from the various bottom composition and inhomogeneities, distortion due to thermal discontinuities in the medium, and losses due to absorption in the ocean bottom, clutter and other biological sources. This detection scenario is much more complicated than in open water (Moore 1997; Houser et al. 2005; Martin et al. 2005; see Fig. 8.11).

Resolving sonar targets in high density clutter, such as fish camouflaged by inshore kelp forests, or the detection of moving prey, either down range of the main sonar beam or cross-range of the beam, is an important capability demonstrated by dolphins. It is noteworthy that even when the target is coplanar with the cluttering objects the dolphin can detect the target when the target echo to clutter backscatter ratio is approximately 0 .25 dB_{pkpk} (the subscript pkpk refers to the peak-to-peak value of the emitted signal) (Au and Moore 1984). Altes et al. (2003) carried out a psychophysical study to test the hypothesis that dolphins combine echoes to improve signal detectability in noisy, cluttered, and reverberant conditions and to determine the best receiver model accounting for the dolphin's performance. They point out that if a moving echolocating animal has the ability to sum echo samples from the same point (target) using different pulse-echo pairs, this ability could be electronically simulated as a synthetic aperture sonar process. In this study, an

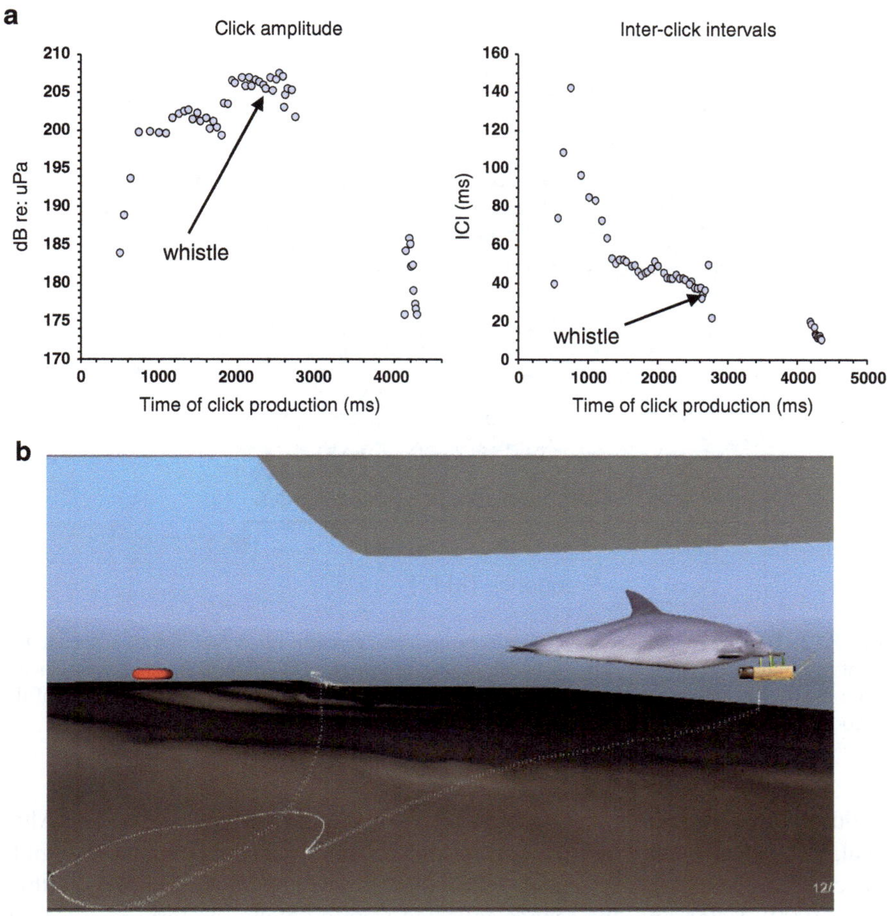

Fig. 8.11 The search strategy of a dolphin as it swims an underwater path around a red float located several hundred meters from a boat. (**a**) Source level in dBre:1 μPa and interclick interval in milliseconds for a target search. The *arrows* indicate the point at which the dolphin whistles, indicating detection of the target (**b**) virtual reality rendering, viewed from animal depth, of the target search path as a series of *white dots*. Data sensors included full three-dimensional position data (heading, pitch, roll, acceleration, angular rates, depth, and velocity). Data collected from the sensor pack are geo-located with a virtual rendering created after the search. (See Martin et al. 2005, and Houser et al. 2005)

echolocating dolphin detected a 50-kHz, 80-μs sinusoid pulse presented at 7 m range in noise. A pulse was delivered for each emitted echolocation click generated by the animal. The detection threshold in noise as a function of the number of delivered pulses was determined for $N=1$, 4, 8, and 16 pulses. They found that for the dolphin's acoustic environment, the binary M-out-of-N detection model closely matched the dolphins' detection performance, but the data were a poor fit for a linear or energy summation model. The binary M-out-of-N detector model is a basic

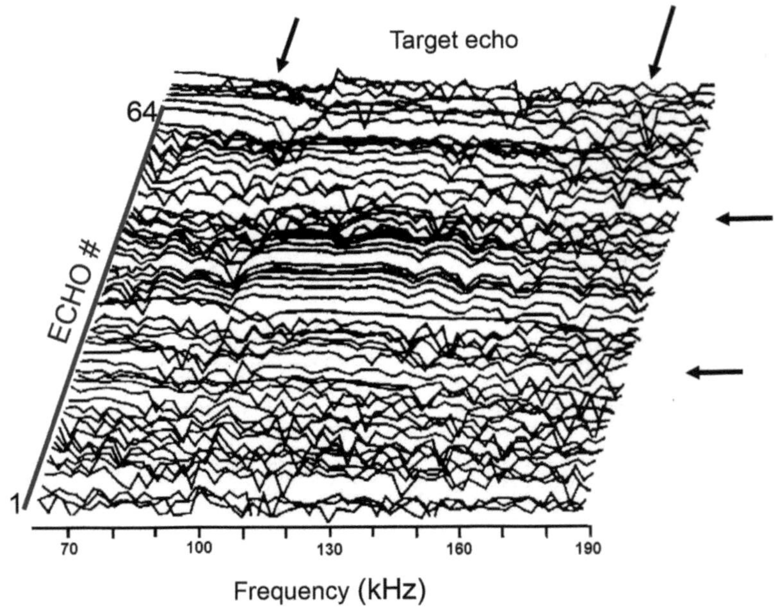

Fig. 8.12 Time-aligned successive target echo spectra (glycerin-filled fuel bottle) suspended in front of blindfolded dolphin (time between echoes removed). The target echo is seen to emerge from the background noise as the train of echoes progresses and can be seen in the center of the echo-train (*arrows* indicate both echo number and frequency range; see Moore et al. 1991)

building block for neural all-or-none signals (binary action potentials). The Altes et al. (2003) study supported an earlier premise that dolphins may use the ensemble of echo returns to discriminate target attributes (Moore and Pawloski 1990; Roitblat et al. 1990). Using only the echoes collected from a dolphin performing a DMTS task in the noise of Kaneohe Bay, Hawaii, these investigators found that a neural network that performed an averaging of returned echo spectra (see Fig. 8.12) could classify simple targets shapes as well as the dolphin when echoes had good signal-to-noise ratio.

Au (1994) and Au et al. (1995) advanced other models that combined temporal and spectral information over an echolocation train in a neural network model with noisy echoes as exemplars. Branstetter et al. (2007) advanced these models of dolphin echo representation with one that incorporates both spectral and temporal resolution of the dolphin auditory system based on processes which have been demonstrated from dolphin psychoacoustic results. Although it is unreasonable to suggest that these models fully represent the underlying processes occurring in dolphin echo recognition, these investigators have applied these models to motivate further psychoacoustic investigations and to better understand the processes that may be at play in the wider arena of auditory scene analysis.

8.3.2.2 Tracking Prey in the Presence of Conspecifics

Dolphins are cooperative foragers which sometime employ scout animals to locate prey. Different species use varying techniques to herd or crowd fish into a small confined area to allow individuals to catch them more easily. Field observations of spinner dolphins indicate a high degree of synchronization with almost immediate transitions of discrete hunting behaviors between animals over large distances where water clarity and light levels (hunting at dusk or night) would make visual cues highly unlikely (Benoit-Bird and Au 2009). Although the capture of individual prey by feeding animals may be aided by bioluminescence (produced by living organisms in the water), recordings (of hunting and feeding) bouts suggested that between feeding bouts these dolphins were using clicks signals, not whistles, to coordinate behaviors between animal groups. Click signals (assumed to be biosonar related) would meet the requirements for this kind of inter-animal coordination of behavior, they are highly directional, have a wide bandwidth (and thus large information-carrying capacity), and allow selective communication between individual animals or groups of animals (Lammers et al. 2003). In addition, the dolphin's impressive passive listening capability surely must play a pivotal role in monitoring conspecifics during social and cooperative hunting behaviors. Xitco and Roitblat (1996) conducted an experiment that demonstrated that echolocating dolphins could "eavesdrop" on conspecifics returning target echoes. They showed that the listening dolphin could perform a matching-to-sample task above chance by using a "champions" echolocation signals. Until more studies like the Benoit-Bird and Au (2009) and the Xitco and Roitblat (1996) investigations are conducted, we must continue to speculate on the auditory scene of echolocating dolphins and the behavioral dynamics of their hunting behavior in the wild.

8.4 Challenges and Future Direction for the Study of Auditory Scene Analysis in Bats and Dolphins

Although bats and dolphins are both echolocators, they operate under vastly different conditions, aerial and aquatic (see Madsen & Surlykke, Chap. 9), and the challenges we face in designing experiments to study their auditory perception are not the same. These two groups of animals exhibit different natural behaviors, which should be taken into consideration when designing studies of auditory scene analysis. Some dolphins and bats are cooperative foragers, and some search for food individually. While foraging in groups, both bats and dolphins face the challenge of sorting out their own calls from others. Echolocating bats employ several strategies, including shifts in call frequency, changes in call duration, and increased pulse intervals, to avoid signal jamming with conspecifics. Dolphins, animals that forage in large groups, produce highly directional, wide bandwidth click signals to hunt for fish, but the strategy they use to avoid signal interference is not yet clear. It is possible that group foragers may use the sonar signals produced by neighbors to guide their prey search and orientation, rather than develop a strategy to avoid interference between their own signals and those of others. Cooperative sonar behavior is a field with many open questions and opportunities for future research.

Both bat and dolphin researchers have studied echoes reflecting from a variety of objects and conducted behavioral experiments to investigate the animal's ability to discriminate among different objects by echolocation. It has been demonstrated that both bats (Simmons et al. 1974; Habersetzer and Vogler 1983; Falk et al. 2011) and dolphins (Moore et al. 1984; Au and Pawloski 1992; Au 1993) can discriminate object structure and they can both integrate echo information over time (Moss and Surlykke 2001; Moore and Finneran 2011). Some researchers also have built models to understand how echolocators use biosonar to perceive their environment (Moore et al. 1991; Branstetter et al. 2007; Yovel et al. 2008). Modeling efforts in the field of echolocation present exciting challenges.

The sonar beam patterns produced by echolocating animals are directional, and therefore bats and dolphins can direct their sonar beam to inspect objects of interest. Studies of target range discrimination in bats suggest that resolution is highest along the central axis. Therefore, bats can maximize information from objects of interest and minimize clutter interference through directional control of the sonar beam (Bates et al. 2011). Although it has been shown that dolphins can detect objects in reverberation and heavy clutter (Au and Turl 1983), an open question remains as to exactly how dolphins use their highly directional sonar beam to minimize masking by interfering echoes.

In our review of psychophysical studies of sonar perception, we note the limitations of understanding auditory scene analysis by echolocation when the animal may be constrained to a limited repertoire of emitted signals, and listening to a mix of simulated and real echoes that can compromise the perceptual salience of the experimental setting. In addition, changes in the animal's head aim that would normally result in large changes in echo features may not be represented in playback echoes and therefore fail to fully capture 3D elements of natural sonar objects. We have learned from psychophysical studies a great deal about the limits of the echolocating animal's echo processing, but there remains much to understand about higher level perceptual processes that contribute to auditory scene analysis.

By contrast, adaptive motor studies are better suited to fully engage the animal in more natural behaviors in which they dynamically modulate their sonar calls in response to echo returns from the environment. Adaptive sonar behaviors are an integral component of echolocation systems that would be expected to feed into auditory scene analysis processes. However, adaptive motor studies have not provided a direct measure of the animal's perception of a complex, natural environment. Instead, auditory perception can only be inferred from the animal's adaptive motor behaviors. Future research on auditory scene analysis by echolocation must embrace the challenge of marrying the advantages of psychophysical and adaptive motor studies, taking creative new approaches to tap into an animal's perception of its complex, 3D auditory world, while allowing it to engage in its natural behaviors.

Acknowledgments National Science Foundation, National Institute of Mental Health, and National Institute on Deafness and Other Communication Disorders provided support for some of the bat echolocation research summarized in this review. Portions of the dolphin research reported here were funded by Office of Naval Research and the Defense Advanced Research Projects Agency. Annemarie Surlykke, Kaushik Ghose, and Ben Falk contributed to the data collection, analysis, and interpretation of several laboratory studies of adaptive echolocation behavior in big brown bats, summarized here. We thank Janna Barcelo and Wei Xian for proofreading the manuscript and Wei Xian for assistance with the figures.

References

Altes, R. A., Dankiewica, L. A., Moore, P. W., & Helweg, D.A. (2003). Multiecho processing by an echolocating dolphin. *Journal of the Acoustical Society of America*, 114, 1155–1166.

Au, W. W. L. (1993). *The sonar of dolphins*. New York: Springer-Verlag.

Au, W. W. L. (1994). Comparison of sonar discrimination-dolphin and an artificial neural-network. *Journal of the Acoustical Society of America*, 95, 2728–2735.

Au, W. W. L., & Turl C.W. (1983). Target detection in reverberation by an echolocating Atlantic bottlenose dolphin (*Tursiops truncatus*). *Journal of the Acoustical Society of America*, 73, 1676–1681.

Au, W. W. L., & Moore, P. W. (1984). Receiving beam patterns and directivity indexes of the Atlantic bottlenose dolphin *Tursiops truncates*. *Journal of the Acoustical Society of America*, 75, 255–262.

Au, W. W. L., & Pawloski, D. A. (1989). A comparison of signal-detection between an echolocating dolphin and an optimal receiver. *Journal of Comparative Physiology A: Sensory Neural and Behavioral Physiology*, 164, 451–458.

Au, W. W. L., & Pawloski, D. A. (1992). Cylinder wall thickness difference discrimination by an echolocating Altantic bottle-nosed-dolphin. *Journal of Comparative Physiology A: Sensory Neural and Behavioral Physiology*, 170, 41–47.

Au, W. W. L., & Banks, K. (1998). The acoustics of the snapping shrimp *Synalpheus parneomeris* in Kaneohe Bay. *Journal of the Acoustical Society of America*, 103, 41–47.

Au, W. W. L., Andersen, L. N., Rasmussen, A. R., Roitblat, H. L., & Nachtigall, P. E. (1995). Neural-network modeling of a dolphin's sonar discrimination capabilities. *Journal of the Acoustical Society of America*, 98, 43–50.

Au, W. W. L., Benoit-Bird, K. J., & Kastelein, R. A. (2007). Modeling the detection range of fish by echolocating bottlenose dolphins and harbor porpoises. *Journal of the Acoustical Society of America*, 121, 3954–3962.

Barber, J. R., Razak, K. A., & Fuzessery, Z. M. (2003). Can two streams of auditory information be processed simultaneously? Evidence from the gleaning bat *Antrozous pallidus*. *Journal of Comparative Physiology A: Sensory Neural and Behavioral Physiology*, 189, 843–855.

Bates, M. E., Stamper, S. A., & Simmons, J. A. (2008). Jamming avoidance response of big brown bats in target detection. *Journal of Experimental Biology*, 211, 106–113.

Bates, M. E., Simmons, J. A., & Zorikov, T. V. (2011). Bats use echo harmonic structure to distinguish their targets from background clutter. *Science*, 333, 627–630.

Bee, M. A., & Micheyl, C. (2008). The cocktail party problem: What is it? How can it be solved? And why should animal behaviorists study it? *Journal of Comparative Psychology*, 122, 235–251.

Benoit-Bird, K. J., & Au, W. W. L. (2009). Phonation behavior of cooperatively foraging spinner dolphins. *Journal of the Acoustical Society of America*, 125, 539.

Branstetter, B. K., Mercado, E., & Au, W. W. L. (2007). Representing multiple discrimination cues in a computational model of the bottlenose dolphin auditory system. *Journal of the Acoustical Society of America*, 122, 2459–2468.

Bregman, A. S. (1990). *Auditory scene analysis: The perceptual organization of sound.* Cambridge, MA: Bradford Books, MIT Press.

Chiu, C., Xian, W., & Moss, C.F. (2008). Flying in silence: Echolocating bats cease vocalizing to avoid sonar jamming. *Proceedings of the National Academy of Sciences of the USA*, 105, 13115–13120.

Chiu C., Xian, W., & Moss, C. F. (2009). Adaptive echolocation behavior in bats for the analysis of auditory scenes. *Journal of Experimental Biology*, 212, 1392–1404.

Falk, B., Williams, T., Aytekin, M., & Moss, C. F. (2011). Adaptive behavior for texture discrimination by the free-flying big brown bat, *Eptesicus fuscus. Journal of Comparative Physiology A: Sensory Neural and Behavioral Physiology*, doi: 10.1007/s00359-010-0621-6.

Finneran, J. J. (2013) Dolphin "packet" use during long-range echolocation tasks. *Journal of the Acoustical Society of America, 113,1796.*

Finneran, J. J., Houser, D. S., Moore, P. W., Branstetter, B. K., Trickey, J. S., & Ridgway, S. H. (2010). A method to enable a bottlenose dolphin (*Tursiops truncatus*) to echolocate while out of water. *Journal of the Acoustical Society of America*, 128, 1483.

Ghose, K., & Moss, C. F. (2003). The sonar beam pattern of a flying bat as it tracks tethered insects. *Journal of the Acoustical Society of America*, 114, 1120–1131.

Gillam, E. H., Ulanovsky, N., & McCracken, G. F. (2007). Rapid jamming avoidance in biosonar. *Proceedings of the Royal Society of London B: Biological Sciences*, 274, 651–660.

Griffin, D. R. (1958). *Listening in the dark.* New Haven, CT: Yale University Press. 2nd ed. (1986). Ithaca, NY: Cornell University Press.

Griffin, D., Webster, F. A., & Michael, C. R. (1960). The echolocation of flying insects by bats. *Animal Behaviour*, 8, 141–154.

Habersetzer, J., & Vogler, B. (1983). Discrimination of surface-structured targets by the echolocating bat *Myotis myotis* during flight. *Journal of Comparative Physiology A: Neuroethology, Sensory, Neural, and Behavioral Physiology*, 152(2), 275–282.

Harley, H. E., Putman, E. A., & Roitblat, H. L. (2003). Bottlenose dolphins perceive object features through echolocation. *Nature*, 424 (6949), 667–669.

Hartley, D. J., & Suthers, R. A. (1989). The sound emission pattern of the echolocating bat, *Eptesicus fuscus. Journal of the Acoustical Society of America*, 85, 1348–1351.

Helweg, D. A., Moore, P. W., Dankiewica, L. A., Zafran, J. M., & Brill, R. L. (2003). Discrimination of complex synthetic echoes by an echolocating bottlenose dolphin. *Journal of the Acoustical Society of America*, 113(2), 1138–1144.

Houser, D., Martin, S. W., Bauer, E. J., Phillips, M., Herrin, T., Cross, M., Vidal, A., & Moore, P.W. (2005). Echolocation characteristics of free-swimming bottlenose dolphins during object detection and identification. *Journal of the Acoustical Society of America*, 117, 2308.

Johnson, R. A., Moore, P. W. B., Stoermer, M. W., Pawloski, J. L., & Anderson, L. C. (1989). Temporal order discrimination within the dolphin critical interval. In P. E. Nachtigall & P. W. B. Moore (Eds.), *Animal sonar: Processes and performance*. New York: Plenum Press.

Kalko, E. K. V., Schnitzler, H. U., Kaipf, I., & Grinnell, A. D. (1998). Echolocation and foraging behavior of the lesser bulldog bat, *Noctilio albiventris*: Preadaptations for piscivory? *Behavioral Ecology and Sociobiology*, 42(5), 305–319.

Kloepper, L., Nachtigall, P. E., & Breese, M. (2010). Change in echolocation signals with hearing loss in a false killer whale (*Pseudorca crassidens*). *Journal of the Acoustical Society of America*, 128, 2233–2237.

Lammers, M. O., Au, W. W. L., & Herzing, D. L. (2003). The broadband social acoustic signaling behavior of spinner and spotted dolphins. *Journal of the Acoustical Society of America*, 114, 1629–1639.

Long, G. R., & Schnitzler, H. U. (1975). Behavioral audiograms from bat, *Rhinolophus ferrumequinum. Journal of Comparative Physiology A: Sensory Neural and Behavioral Physiology*, 100(3), 211–219.

Martin, S. W., Phillips, M., Bauer, E. J., Moore, P. W., & Houser, D. S. (2005). Instrumenting free-swimming dolphins echolocating in open water. *Journal of the Acoustical Society of America*, 117(4), 2301–2307.

Moore, P. W. (1997). Dolphin psychoacoustic. *Bioacoustics*, 8, 61–78.
Moore, P. W. B., & Pawloski, D. A. (1990). Investigations on the control of echolocation pulses in the dolphin. In J. A. Thomas & R. Kastelein (Eds.), *Dolphin sensory processes* (pp. 305–316). New York: Plenum Press.
Moore, P. W., Finneran, J., and Houser, D. S. (2004). Hearing loss and echolocation signal change in dolphins. *Journal of the Acoustical Society of America* 116, 2503-2503.
Moore, P. W., & Finneran, J. J. (2011). Auditory scene analysis in the echolocating dolphin. *Journal of the Acoustical Society of America*, 129(4), 2469–2469.
Moore, P. W., Dankiewicz, L. A., & Houser, D. S. (2008). Beamwidth control and angular target detection in an echolocating bottlenose dolphin (*Tursiops truncatus*). *Journal of the Acoustical Society of America*, 124, 3324.
Moore, P. W. B., Hall, R. W., Friedl, W. A., & Nachtigall, P. E. (1984). The critical interval in dolphin echolocation: What is it? *Journal of the Acoustical Society of America*, 76, 314–317.
Moore, P. W. B., Roitblat, H. L., Penner, R. H., & Nachtigall, P. E. (1991). Recognizing successive dolphin echoes with an integrator gateway network. *Neural Networks*, 4, 701–709.
Moss, C. F., & Zagaeski, M. (1994). Acoustic information available to bats using frequency-modulated sounds for the perception of insect prey. *Journal of the Acoustical Society of America*, 95, 2745–2756.
Moss, C. F., & Schnitzler, H. U. (1995). Behavioral studies of auditory information processing. In A. N. Popper & R. R. Fay (Eds.), *Hearing by bats* (pp. 87–145). New York: Springer-Verlag.
Moss, C. F., & Surlykke, A. (2001). Auditory scene analysis by echolocation in bats. *Journal of the Acoustical Society of America*, 110, 2207–2226.
Moss, C. F., & Surlykke, A. (2010). Probing the natural scene by echolocation in bats. *Frontiers in Behavioral Neuroscience*, 4, 1–16.
Moss, C. F., Chiu, C., & Surlykke, A. (2011). Adaptive vocal behavior drives perception by echolocation in bats. *Current Opinion in Neurobiology*, 21, 645–652.
Nachtigall, P. E., & Moore, P. W. B. (1988). *Animal sonar: Processes and performance*. New York: Plenum Press.
Neuweiler, G. (1989). Foraging ecology and audition in echolocating bats. *Trends in Ecology and Evolution*, 4, 160–166.
Roitblat, H. L., Penner, R. H., & Nachtigall, P. E. (1990). Matching-to-sample by an echolocating dolphin (*Tursiops truncatus*). *Journal of Experimental Psychology: Animal Behavior Processes*, 16(1), 85–95.
Schnitzler, H. U. (1968). Ultrasonic sounds of horseshoe bats (Chiroptera Rhinolophidae) in different orientation situations. *Zeitschrift fur Vergleichende Physiologie*, 57, 376–408.
Schnitzler, H. U., & Grinnell, A. D. (1977). Directional sensitivity of echolocation in horseshoe bat, *Rhinolophus ferrumequinum*: 1. Directionality of sound emission. *Journal of Comparative Physiology A: Sensory Neural and Behavioral Physiology*, 116(1), 51–61.
Schnitzler, H. U., & Flieger, E. (1983). Detection of oscillating target movements by echolocation in the greater horseshoe bat. *Journal of Comparative Physiology A: Sensory Neural and Behavioral Physiology*, 153(3), 385–391.
Schnitzler, H. U., Menne, D., Kober, R., & Heblich, D. (1983). The acoustical image of fluttering insects in echolocating bats. In F. Huber & H. Markl (Eds.), *Neuroethology and behavioral physiology: Roots and growing points* (pp. 235–250). Berlin, Germany: Springer-Verlag.
Schnitzler, H. U., Kalko, E. K. V., Kaipf, I., & Grinnell, A. D. (1994). Fishing and echolocation behavior of the greater bulldog bat, *Noctilio leporinus*, in the field. *Behavioral Ecology and Sociobiology*, 35(5), 327–345.
Schnitzler, H. U., Moss, C. F., & Denzinger, A. (2003). From spatial orientation to food acquisition in echolocating bats. *Trends in Research in Ecology and Evolution*, 18, 386–394.
Simmons, J. A., Lavender, W. A., Lavender, B. A., Doroshow, C. A., Kiefer, S. W., & Livingston, R. (1974). Target structure and echo spectral discrimination by echolocating bats. *Science*, 186(4169), 1130–1132.
Simmons, J. A., Moss, C. F., & Ferragamo, M. (1990). Convergence of temporal and spectral information into acoustic images of complex sonar targets perceived by the echolocating bat,

Eptesicus fuscus. Journal of Comparative Physiology A- Sensory Neural and Behavioral Physiology, 166, 449–470.

Starkhammar, S., Dankiewicz-Talmadge, L. A., Houser, D. S., & Moore, P. W. (2010). Frequency-dependent echolocation beam pattern of the bottlenose dolphin. *Journal of the Acoustical Society of America*, 128, 2484.

Stilz, P. (2004). Akustische untersuchungen zur echoortung bei fledermauesen. Tier Phisiologie. University of Tübingen. Retrieved from http://www.biosonarlab.uni-tuebingen.de/public/diss-stilz/diss.pdf

Surlykke, A., & Moss, C. F. (2000). Echolocation behavior of big brown bats, *Eptesicus fuscus*, in the field and the laboratory. *Journal of the Acoustical Society of America*, 108, 2419–2429.

Surlykke, A., Ghose, K., & Moss, C. F. (2009). Acoustic scanning of natural scenes by echolocation in the big brown bat, *Eptesicus fuscus. Journal of Experimental Biology*, 212, 1011–1020.

Ulanovsky, N., Fenton, M. B., Tsoar, A., & Korine, C. (2004). Dynamics of jamming avoidance in echolocating bats. *Proceedings of the Royal Society of London B: Biological Sciences*, 271, 1467–1475.

Urick, R. J. (1975). Rician distribution and amplitude fluctuation of transmitted signals in sea. *Journal of the Acoustical Society of America*, 57, S59–860.

Urick, R. J. (1983). *Principles of underwater sound*. New York: McGraw-Hill.

Vanderelst, D., De Mey, F., Peremans, H., Geipel, I., Kalko, E. K. V., & Firzlaff, U. (2010). What noseleaves do for FM bats depends on their degree of sensorial specialization. *Public Library of Science: One*, 5(8), e11893.

Vel'min, V. A., & Dubrovskiy, N. A. (1976). The critical interval of active hearing in dolphins. *Soviet Physics: Acoustics*, 2, 351–352.

von der Emde, G., & Menne, D. (1989). Discrimination of insect wingbeat frequencies by the bat *Rhinolophus ferrumequinum. Journal of Comparative Physiology A: Sensory Neural and Behavioral Physiology*, 164(5), 663–671.

von der Emde, G., & Schnitzler, H. U. (1990). Classification of insects by echolocating greater horseshoe bats. *Journal of Comparative Physiology A- Sensory Neural and Behavioral Physiology*, 167(3), 423–430.

Xitco, M. J., Jr., & Roitblat, H. L. (1996). Object recognition through eavesdropping: Passive echolocation in bottlenose dolphins (*Tursiops truncatus*), *Animal Learning & Behavior* 24, 355–365.

Yovel, Y., Franz, M. O., Stilz, P., & Schnitzler, H. U. (2008). Plant classification from bat-like echolocation signals. *Public Library of Science: Computational Biology*, 4, e1000032.

Yovel, Y., Stilz, P., Franz, M. O., Boonman, A., & Schnitzler, H. U. (2009). What a plant sounds like: The statistics of vegetation echoes as received by echolocating bats. *Public Library of Science: Computational Biology*, 5, e1000429.

Yovel, Y., Franz, M.O., Stilz, P., & Schnitzler, H. U. (2011). Complex echo classification by echo-locating bats: A review. *Journal of Comparative Physiology A: Sensory Neural and Behavioral Physiology*, 197, 475–490.

Chapter 9
Echolocation in Air and Water

Peter Teglberg Madsen and Annemarie Surlykke

Keywords Bat • Biosonar • Buzz • Call • Chiroptera • Click • Echolocation • Odontocete • Toothed whale

9.1 Introduction

Echolocating animals listen for, and process, echoes from objects ensonified by self-emitted sound pulses to gather information from their environment. Many animals use echo information, and the capability to echolocate has been ascribed to members of several orders among both mammals and birds (Griffin 1958). However, of those, only species from three mammalian suborders, the laryngeal echolocating bats (Yangochiroptera and Yinpterochiroptera, Chiroptera) and the toothed whales (Odontoceti, Cetacea), can orient and locate, track, discriminate, and catch small moving prey objects in a three-dimensional world using just self-generated echo-information.

The independent evolution of echolocation as the primary sensory modality for navigation and foraging has, along with the capabilities to fly or perform long breath-hold dives, allowed bats and toothed whales to exploit dark foraging niches where there is little competition from other predators. This had led to a successful radiation of echolocating species adapting to a large range of different habitats from

P.T. Madsen (✉)
Zoophysiology, Department of Bioscience, Aarhus University,
CF Møllers Alle, 8000 Aarhus C, Denmark
e-mail: peter.madsen@biology.au.dk

A. Surlykke
Department of Biology, University of Southern Denmark,
Odense M 5230, Denmark
e-mail: ams@biology.sdu.dk

A. Surlykke et al. (eds.), *Biosonar*, Springer Handbook of Auditory Research 51,
DOI 10.1007/978-1-4614-9146-0_9, © Springer-Verlag New York 2014

gleaning prey from surfaces to locating high-altitude insect swarms in bats, and from detecting prey at mesopelagic depths to shallow rivers in toothed whales.

In fact, the independent evolution of biosonar in toothed whales and bats is so successful that almost one in four mammalian species echolocate—:around 1,100 species of echolocating bats and some 80 species of echolocating toothed whales. Bats probably evolved from small gliding insect-eating predators to echolocate prey on the wing more than 50 million years ago (Simmons et al. 2008; Veselka et al. 2010), and toothed whales evolved from artiodactyle ancestors to echolocate in water as part of a series of secondary adaptations to a life in water more than 30 million years ago (Nummela et al. 2004; Fordyce 2009; Fahlke et al. 2011). The vast differences in ancestry and properties of the media in which these two groups of mammals have evolved the same sensory means, to locate and catch prey using ultrasound under poor lighting conditions, present a fascinating area of research in comparative sensory physiology and behavioral ecology (Madsen and Surlykke 2013). Most echolocating animals that use sonar to home in on a prey item go through the three phases of search, approach, and capture as defined by Griffin et al. (1960), but within that general framework, there is considerable variation both within and between species of bats and toothed whales.

Since the discovery of echolocation in bats in the late 1930s (Griffin and Galambos 1941; Griffin 1958) and in toothed whales in the late 1950s (Kellogg et al. 1953; Norris et al. 1961), there has been considerable basic and applied research into understanding the operation of biosonar systems in air and water. As with any research on complex systems, these studies have often taken a reductionist approach to uncover the function and performance of biosonar systems: how sounds are produced and radiated, how the sounds propagate and reflect off targets, and how the echoes are detected and processed in the auditory systems to guide motor patterns for prey interception. That has led to a firm understanding of how echolocating animals produce, receive, and process ultrasound. Though these important studies are fundamental for understanding how biosonar systems operate, it is nevertheless unlikely that the combined knowledge from such dedicated laboratory studies provides the full picture of how free-ranging animals use echolocation for navigation and foraging in the habitats in which this sense evolved. A natural environment will often comprise an extremely complex soundscape compared to the controlled, simple surroundings of a lab setting. Besides, stimuli in the lab are often simplified or isolated to control one experimental variable at a time, such as range, target strength, frequency, delay, noise, clutter, or intensity, whereas the number, properties, and behavior of live echoic targets in the wild are unpredictable, and vary along all acoustic parameter axes. Also, signals and echoes in the wild may be buried in clutter, background noise, or even similar signals and echoes from other echolocators. Furthermore, laboratory studies are often focused on testing a certain aspect of biosonar operation, thereby driving the animal to optimize its payoff matrix by focusing exclusively on solving the defined task at hand, while ignoring other parameters it would have to negotiate in the wild.

For example: Do bats flying at 3 m s^{-1} in dense clutter toward a prey that they have less than 500 ms to detect, track, and capture employ the same discrimination and ranging resolution as stationed bats with much more time on a platform in an anechoic lab chamber? Will a breath-holding toothed whale at 1,000 m depth dedicate the same amount of clicks to a moving elusive target as when stationed at 1 m depth in a fixed target detection experiment? To get at these questions and obtain a fuller picture of how echolocating animals function and evolved to use ultrasound for navigation and foraging, it is important to study animals in their natural habitat and design lab experiments on the basis of realistic conditions encountered by the animals in the wild (Madsen and Surlykke 2013).

An increasing number of techniques and new technologies have emerged with more and more power to study how echolocating animals operate their sonar systems in the wild or under more natural conditions in the lab. Developments, especially over the last 10 years have provided the advent of small electronic recording devices that are easy to bring to the field and even in some cases to deploy on free-ranging animals in the form of archival tags (see Johnson, Chap. 7). These developments have provided us with an unprecedented insight into the sensory and behavioral ecology of echolocating animals in the wild. This chapter is about that: Given the physical properties of their environment, how do echolocating bats and toothed whales use ultrasound to guide motor patterns during detection, approach, and capture of prey while negotiating the challenges of noise, clutter, eavesdropping, predation, and prey avoidance?

Previous reviews by Au (1993, 2004) and Au and Simmons (2007) compare echolocation in bats and toothed whales with a focus on captive studies. The aim of this chapter is to complement such previous reviews and a recent one (Madsen and Surlykke 2013) with a comparative overview of what has been learned from the new field techniques developed over recent years on how bats and toothed whales use their sonars in the wild to catch live prey in the two very different media of air and water. This aim is pursued by (1) providing an introduction to the physical framework of biosonar operation in air and water; (2) briefly introducing recent developments in field techniques; (3) outlining the general patterns of how bats and toothed whales use sound to echolocate with examples of the constraints from their natural habitats, and how comparable challenges are solved in different media; (4) providing an overview of acoustic predator–prey interactions of bat and toothed whales and their food; and (5) comparing biosonar operation in air and water with a discussion of functional divergence and convergence in an evolutionary perspective.

9.2 The Physical Framework of Operating Biosonars in Air and Water

9.2.1 Background

Echolocating animals actively update their auditory scene through discrete acoustic sampling at a rate given, at best, by the pulse interval of their sonar emissions. If they emit pulses too fast they may emit a new biosonar pulse before echoes generated from the previous emission have arrived, which will lead to range ambiguity problems and spatial aliasing. On the other hand, if they use too long pulse intervals with respect to the speed at which they or their prey move, they may miss potential prey targets or fail to provide frequent enough feedback to avoid obstacles or track targets in time and space. The longer the detection range of the biosonar system and the slower the sound speed, the longer the pulse intervals must be to avoid range ambiguity. Hence, because of the roughly 4.5 times slower sound speed in air compared to water, bats must wait about 4.5 times longer for the echo to return compared to a whale ensonifying a target at the same range (Madsen and Surlykke 2013). On the other hand, the detection ranges of toothed whale sonars are considerably more than five times longer than those of bats, resulting in potentially much longer two-way travel times to the targets. So overall, the pulse emission rates of bats and whales are fairly similar, on the order of 2–40 Hz during the search phase. Thus, in general for biosonar systems, the pulse emission rate determines the update rate of the control system that informs changes in motor patterns to track or avoid targets (Madsen and Surlykke 2013).

The detection range of a sonar system for a single target is determined by the sound level emitted from the source, the transmission loss to and from the target, the backscattering properties of the target, and the echo to noise or clutter ratio required by the auditory system to detect the echo. This process is expressed in the sonar equation (Møhl 1988; Au 1993; Urick 1995), which in a simple form for a point target may be written on a decibel scale as:

$$\mathrm{EL} = \mathrm{SL} - 2 \times \mathrm{TL} + \mathrm{TS}$$

where EL is the echo level received back at the auditory system of the echolocating animal; SL is the source level, defined as the sound level at a standard range ahead of the animal on the acoustic axis; TL is the transmission loss between the source and the target; and TS is the target strength, defined as the ratio between the sound level impinging on the target and the echo level on the same axis at a standard range back toward the source (Fig. 9.1). Detection happens when the EL on a statistical basis is larger than the noise level in the receiving part of the sonar system; this noise level may be set by internal noise of the auditory system (the hearing threshold) or, if strong enough, by masking noise or clutter (echoes from objects close to the target) from the environment. The reference distances for the sonar equations when dealing with echolocating animals are 10 cm for bats and 1 m for toothed whales, which should be kept in mind when comparing sonar operation in these two groups (Fig. 9.1).

Though biosonars in air and water have the same objective, namely to process an actively generated auditory scene to navigate and detect, classify, and track prey items for capture using echo information, the two media offer very different conditions for the formation, propagation, and reflection of ultrasound. The densities and sound speeds in water and air are very different and will hence pose different challenges and advantages that echolocating animals must negotiate or exploit. Water is about 830 times denser than air and has a sound speed of around 1,500 m s^{-1} compared to a sound speed of around 340 m s^{-1} in air. In the following Sect. 9.2.2 we explore the consequences for each of the elements of the sonar equation of using ultrasound in air and water.

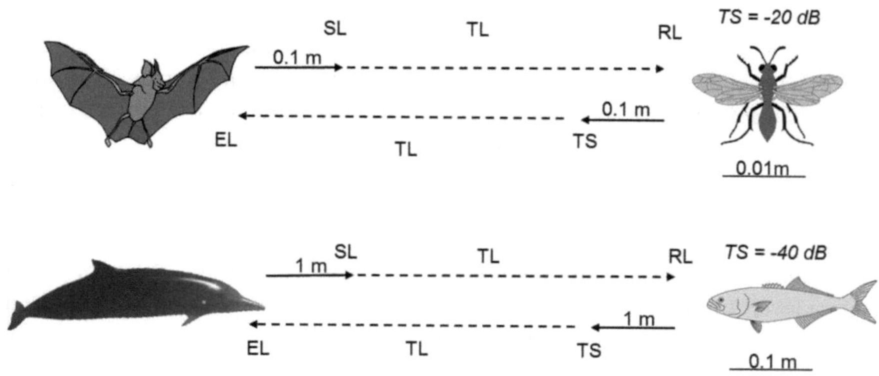

Fig. 9.1 Graphic display of the parameters of the active sonar equation for a bat and a toothed whale. *EL* echo level, *RL* received level, *SL* source level, *TL* transmission loss, *TS* target strength. Note the relatively high TS for the small insect prey compared to the larger fish prey, when taking the different reference ranges into account

9.2.2 Source Levels and Acoustic Outputs

The acoustic resistance or impedance of a medium is given by the product of sound speed and density. The impedance in an acoustic free field is a measure of the relationship between the sound pressure ($Pa = N\ m^{-2}$) and the particle velocity of the medium. In an acoustic free field, the impedance is about 3,700 times higher in water compared to air, which means that an animal in water can generate a pressure that is 3,700 times higher than in air for the same energy and sound production efficiency. The impedance difference also means that the sound pressure limit, at which there is diminishing pressure change for increasing energy input, will be much lower in air compared to water. The impedance differences thus mean that it is relatively easy to generate high sound pressures in water, and rather difficult in air.

Echolocating animals must produce high source levels to be able to generate detectable echoes from small targets at ranges great enough to, ultimately, allow them to find and acquire enough food for survival. Source levels have been estimated only for a few bat species in the wild so far. The available data suggest that most bat species can emit at least around 110 dB re 20 μPa (pp) at 10 cm (Brinkløv et al. 2009; Schuchmann and Siemers 2010), and the highest source sound pressure levels reported for bats may reach more than 140 dB re 20 μPa (pp) at 10 cm (Surlykke and Kalko 2008). Free-ranging toothed whales normally generate click source levels between 180 and 226 dB re 1 μPa (pp) at 1 m (Au 1993; Madsen and Wahlberg 2007), and in the case of the sperm whale (*Physeter macrocephalus*) up to 240 dB re 1 μPa (pp) at 1 m (Møhl et al. 2003).

In the following we compare a bat making a 2-ms call with a source level of 138 dB re 20 μPa (pp) at 0.1 m with a dolphin making a 50-μs click with a source level of 226 dB re 1 μPa (pp) at 1 m. At first glance, it would seem that there is an 88 dB (25,000 times) difference between these two source levels, but there are, however, several issues that make such direct comparison erroneous. For starters, the reference distances and the reference pressures are different, so the sound pressure of a bat emitting a source level of 138 dB re 20 μPa (pp) at 10 cm is 20 dB lower at the 1 m reference distance of toothed whales. In combination with a 20 times (26 dB) higher reference unit in air (20 vs. 1 μPa), the bat source level of 138 dB re 20 μPa (pp) at 0.1 m will, on a linear scale, be 16 Pa (pp) at 1 m, which can then be compared to 200,000 Pa (pp) at 1 m for a dolphin emitting a source level of 226 dB re 1 μPa (pp) at 1 m.

So the peak-to-peak (pp) pressure ratio is rather 12,500 (82 dB) between the source levels of these two biosonar systems in air and water. This very large difference in source sound pressures is, however, not a meaningful parameter for comparison in the light of how sounds are detected in mammalian auditory systems. The ears of both bats and toothed whales operate as energy detectors that integrate sound intensity (given by the pressure squared divided by the impedance) over time up to a certain maximum integration time. Therefore, the most meaningful way of comparing source levels for sonar operation in bats and toothed whales is by using the energy of the sonar pulses and not the pressure, despite that most biosonar researchers report peak-to-peak pressure; likely because it is easier to measure.

The energy of a sound pulse increases linearly with the duration and with the square of the pressure. Thus, to quadruple the energy of a sound pulse, an animal must either double the sound pressure or make the pulse four times longer. Given the relative ease with which high sound pressures can be generated in water, echolocating toothed whales can generate high energy source levels (J m^{-2}) even when producing very short clicks (Fig. 9.2). The energy source level of a 50-μs dolphin click with a source pressure level of 226 dB re 1 μPa (pp) at 1 m in water will be around 0.04 J m^{-2} (–14 dB re 1 J m^{-2}) (Fig. 9.2b). If a bat was to echolocate with a 50 μs long cry with a source level of 138 dB re 20 μPa (pp) at 10 cm in air, the energy source level at 1 m would be around 1.33×10^{-6} J m^{-2} (–59 dB re 1 J m^{-2}), which is about 30,000 times less energy per m^2 on axis than the toothed whale. The most powerful echolocating bats are close to the limit for sound pressure generation in air, but they emit much longer signals than toothed whales, hence dramatically increasing the energy of the pulse for the same pressure (Fig. 9.2a). For example, the 2 ms long search signal of a bat considered in Fig. 9.2a with a source level of 138 dB re 20 μPa (pp) at 10 cm (=118 dB re 20 μPa pp at 1 m) will have an root mean square (rms) source level over its duration of 104 dB re 20 μPa rms at 1 m. Integration of that rms pressure squared over the duration and divided by the impedance in air gives an energy source level at 1 m of 5.45×10^{-5} J m^{-2}, which is the expected 40 times higher than if emitting a 50-μs transient, and now only about 750 times lower in energy than the 50-μs click of the toothed whale (Fig. 9.2) (see also Wahlberg & Surlykke, Chap. 4, Table 4.2). Clearly, such ratios in energy source levels critically hinge on the chosen examples of bat and toothed whale signals, but it is generally a good rule of thumb to consider that the energy ratios between bat and toothed whales biosonar pulses are between two and three orders of magnitude.

Hence it follows that bats theoretically could reach the same energy source levels as toothed whales if they made their signals 1,000 times longer. However, FM bats,

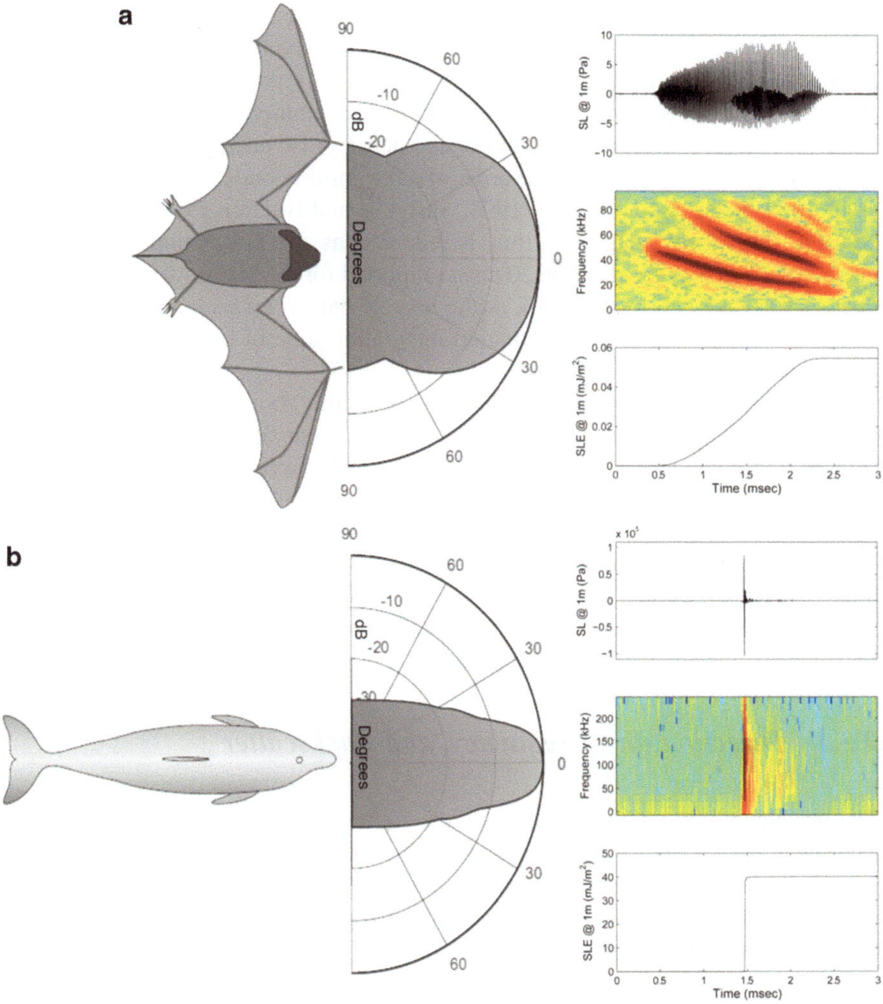

Fig. 9.2 (**a**) Sound radiation from a bat (*Myotis*) producing a sound pulse with a directionality index (DI) of 16 dB along with the waveform (assumed SL of 138 dB re 20 μPa at 0.1 m, pp), spectrogram, and energy accumulation of the emitted call. (**b**) Sound radiation from a bottlenose dolphin (*Tursiops*) with a DI of 26 dB along with the waveform (assumed SL of 226 dB re 1 μPa at 1 m, pp), spectrogram, and energy accumulation of the emitted click. Note that SLs are normalized to the same reference distance of 1 m. (Modified from Madsen and Surlykke 2013)

emitting frequency-modulated signals, cannot just keep increasing the duration of their calls. They generally avoid emitting a new pulse before the echo from the previous cry has returned, and they also try to end the emission before the echo of that pulse returns (Schnitzler and Kalko 2001; Wilson and Moss 2004; Moss et al. 2006). The only exceptions to this rule are the CF bats emitting long calls (up to 80 ms) of constant frequency (CF) and with a high duty cycle (% call emission time out of total time) generating considerable overlap between emitted calls and returning echoes (Fenton et al. 2012). For such CF bats, it may be their acute frequency discrimination abilities that help them to solve ambiguity problems (Kössl and Vater 1995). However, based on current knowledge, detectability of echoes will not keep increasing linearly with pulse duration, and hence pulse energy, owing to the relatively short integration time on the order of a few milliseconds in bat ears (Simmons et al. 1989; Schmidt and Thaller 1994; Surlykke and Bojesen 1996). It remains to be seen if this 2-ms integration time is indeed constant. For example, in FM bats pulse durations may reach some 20 ms, as reported for the big brown bat (*Eptesicus fuscus*: Surlykke and Moss 2000), or the noctule bat (*Nyctalus noctula*: Miller and Degn 1981), and it would seem that an integration time of only 2 ms would be a mismatch to such long signals.

Thus, on an absolute scale, echolocating bats emit about two to three orders of magnitude less energy per unit of area on-axis per sonar signal than toothed whales (Fig. 9.2), but about one to two orders of magnitude more of total acoustic energy per unit of body mass (3–100 g for bats and 0.04–70 tons for toothed whales) when correcting for the different directionalities and body weights. However, the physical conditions for propagation of sound in air and water are also quite different, as outlined in the next section, so even *if* bats could emit as much total energy as toothed whales their sonar operation range would not be anything close to that of the whales.

9.2.3 Directionality, Frequency, and Backscatter

The wavelength of a sound is given by the sound speed divided by the frequency. The spectral and hence wavelength composition of a sound pulse will affect how it is radiated from a source, reflected off a target, attenuated through the medium, and to what degree it is masked by ambient noise. Despite their vast size differences and the almost five times longer wavelengths in water compared to air for the same frequency, bats and toothed whales produce sonar signals in a surprisingly similar frequency range between 10 and 200 kHz. That then begs the question of why have bats and toothed whales converged on using ultrasound in this frequency range for echolocation?

For a sonar system to be effective in detecting a small target, it must (1) produce relatively high source levels, (2) reduce masking noise and unwanted echoes in the form of clutter that can interfere with detection of faint echoes, (3) maximize the proportion of energy reflected off the target, and (4) keep the transmission loss low (Fig. 9.1). High source levels can, as outlined in the previous section, be achieved by putting more energy into the sound pulses. However, biosonar systems do not draw from an infinite power source and the key to generating high source levels given limited power is to make them directional. High directionality by which the sound is emitted in a forward directed beam also reduces clutter by only ensonifying a limited area in the general direction that echolocating animal is pointing its head (Figs. 9.2 and 9.8). The conventional measures of directionality of sonars are the half power beam width (–3 dB) and the directionality (or directivity) index (DI), while the half amplitude (–6 dB) beam measure has been used for bats, rendering comparison between bats and toothed whales difficult when using older literature. The half power beam gives the width of the sound beam in degrees where it has dropped to half the power (–3 dB) compared to the level on the acoustic axis. The DI gives the decibel difference between the source level of a directional source and the source level of an omnidirectional source radiating the same power (Fig. 9.2).

The sound radiation from bats (Strother and Mogus 1970; Mogensen and Møhl 1979; Surlykke et al. 2009) and toothed whales (Au et al. 1986, 1995; Beedholm and Møhl 2006) have both been modeled as flat pistons oscillating in an infinite baffle, not because bat and toothed whale sound emission structures indeed are flat pistons, but because of the relative ease with which flat piston models can be applied to explain sound radiation patterns from toothed whales and bats reasonably well (Strother and Mogus 1970; Au 1993; Madsen and Wahlberg 2007). The directionality of a flat piston goes up with the number of wavelengths across its transmitting aperture as given by the wave number k times the equivalent aperture radius a: $ka = 2\pi a/\lambda$, where λ is the wavelength. The DI for a flat piston can subsequently be estimated from $\mathrm{DI} = 20\log_{10}(ka)$ and the half power beam width can be estimated by $185°\ ka^{-1}$ (Zimmer et al. 2005). Toothed whales have measured DIs between 22 and 32 dB (Au 1993), meaning that there are some 3–12 wavelengths across their effective transmitting apertures. For example, for the bottlenose dolphin, the average DI is around 26 dB for high source level clicks (Au 1993), which with a centroid frequency of 100 kHz gives an equivalent circular transmitting aperture of around 10 cm in diameter (Fig. 9.2b). So the larger the effective transmitting structure, the lower the frequency the animal can use and still maintain a high directionality, and there is indeed a strong negative correlation between echolocation frequency and size in both toothed whales and bats (Barclay and Brigham 1991; Au et al. 1999; Jones 1999; Madsen et al. 2002b). That also means that echolocating animals can change their directionality by changing either the size of the effective transmitting aperture or the frequency of the transmitted sound. A doubling in either frequency or effective aperture will lead to a DI that is 6 dB higher all other things equal. Both toothed whales and bats change their directionality by changing the frequency (Au et al. 1995; Madsen et al. 2004; Jakobsen and Surlykke 2010; Jakobsen et al. 2013). Bats emitting signals through the mouth may also change the directionality of their calls by adjusting the size of their mouth opening. Daubenton's bat (*Myotis daubentonii*) changes directionality from a DI of some 11 dB in the lab to 16 dB in the field at 55 kHz (Fig. 9.2a), probably simply by opening the mouth wider when making more powerful calls in the field (Surlykke et al. 2009). The anatomical basis for large aperture changes in toothed whales is less evident, but there are some indications that toothed whales can change both the direction and width of their sonar beams for pulses of the same centroid frequency (Moore et al. 2008; Au et al. 2009; Madsen et al. 2010), but that intriguing notion calls for further testing.

Bats are much smaller than toothed whales, and correspondingly their mouths or nose leaves in general provide equivalent transmitting apertures in the order of a few centimeters or less in diameter (Mogensen and Møhl 1979). However, the sound speed in air of around 340 m s^{-1} means that the wavelength is about five times shorter in air for the same frequency. This compensates for some of the difference in aperture, such that the *ka* and hence DI of a bat with a 1 cm diameter of the equivalent aperture of the emitter is only 10–16 dB lower than those of a bottlenose dolphin with a 10 cm diameter of the equivalent aperture of the emitter for the same frequency (Fig. 9.2). However, to achieve the directionality of a toothed whale, a bat would nevertheless have to echolocate at frequencies between 200 and 300 kHz, which very few if any species do. Rather, most bats operate their sonars in the same frequency range as toothed whales from 10 to 200 kHz, and most species between 20 and 100 kHz, depending on their size. That means that the directionalities of bat sonar beams in general are between 2 and 20 times lower than those of toothed whales (Fig. 9.2), and the estimated DIs of bats fall between 6 and 18 dB (Brinkløv et al. 2011; Jakobsen et al. 2013), meaning that there are some one to two wavelengths across their effective transmitting apertures. In other words; the beam widths or the acoustic fields of view used by bats, relative to whales, are 2–20 times wider. However, within each group there is considerable convergence on a relatively narrow range of DI values that seems to drive the evolution of biosonar frequencies so that smaller animals use higher frequencies than larger species to achieve the same directionality (Jakobsen et al. 2013; Madsen and Surlykke 2013). Nevertheless, directionality is not the only evolutionary driving force on the frequencies of sonar pulses emitted by echolocators. Target strength and transmission loss (Fig. 9.1) also play important roles, as discussed below.

Target strength is defined as the difference between the energy level that ensonifies the target and the energy level of the echo reflected off the target one reference distance back towards the source (Fig. 9.1). Target strength increases with size of the target and with the impedance difference with respect to the medium in which the sound wave propagates. The body tissues of fish and squid have densities and sound speeds close to that of water and hence displays small impedance differences with respect to water. That means that the typical prey of toothed whales have low target strengths for their size, typically between –30 and –50 dB re 1 m at toothed whale click frequencies (Benoit-Bird and Au 2001; Au et al. 2007; Madsen et al. 2007; Au and Benoit-Bird 2008). Many bony fishes have air-filled swim bladders that greatly increase the impedance difference and thus their target strength by 5–10 dB (Foote 1980), making them easier to detect than similar sized fish with no swim bladder (Fig. 9.1). Bats, on the other hand, echolocate in air, where all targets (insects, vertebrates, plants, ground) have large impedance differences compared to air. Insects have target strengths of –10 to –40 dB re 10 cm (equal to –30 to –60 dB re 1 m) at bat call frequencies (Møhl 1988; Surlykke et al. 1999; Jung et al. 2007). So even though bat prey are one to two orders of magnitude smaller than the typical prey of toothed whales, they have comparable target strengths (Fig. 9.1).

The target strength is dependent of the wavelength of the sound that ensonifies the target. If the wavelength is too long compared to the target, the proportion of reflected sound energy will be very low owing to so-called Raleigh scatter. To be in the so-called geometric scatter zone where there is effective backscatter and where the target strength is relatively independent of the frequency, the wavelengths used for ensonification of solid targets must be shorter than the effective circumference of the target. Fig. 9.3a and b displays the minimum effective target diameters for

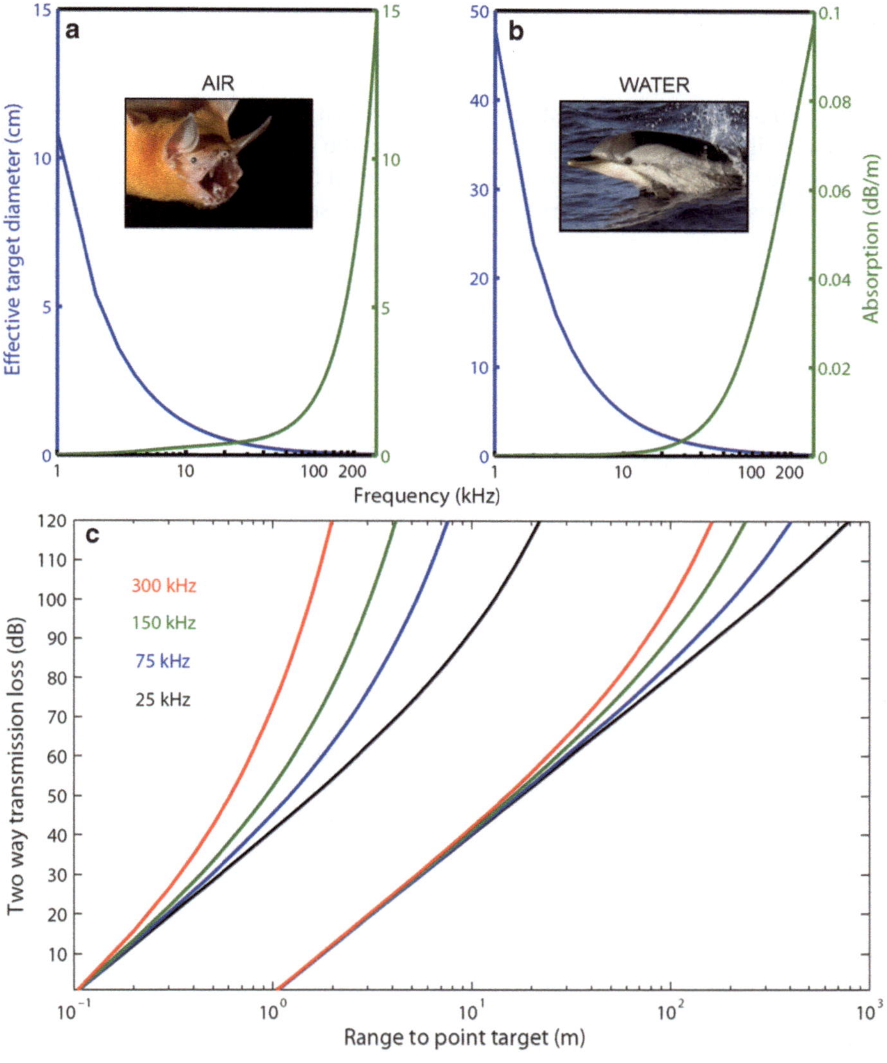

Fig. 9.3 Target size that generate geometric backscatter (*blue*, left *y*-axis) and absorption (*green*, right *y*-axis) as a function of frequency in air (**a**) and water (**b**). Note how the frequency span between 10 and 150 kHz offers a window of geometric backscatter and relatively low absorption for both bats and toothed whales, although the absorption per meter is much higher in air for a given frequency. Absorption in air is calculated in accordance with ISO 9613 at 25 °C and 60 % rel. humidity. Absorption in water is calculated at 10 m depth, 3.5 % salinity, and 25 °C. (**c**) The two-way transmission loss from spherical spreading and absorption as a function of frequency for reference distances of 0.1 m (bats) and 1 m (toothed whales). (Modified from Madsen and Surlykke 2013.) Note how the transmission loss increased dramatically with frequency in air, even for short ranges

which a target will enter the geometric scatter zone as a function of frequency in air and water. Toothed whales generally search for prey with effective diameters of more than 5 cm, whereas the prey of most bats will have effective diameters in the order of 0.5–4 cm. In general it therefore seems that both bats and toothed whales use sonar signals at frequencies that are well into the geometric scatter zone of their typical prey targets. It has been suggested that small bat species use high frequency sonar pulses to obtain geometric scatter from their small prey (Pye 1993), but this notion is hard to decouple from the fact that small bats, targeting small insects, have smaller vocal cords and vocal tracts and hence produce high frequencies for biophysical reasons (Jones 1999). Higher frequencies result in higher directionality for the same transmitter size, and if a certain DI is required to obtain sufficient clutter reduction to operate biosonars, higher frequencies in smaller animals may in part be driven by a need for a certain directionality (Koblitz et al. 2012; Jakobsen et al. 2013; Madsen and Surlykke 2013). Most likely, the correlation between size and emitted frequency in bats is due to the combined effect of scaling with body size, frequency-dependent backscatter of their small prey, and keeping up the directionality of their sonar beams.

The spectral content of biosonar signals affects not only radiation from the head and reflection off targets, but also the amount of information that can be encoded in the echo. Echolocating animals likely use spectral information to discriminate between targets, and to do that they must use wavelengths short enough to generate interference patterns from reflections off thinly spaced tissue barriers within the body of their ensonified prey. Accordingly, if a whale cannot produce or hear high-frequency energy, its discrimination capabilities deteriorate (Kloepper et al. 2010). Toothed whales with normal hearing abilities have acute discrimination capabilities (Au 1993), and seem to employ fine scale echo-analysis in a selective foraging scheme where many potential prey items are ensonified, but not selected for capture attempts (Madsen et al. 2005a; Au et al. 2009).

To the contrary, though bats can discriminate minuscule differences between targets in the lab (Simmons et al. 1974; Genzel and Wiegrebe 2008; Ulanovsky and Moss 2008), they generally seem to hunt more opportunistically in the wild, and active prey discrimination has not yet to our knowledge been demonstrated unequivocally (Barclay and Brigham 1994; Houston et al. 2004; Safi and Siemers 2010). However, gleaning bats and bats that hunt in clutter close to vegetation must be able to discriminate between the prey and clutter echoes from the background (Moss et al. 2006). The gleaning phyllostomid bat *Micronycteris microtis*, for example, can discriminate motionless insects sitting on leaves against empty leaves (Geipel et al. 2013). Owing to the short delay difference between prey and leaf echoes, the echoes will overlap and cannot be processed separately. Thus, this bat species likely provides an example of target discrimination by spectral cues in the wild, emphasizing the need for high bandwidth.

The vast number of bat species and their wide range of habitats and food are reflected in the large diversity in call types, where some calls are adapted for detection in the open (in general longer, narrow band calls at lower frequencies), whereas bats hunting closer to clutter may employ different strategies to discriminate between insect prey and background clutter (by emitting very short, broad band calls) (Fenton, Jensen, Kalko, & Tyack, Chap. 2; Schnitzler and Kalko 2001). A broad bandwidth can be achieved by a broad sweep like that of, for example, *Myotis nattereri* (Siemers and Schnitzler 2000, 2004) or by emitting a multiharmonic call (e.g., *Carollia perspicillata*: Brinkløv et al. 2011). However, other bats may employ a completely different strategy fighting clutter by emitting long CF calls and discriminating prey against a more stationary background by the Doppler shifts from prey wing movements (Schnitzler and Henson 1980; Fenton et al. 2012).

Thus, bats and all toothed whales use short wavelengths at ultrasonic frequencies to make their sound beams directional (to decrease clutter levels and produce higher source levels), to be in the geometric scatter zone of their prey targets, and also to extract spectral information about the ensonified targets. However, as outlined earlier in this section, bats use the same frequency range for echolocation as toothed whales, which is surprisingly low given their much smaller sizes of transmitting apertures and prey. That leads us back to the initial question: Why do bats not use higher frequencies for echolocation? The short answer is likely: absorption (Griffin 1971) as discussed in Sect. 9.2.4.

9.2.4 Transmission Loss and Masking Noise

Transmission loss, TL (dB re 1 reference distance), for sound propagation over a range, R, in an acoustic free field is the sum of geometrical spreading loss (20 log(R)) and absorption loss (αR) in length units of reference range (R):

$$\mathrm{TL} = 20\log(R) + \alpha R$$

Geometrical spreading is due to distribution of sound energy over an ever increasing surface area with distance away from the source, and that spreading model generally holds for the distances relevant for biosonars. That implies that the sound pressure is halved for every doubling in range, and in the case of a sonar where the sound has to travel out to a point target and back again (Fig. 9.1), the pressure level will go down by four times, and the energy by 16 times, for every doubling of target range (a decrease of 12 dB in pressure and energy). On top of that comes absorption, α, expressing the frequency dependent energy loss that, for a given frequency, is directly proportional to the distance traveled in the medium. The absorption per distance goes up with frequency in both air and water, but by about two orders of magnitude more so in air. In Fig. 9.3a and b the transmission loss from absorption as a function of frequency in air and water have been plotted, along with the total two way transmission loss as a function of range for four different frequencies in Fig. 9.3c. The absorption increases dramatically with frequency in air and much less so in water. Bats can therefore not effectively use frequencies that would provide them with the same directionality as toothed whales, that is, around 300 kHz. At 300 kHz the total (spherical + absorption) atmospheric two-way transmission loss to a point target would be 120 dB at a short target range of 2 m (Fig. 9.3c). Such a transmission loss is together with the TS of a normal prey item larger than the dynamic range between the SL and the hearing threshold, hampering the use of such high-frequency echolocation in foraging significantly. Thus, the high absorption in air at high frequencies, apparently forces bats to operate their sonars at relatively low frequencies with broad beams (Fig. 9.3) to maintain useful prey detection ranges, indicating that the few bats emitting around 200 kHz (Fenton and Bell 1981; Fenton, Jensen, Kalko, & Tyack, Chap. 2) operate with very short detection ranges.

Absorption in water at ultrasonic frequencies is much smaller than in air (Fig. 9.3a, b), which in combination with much higher source levels provide whales with target detection ranges that are 1–2 orders of magnitude larger than for bats, meaning toothed whales in general can echolocate their prey at ranges between 20 and 500 m compared to the 2–10 m of bats. Therefore, echolocating animals face a trade-off: They must use ultrasonic frequencies to generate directional sound pulses with high resolution and geometric backscatter off their prey targets, but on the other hand keep the transmission loss down to a level that allows them to detect prey at large enough ranges to find enough food to meet their energetic requirements. Physics therefore predicts that larger echolocating animals targeting large prey can use lower frequencies to achieve the same directionality, and relative spectral resolution and geometric backscatter of their prey, as small echolocating animals using higher frequencies to detect smaller prey. Such a size correlation seems indeed to be supported at least in general for both mouth-emitting bats and toothed whales, where the biggest echolocators in both media operate around 15 kHz and the smallest beyond 130 kHz.

Under very quiet conditions, the hearing threshold (i.e., neural noise in the auditory system) will set the detection threshold for detection of echoes, but in water in particular, the ambient noise will often limit the detection of echoes. For both air and water, the increasing absorption at high frequencies will overall generate pink noise characteristics (where the spectral noise reduces by 3 dB per octave) of the ambient noise profiles. In water, the thermal noise starts to dominate over the wind and wave generated noise between 30 and 140 kHz depending on sea state, which means that the spectral noise in deep oceanic water will reach a minimum between 30 and 130 kHz. That, in combination with the bandwidth of the detected echo and the receiving directionality of the auditory system (Au and Moore 1984), likely sets the lower bound on echo hearing threshold for toothed whales, as there will likely be no evolutionary driving force to increase the hearing sensitivity below those levels. Hence in general, the ambient noise falls off at higher frequencies, while the receiving directionality increases (Au and Moore 1984), which in combination provides lower masking noise levels in the direction of incoming echoes. However, ambient noise levels at ultrasonic frequencies may at times be much higher because of rain or biotic noise sources such as snapping shrimps that in tropical, shallow waters can raise the high-frequency noise levels by one or more orders of magnitude. Few studies have addressed how such elevated high-frequency ambient noise levels affect echolocation, but it seems that at least the beluga whales may adjust to increased ambient noise by producing higher source levels and higher frequencies (Au et al. 1985). In other habitats, such as the deep waters off El Hierro in the Canary islands (Johnson et al. 2006), the very low ambient noise levels means that it is likely the hearing threshold that sets the detection thresholds of echolocating beaked whales. One preliminary study suggests that toothed whales may in fact abort echolocation efforts to find food if the masking noise is too high (Aguilar de Soto et al. 2006).

In air, both biotic (calling insects and con-specific bats) as well as abiotic (waterfalls, running water) and anthropogenic noise may contribute to increased noise levels at bat echolocation frequencies and reduce the detection ranges of their sonars. The insect-generated noise may increase dramatically around sunset (Lang et al. 2006); calling of an individual insect may increase the chance of it being located by a bat using passive acoustics (Belwood and Morris 1987; ter Hofstede and Fullard 2008), but will when many insects are calling in unison make it more difficult for a bat to hunt due to masking. As a consequence, echolocating bats as well as bats relying also on passive listening to detect their prey have been shown to avoid areas with high noise levels, natural as well as anthropogenic (Gillam and McCracken 2007; Schaub et al. 2008). Wind noise when flying may also play an important but largely overlooked role, in particular for bats with large ears (Surlykke et al. 2013). Overall, little is known about how different ambient noise levels in the ultrasonic range affect echolocating animals and how, on an evolutionary scale, it may have played a role in determining the frequency range at which different echolocators emit their sonar pulses.

9.3 Methods for Studying Echolocation in the Wild

9.3.1 Historical Background

Griffins' discovery of echolocation in bats (Pierce and Griffin 1938; Griffin and Galambos 1941) was one of the most important breakthroughs in studies of bioacoustics, and impetus for much research on echolocation and animal behavior in general in the years to follow by Griffin himself as well as many others. Not only did Griffin discover echolocation, but he also uncovered many of the basic features of biosonar and the nomenclature and measures to describe and quantify them (Griffin 1958). Griffin's studies on active sonar operation in captive bats also inspired a large number of important studies on captive toothed whales, and set the stage for designing field studies to understand echolocation in the environment where biosonars evolved. A major leap forward to understand echolocation in the wild has been made possible by the recent miniaturization of electronics, allowing for ground breaking investigations of natural echolocation behavior in whales and bats.

9.3.2 Bats

Griffin emphasized the importance of studying animals in their natural environment, but was limited by the constraints of equipment that he, to a large extent, had to develop himself. Subsequent developments of handheld mobile bat detectors or heavy (stationary), high frequency tape recorders combined with careful observations of flight and capture behavior have greatly advanced our understanding of bat biology and the adaptation of echolocation call features to different acoustic niches and feeding behaviors (Griffin 1958; Neuweiler 1989; Fenton 1990; Schnitzler and Kalko 2001).

The recent development of small, high sampling rate digital recorders has allowed for high-quality, multichannel recordings of echolocation signals in the field using microphone arrays. The array technique is useful for recording sound in air, as the relatively slow speed of sound gives rise to quite large differences in arrival times of the same sound on different receivers even with an array of moderate dimensions (0.5–1 m between microphones), which allows for acoustic localization of the bat at each call emission (Surlykke et al. 1993; Jensen and Miller 1999) (Fig. 9.4). Another array technique involves two arrays of four microphones, with each microphone arranged in a symmetrical star, to reconstruct flight paths accurately (Holderied and von Helversen 2003). Infrared video techniques have also been applied to track bats in both two and three dimensions (Jones and Holderied 2007). When the bat's location relative to the microphones is known at the time of emission, the source level and directionality can be estimated for freely flying bats in the wild (Jensen and Miller 1999; Holderied and von Helversen 2003; Surlykke and Kalko 2008) to reveal adjustments of call features relative to distance to the target (the array in many cases), the echoic background, and prey behavior.

The range over which a target may be echolocated by bats has been estimated based on behavioral observations as well as on source levels compared to realistic target strengths of insects. All evidence suggest that the range of bat sonar is short, probably below 10 m for most species, and perhaps more typically 2–3 m (Fenton et al. 1998; Wahlberg & Surlykke, Chap. 4). Detection ranges for large objects as for example trees or the ground are larger (Jensen and Miller 1999; Jung et al. 2007), but details of the echoscape perceived by bats in the wild must await development of archival bat tags that can be used in the field (see Johnson, Chap. 7). Much effort is currently dedicated to the development of ultrasonic tags, but the small size of bats makes it a challenge to achieve sensitivity and tele-range and/or storage capability anywhere near what is presently possible with tags for toothed whales (Johnson et al. 2009; Johnson, Chap. 7). So far tele-mikes have been used with success only over short ranges in the lab (Hiryu et al. 2007, 2008; Johnson, Chap. 7), and the low transmission power of small tele-mikes will make them very difficult to use in the wild. Radio tracking of bats (Kalko et al. 1999; Weinberg et al. 2006; Holland and Wikelski 2009) has provided interesting results on habitat use and home range, but correlation with acoustic data on adaptive control and flexibility of

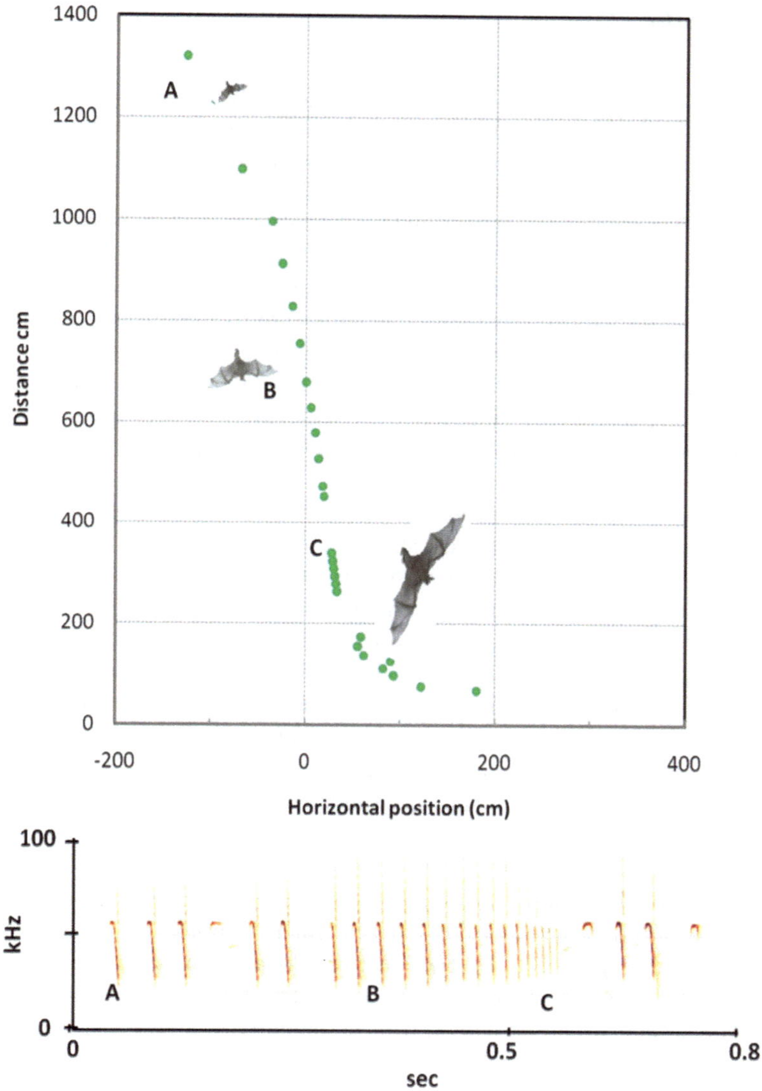

Fig. 9.4 Pursuit of prey by *Noctilio leporinus* recorded in the field. In the search phase (**a**) the call rate is slow (10–15 Hz) and the frequency-modulated part is occasionally omitted. After detection, which appears to happen 3–4 m from the prey, the approach phase (**b**) starts with gradual increase of call rate and decrease of call duration, until the terminal buzz phase (**c**) where the call rate is 150 Hz and call duration around 1 ms

echolocation has been made only by inference. Recently, another promising new technique to close the gap between the short range acoustic methods and the needed range is the thermocamera, allowing for long range observation of bats as well as their prey (Hristov et al. 2008).

9.3.3 Toothed Whales

Studies of toothed whales in the wild started by a series of qualitative studies of free-ranging toothed whales species where their click sounds were described from single hydrophone recordings, but with only a few attempts to quantify source. That changed in the late 1980s, when Møhl et al. (1990) used a vertical hydrophone array to show that free-ranging narwhals could produce source levels comparable to those reported from the long-range target detection experiment of Au et al. (1974) for bottlenose dolphins in open water pens. Since then, a number of studies have used hydrophone arrays with an ever increasing number of hydrophones to report on the source parameters of free ranging toothed whales (e.g., Møhl et al. 2003; Au and Würsig 2004; Madsen et al. 2004). Such source parameter data have on the basis of target detection experiments with captive animals (Au 1993) been used to assess the prey detection for free-ranging toothed whales (Au et al. 2004, 2007; Madsen et al. 2004, 2007).

Lab studies have provided a good understanding of how toothed whales use their sonars when echolocating on station for a fixed target in captivity (see Wahlberg & Surlykke, Chap. 4; Fenton, Jensen, Kalko, & Tyack, Chap. 2). The general picture from these studies is that the animals click gradually faster and with lower outputs when they are close to the target. However, there are several important aspects of biosonar operation that such studies do not address. An obvious one is: How do free ranging toothed whales operate their sonars when they search, approach, and capture live prey in a natural environment? Hydrophone arrays in the field may in part speak to that question (Madsen et al. 2002b; Wahlberg 2002), in particular if the acoustic recordings are supplemented by concomitant visual or video sampling of the behavior. However, owing to the very narrow sonar beam of toothed whales, it is for most species difficult to record echolocation clicks from animals engaged in foraging while ensuring that their sonar output is in fact not adjusted to the recording devices rather than their prey (but see Miller et al. 1995; Madsen et al. 2002b). Therefore echolocation in the wild should ideally be studied in a way where the acoustic output and the foraging behavior of the animal is recorded while the animal behave naturally in a way that is decoupled from the presence of recording gear in the water and researchers on nearby and often noisy platforms.

Over the last 10 years our ability to study the echolocation behavior of free-ranging toothed whales has improved dramatically through the advent of acoustic tags that to some extent meet the requirements of unperturbed power to see (Johnson, Chap. 7; Johnson et al. 2009). With acoustic tags it has become possible to sample the acoustic out- and inputs of echolocating whales along with their overall behavior as measured by depth and inertial sensors of the tags (Madsen et al. 2002a; Johnson and Tyack 2003; Akamatsu et al. 2005; Johnson et al. 2009). The following section provides an overview of the current knowledge of echolocation in the wild of both bats and toothed whales as uncovered over the last 10 years.

9.4 Echolocation in the Wild

The large number of echolocating species show adaptations to a diverse range of foraging niches in which echolocation may be used to navigate and find food. Most of our present knowledge stems from a limited number of studies on relatively few species in the wild. It is therefore inherently difficult to outline how *A* bat and *A* toothed whale echolocate in the wild. It is nevertheless tempting to generalize if the limited data at hand suggest the same pattern of acoustic behavior, but such generalizations should be made with caution. The following is an attempt to try to draw some general inferences that can form the basis for a comparison in the last section of this chapter.

9.4.1 Bat Echolocation in the Wild

There are a vast number of field studies of bats. However, most of these report only temporal and spectral features of bat echolocation in unknown behavioral contexts. Only recently have microphone array and video techniques enabled estimation of the emitted intensity and directionality, which are also acoustic features that are dynamically adapted to habitat and behavioral context by bats in the wild. In spite of the technological problems with bulky heavy equipment with limited bandwidth, dynamic range, and recording channels, Griffin pioneered studies of echolocation behavior of bats in their natural habitats (Griffin 1958). He described the typical temporal pattern of calls when aerial insectivorous bats hunt for prey on the wing in a natural situation. He divided the pursuit into three task-related phases coined "search," "approach," and "terminal" or "buzz" phase (Fig. 9.4). The search phase is characterized by relatively long narrowband signals well suited for detection of prey, the approach phase by shortening of the call durations but with increasing band width to facilitate localization, and the buzz phase by very short calls at high repetition rates of up to 200 calls per second just before capture (Figs. 9.4 and 9.5) (Griffin 1953, 1958; Kalko 1995). The same general temporal pattern is found in trawling bats hunting for prey on or above water surfaces, a hunting strategy that has evolved in several distantly related bat families (*Vespertilinoidae*, *Noctilionidae*, *Phyllostomidae*, *Emballonuridae*) emitting quite diverse echolocation signals (Fenton, Jensen, Kalko, & Tyack, Chap. 2) (Fig. 9.5).

Temporal features of search phase calls scale with size such that larger bats emit longer calls with longer intervals in accordance with their higher flight speed and slower wing beat rate. However, the buzz phase shows only little scaling with bat size, with almost identical high repetition rates (180–200 calls per second) and short call durations (approximately 0.5 ms) across families and species of trawling and aerial insect hunting bats (Fig. 9.5). The ubiquitous nature of these temporal changes suggests a high functional significance. Gleaning bats and frugivores taking stationary prey, on the other hand, in general do not produce buzzes (Ratcliffe et al. 2013),

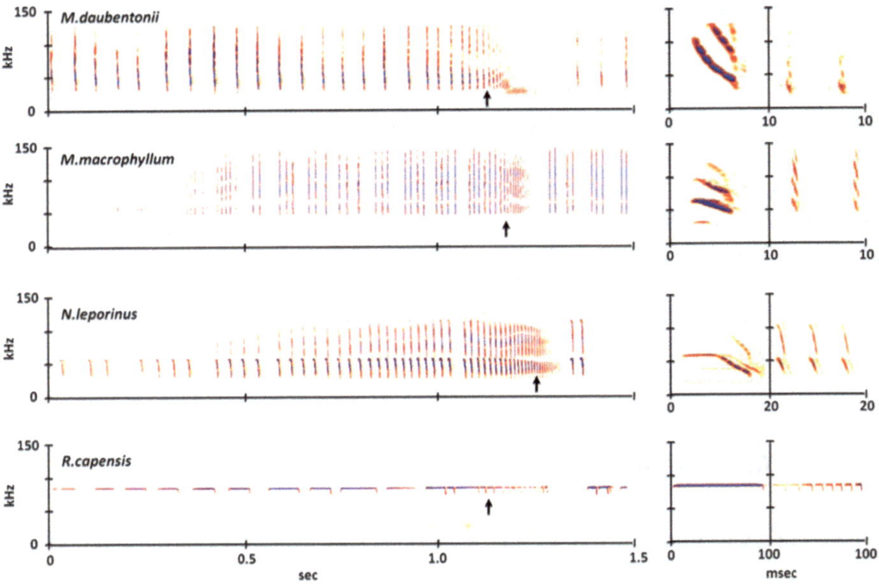

Fig. 9.5 Spectrograms of acoustic pursuit. Acoustic pursuit sequences for four insect eating bats *Myotis daubentonii* (Vespertilio-nidae), *Macrophyllum macrophyllum* (Phyllostomidae), *Noctilio leporinus* (Noctilionidae), and *Rhinolophus capensis* (Rhinolophidae). The first three species trawl for insect prey (and in case of *N. leporinus* also small fishes) at or above water surfaces. *R. capensis* search for flying insects while hanging from a perch and fly off to catch it in mid-air on detection. Across the families and hunting strategies there are striking similarities between temporal and spectral changes through the pursuit sequences. The search calls are longer and repeated slowly and, except the rhinolophid, show relatively narrow band widths with shallow frequency modulation (*left panels*). In the terminal buzz phase the calls are very short, repeated very fast, and frequency sweep rates (*right panels*) are steep. The *arrows* indicate the start of the buzz phases

indicating that the terminal phase is important for last-second update of the location of moving prey.

Bats have been categorized according to emitted echolocation signal intensity as high-intensity bats or whispering bats. Griffin (1958) measured high intensity calls to around 110 dB re 20 μPa (pp) at 10 cm and whispering calls to approximately 70 dB re 20 μPa (pp) at 10 cm. Recent field recordings have added 20–40 dB to both these numbers, but a significant difference remains between powerful and more silent calling bats. So-called whispering bats have now been shown to emit calls around 100–110 dB re 20 μPa (pp) at 0.1 m (Brinkløv et al. 2010, 2011). Aerial insectivores and trawling bats hunting in open and semi-open areas seem to emit intense source levels in the field, up to between 120 and 140 dB re 20 μPa (pp) at 0.1 m, which is much more powerful than anything recorded in the lab for the same or comparable species (Jensen and Miller 1999; Holderied et al. 2005; Surlykke and Kalko 2008). Even so, behavioral observations as well as estimates based on realistic target strengths of insects show that prey detection range of bat sonar is limited to around 10 m at the most, and in most cases maybe only 2–3 m. Normal flight speeds for bats are between 3 and 5 m s^{-1}, meaning that they cover a typical detection distance to their prey per second. When closing in on targets in the field high intensity bats lower the source level, probably to compensate for the increase in received echo level with decreasing distance keeping echo intensity within a reasonable range for auditory processing (Nørum et al. 2012; Wahlberg & Surlykke, Chap. 4). This is in accordance with the observed lower source levels in the confined environment of a flight room.

It should also be kept in mind that all remote acoustic methods for studying bats have a strong bias toward those species with powerful echolocation calls, and it is very likely that a number of bats have not yet been discovered or studied under natural conditions because they use very low intensities and/or very high frequencies. For example, the European long eared bat (*Plecotus auritus*) is almost impossible to detect with a bat detector unless it is very close.

Bats have also been categorized according to the duration of their calls as either high duty cycle or low duty cycle callers (Fenton et al. 2012) or according to the time-frequency structure of their calls as CF or FM bats (Griffin 1958; Schnitzler and Henson 1980; Fenton et al. 1998; Jones and Teeling 2006). Short, broad-band signals are produced by many species of different families, for example, *Myotis nattereri* (Vespertilionidae) (Siemers and Schnitzler 2004) or *Megaderma lyra* (Megadermatidae) (Marimuthu and Neuweiler 1987), that employ this strategy to increase resolution and avoid overlap between prey and background echoes in cluttered habitats. Another strategy is to emit long (up to approximately 80 ms), high duty cycle signals of extremely narrow bandwidth and listen for the rhythmic Doppler shifts of the echoes due to the movement of the wings of the prey to disentangle prey echoes from background clutter. CF calls and Doppler shift detection have been reported for several species of the Old World families Rhinolophidae and Hipposideridae (Schnitzler and Henson 1980), and by convergent evolution in a tropical species in the New World, *Pteronotus parnellii* (Mormoopidae) (Fenton, Jensen, Kalko, & Tyack, Chap. 2). Convergent evolution of echolocation signals across species and families using similar habitats point to the importance of ecological effects on call parameters. Some bats may be limited by their echolocation calls to certain habitats, but a number of bats are very flexible in their acoustic output and able to control and change their calls to adapt them to the constraints of different environments. Molossid bats are perhaps the best examples of bats with a large repertoire of calls (Simmons et al. 1978; Guillén-Servent and Ibáñez 2007).

9.4.2 Toothed Whale Echolocation in the Wild

Studies of source parameters of toothed whales (see also Fenton, Jensen, Kalko, & Tyack, Chap. 2 and Wahlberg & Surlykke, Chap. 4) have revealed overall four different types of echolocation clicks: (1) narrow-band high-frequency clicks produced by porpoises, members of the genus *Cephalorynchus*, and the pygmy and dwarf sperm whales (Møhl and Andersen 1973; Dawson 1991; Madsen et al. 2005b; Kyhn et al. 2010); (2) broad-band transient clicks produced by most delphinids (Au 1993); (3) FM clicks from beaked whales (Zimmer et al. 2005; Johnson et al. 2006;

Wahlberg et al. 2011b); and (4) the multipulsed clicks of sperm whales (Madsen et al. 2002b; Møhl et al. 2003; Fenton, Jensen, Kalko, & Tyack, Chap. 2). The inherent danger of listing source properties is that the conditions under which the sounds are recorded, and possible differences in the methods used, will inflate or miss the actual variation in source parameters (Madsen and Wahlberg 2007; Moore et al. 2008). Laboratory studies have shown that delphinids in particular have a very dynamic sound production that is under active control (Moore and Pawloski 1990; Au 1993; Madsen et al. 2013a), and field studies support that notion (Madsen et al. 2004; Wahlberg et al. 2011a). Echolocating toothed whales can regulate their click intervals and to a certain degree other source parameters such as bandwidth, centroid frequency, directionality, and source level in a form of active acoustic gaze control (Au et al. 1995; Akamatsu et al. 1998; Madsen et al. 2002b; Au and Wursig 2004; Johnson et al. 2008; Moore et al. 2008). The temporal structure and spectral content seem to vary little for narrow-band high-frequency clicks (Kyhn et al. 2010) and FM clicks (Johnson et al. 2006), whereas the spectral content of broad band transient and multipulsed clicks may vary over one or more octaves (Au et al. 1995; Houser et al. 1999; Madsen et al. 2004; 2013a). In general, the bandwidth, centroid frequencies and directionality increase with increasing source level, because higher outputs generate more energy at higher frequencies (Au et al. 1995; Madsen et al. 2002b; 2013a; Wahlberg and Surlykke, Chap. 4).

Using a combination of detection thresholds derived from captive studies and source properties measured in the field, it is estimated under noise limited conditions that the small porpoises may detect typical prey targets at ranges of up to 25 m, whereas the bottlenose dolphins may be able to detect prey targets beyond 100 m (Au et al. 2007). For the sperm whale, producing the highest source levels in the animal kingdom (Møhl et al. 2003), it has been estimated that they may be able to detect individual prey at ranges well in excess of 300 m and prey patches perhaps beyond 1,000 m (Møhl et al. 2003; Madsen et al. 2007). Toothed whales normally swim at speeds between 1 and 3 m s^{-1}, which means that even porpoises cover less than 1/10th of a detection distance per second, and they may potentially at times have many seconds between the time of detecting and classifying a prey item and the time at which it must be subdued (Madsen et al. 2013b).

There is a growing notion suggesting that smaller animals produce clicks at higher frequencies, lower source levels and shorter interclick intervals (ICIs) than larger species (Au 1993; Kyhn et al. 2009), and that larger species have higher DIs than smaller species (Au et al. 1999). Array recordings also suggest that free-ranging animals, like their trained counterparts, in general adjust click intervals to the reducing two-way travel time (TWTT) by clicking faster the closer they are to the recording array (Au and Benoit-Bird 2003; Jensen et al. 2009). Perhaps as a biomechanical consequence of faster clicking with sound generator that acts pneumatic oscillator (Cranford et al. 1996; Au and Suthers. Chap. 3), toothed whales do in many cases lower their source levels following 20*log(range, m) which reduces the increase in received level to a doubling for each halving of range instead of a quadrupling without reduction of output level, a phenomenon that has been coined automatic gain control (AGC) (Au and Benoit-bird 2003; Beedholm and

Miller 2007; Jensen et al. 2009). Other species, such as beaked whales, do not follow this pattern.

One thing that many years of studies of trained animals and the use of hydrophone arrays in the wild has barely addressed (but see Norris et al. 1961; Morozov et al. 1972; DeRuiter et al. 2009; Verfuss et al. 2009) is the issue of how an echolocating toothed whale use sonar to guide its motor patterns during the last few meters of closing in on a live prey item. The process of biosonar-mediated foraging involves a navigation part where the whale after leaving the surface must identify the depth layer or habitat where the prey is found. Second, it must search for targets and subsequently classify them as prey and non-prey, and then start an approach phase that, from the predator's perspective, ideally ends with the prey being close enough to be captured (Madsen et al. 2013b). Given these steps, it is clear that toothed whales must follow the same phases of search, approach, and capture outlined by Griffin (1958). Available data from toothed whales in the wild nevertheless suggest that the acoustic changes during these phases may only in part follow those seen for bats, and that the Griffin model cannot explain the entire sampling behavior of neither bats nor toothed whales (Madsen et al. 2013b).

Despite dedicated use of acoustic tags for nearly a decade (see Johnson, Chap. 7), there is still only echolocation behavior data from multiple acoustic tags on a few toothed whale species in the wild: harbor porpoises (Akamatsu et al. 2005), finless porpoises (Akamatsu et al. 2005, 2010), pilot whales (Aguilar de Soto et al. 2008), Blainville's and Cuvier's beaked whales (Johnson et al. 2004; Madsen et al. 2005a; Johnson et al. 2006, 2008; Madsen et al. 2013b), and sperm whales (Madsen et al. 2002b; Watwood et al. 2006).

Ironically, the most revealing data from free ranging toothed whales stems from tagging some of the most elusive and up until recently basically unknown species from the beaked whales family; the Blainville's beaked whales (*Mesoplodon densirostris*). On those whales, it has been possible for the first time to record both the outgoing click and the returning echoes, and link those data with movements of the whale as revealed by depth and inertial sensors of the D-tag (Johnson et al. 2004; Madsen et al. 2005a, 2013b). Sect. 9.4.3 therefore focuses on the findings from D-tags on Blainville's beaked whales, and ends with attempt to draw a bigger picture for toothed whales by including tagging data from other species.

9.4.3 A Case Study: Blainville's Beaked Whale

Blainville's beaked whales are deep diving predators that target small mesopelagic cephalopods and fish during long, deep foraging dives (Fig. 9.6a) (Arranz et al. 2011). During descent on a deep foraging dive, they start emitting frequency modulated clicks at a depth of 200–500 m using slow ICIs around 300–400 ms (Fig. 9.6a, b) (Madsen et al. 2005a, 2013b). When D-tags recordings were first analyzed from these animals (Johnson et al. 2004; Madsen et al. 2005b), it was discovered that the tags recorded not only the outgoing click, but also the returning echo from

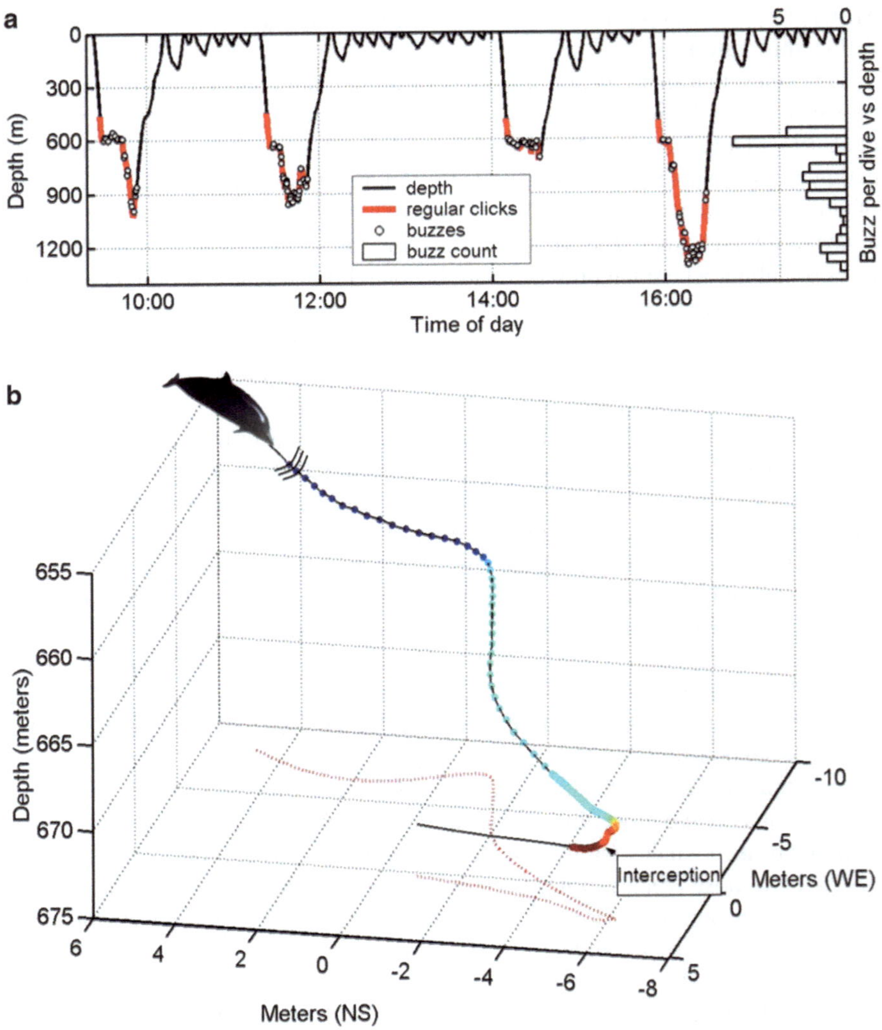

Fig. 9.6 (**a**) Dive profile of Blainville's beaked whale with clicks and buzzes superimposed. (Figure courtesy of M. Johnson.) Note how echolocation clicks are produced only in deep dives. (**b**) 3D pseudo-track of diving Blainville's beaked whale before and during prey capture attempt. Each click is color coded with roll of the animal so that *dark blue* is dorsal side up and *dark red* is ventral side up. Accelerometer data and acoustics indicate a prey capture attempt some two-thirds into the buzz (marked "interception"). (Figure courtesy of M. Johnson)

organisms in the water column (see Johnson, Chap. 7). The discovery of echoes from outgoing clicks has revolutionized studies of echolocation in the wild by allowing researchers to tap into the sensory stream of information back to auditory system of the whale, while concomitantly logging the movements of the whale via inertial sensors (Johnson et al. 2009). It may be imagined that deep diving whales would likely try to target any organisms they happen to locate in a large, deep, three-dimensional world of mesopelagic darkness, but the tag recordings have shown that when the whales reach the deep scattering layer of organisms at depths between 500 and 800 m, they do not try to capture all the many organisms they ensonify, but they rather ignore hundreds of organisms while only catching a small percentage of those ensonified (Madsen et al. 2005a; Arranz et al. 2011).

It is therefore strongly implied that the beaked whales employ selective foraging based on echo discrimination, a notion that has been supported by analysis of selected versus ignored target echoes as recorded on the D-tags (Jones et al. 2008). When the whale has located a target of interest it locks its sonar on the prey and starts an approach phase, however, involving largely the same output and click intervals as during the search phase (Madsen et al. 2005a). The approach phase can therefore not be identified acoustically as in bats and smaller toothed whales (Fig. 9.7), and has accordingly been defined to begin when the first prey echoes are detected on the tag recordings. However, it is likely that the whale has detected and selected the prey target for approach well before that (Madsen et al. 2005a). Echoes are recorded from prey at ranges of more than 10 m (Fig. 9.8), and it is likely that Blainville's beaked whales not only detect, but also classify their prey at much longer ranges by which these toothed whales detect and discriminate their prey at ranges at least an order of magnitude further away than bats (Madsen et al. 2013b).

In contrast to smaller toothed whales (Au and Benoit-Bird 2003), Blainville's beaked whales do not reduce their emitted click levels in accordance with a 20 log(R) fashion. Consequently, the received echo levels increase approximately by 12 dB per halving in target range for the same SL. Second, they maintain ICI's that are an order of magnitude longer than the TWTT to the prey with no or only little reduction in ICIs as they close in on the target. It has been suggested that the long click intervals are maintained to facilitate auditory streaming of a complex auditory scene made of many targets, of which only a few are prey for the beaked whales. Maintaining long ICIs in a situation where the whale is approaching a small prey target surrounded by multiple other targets with approximately the same target strength may thus reduce range ambiguity problems and allow the whale to detect and approach the prey for capture in a complex, self-generated auditory scene (Madsen et al. 2005a). However, when the whales turns sharply to re-access a prey patch, the ICIs are halved to about 200 ms, a phenomenon that likely relates to the high encounter rates of new water volumes and a narrow field of view (Fig. 9.8) (Madsen et al. 2013b) and parallels what some echolocating bats also do when turning (Ghose and Moss 2006). Thus, the Griffin model of search, approach, and capture must be augmented with a phase that involves search for and navigation of patches in which the acoustic behavior may have more to do with what the echolocator is doing rather than the distance to and behavior of individual prey (Madsen et al. 2013b).

When the beaked whale is about a body length away from the prey item (Figs. 9.6b and 9.7), it dramatically changes its output and repetition rate; it switches to a shorter, more broad band click (Johnson et al. 2006), and reduces the ICIs to some 5 ms and the source level by some 20 dB (Johnson et al. 2004; Madsen et al. 2005a). In line with the functional term for bats, this acoustic event is coined a terminal "buzz" (Griffin 1958).

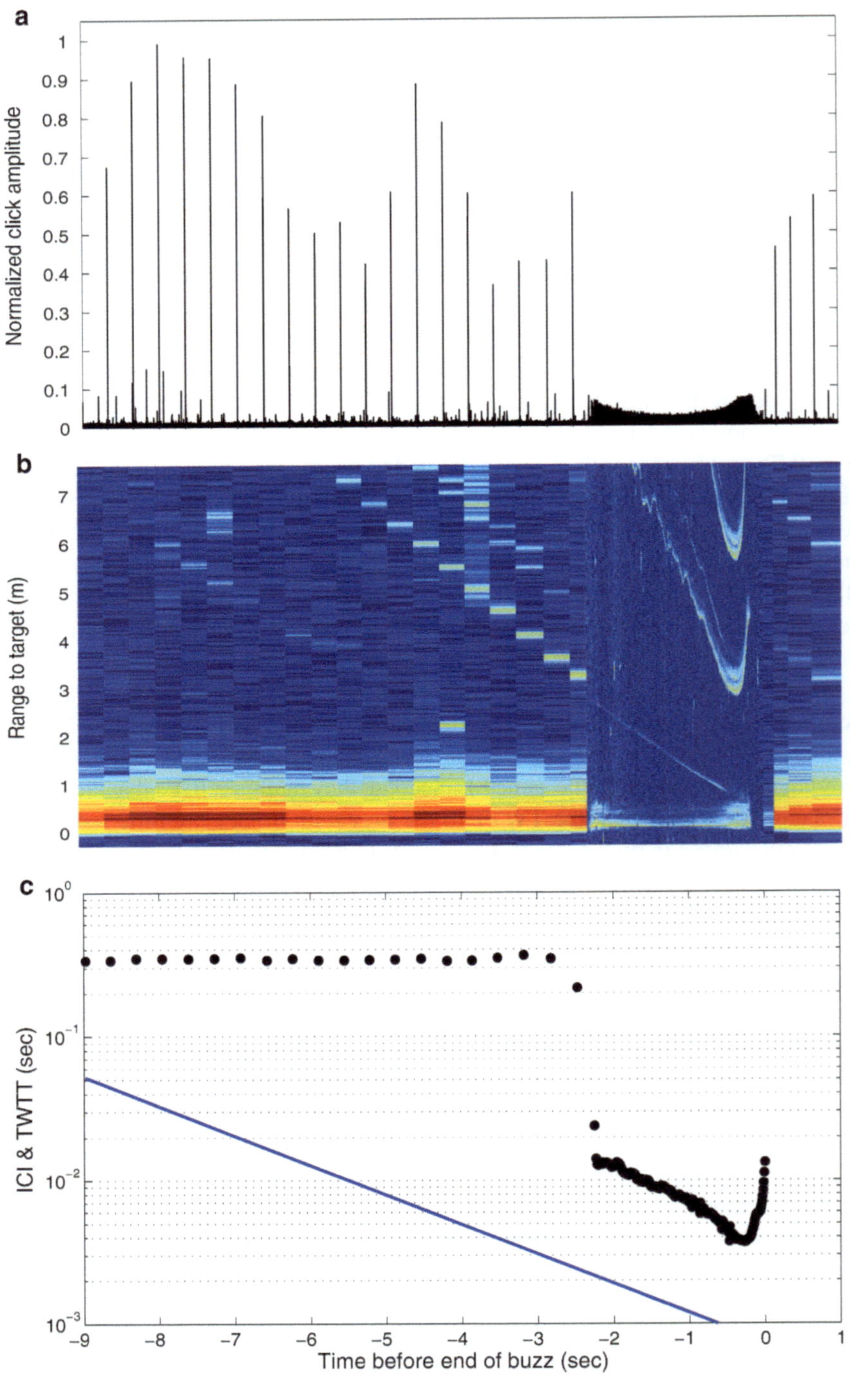

Fig. 9.7 (**a**) Envelopes of clicks from Blainville's beaked whale before and during a buzz. Note how the click amplitude is reduced by an order of magnitude from regular clicking to the buzz. (**b**) Echogram of the same event where the outgoing clicks are seen as a dark band around 0 m range and with the whale moving in on a target that can be seen both before and during the buzz. The two curved energy bands above the target echo in the buzz are the following clicks. (**c**) Interclick interval (ICI) along with the two-way travel time (TWTT, *blue line*) to the target. Note that there is no consistent range (nor output **a**) adjustment in the approach phase whereas the ICI is reduced with reducing TWTT in the buzz. (From Madsen et al. 2013b)

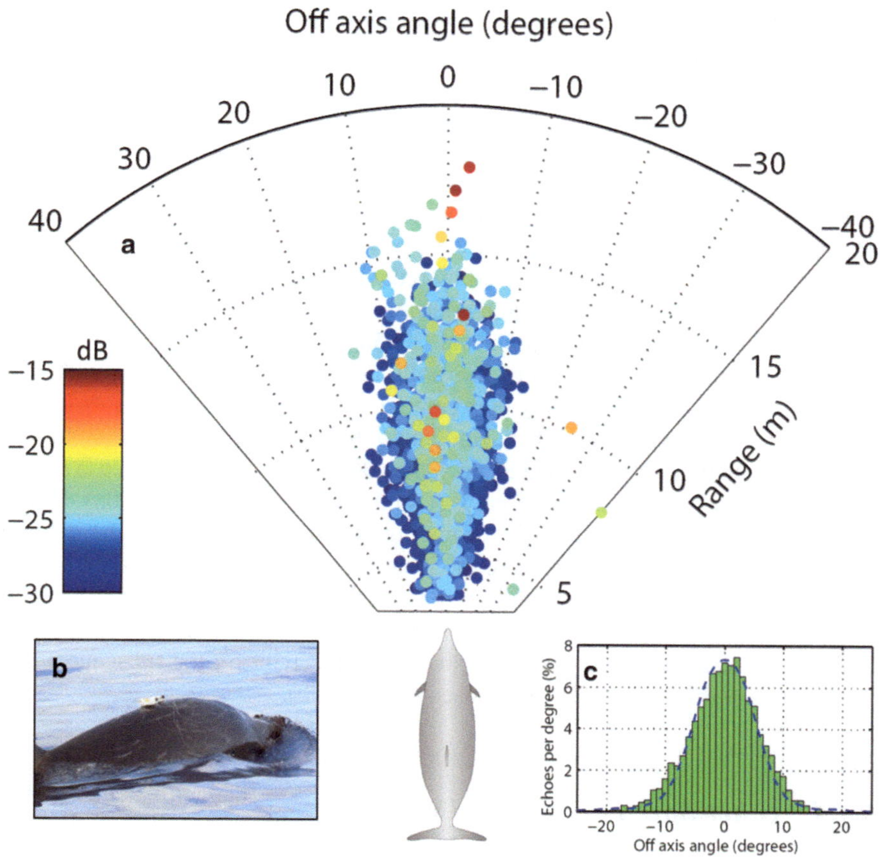

Fig. 9.8 (**a**) 2D projection of targets ensonified by a tagged Blainville's beaked whale (**b**) as a function of range in meters and angles in degrees. Note how only targets within a narrow cone ahead of the tagged animal give rise to detectable echoes (relative echo strengths color coded in decibels). This transmitting directionality provides a narrow acoustic field of view ahead of the whale in which 90 % of echoes recorded by the tag are within ±10° of the acoustic beam center (**c**), suggesting a functional beamwidth of some 20°. The detection vs. angle histogram shown in (**c**) is closely approximated by the beam pattern of a beaked whale FM click radiated from a 25 cm diameter circular piston (*dotted line*), implying a half-power beamwidth of about 9°. (From Madsen et al. 2013b)

In the buzzes of beaked whales, the echo-delays to the prey go toward zero around two-thirds into the 1–3 s duration of a normal buzz (Fig. 9.7), strongly implying that that buzzes are providing fast sonar updates on the location of the prey for capture and that the prey is caught during the buzz (Madsen et al. 2005a). Thus beaked whales use long 200–400 ms click intervals and high outputs of FM clicks emitted in a narrow cone ahead of the animal (Fig. 9.8) when searching for and approaching prey, and when they are about a body length away from the prey, they switch to a buzz in which output level is traded for resolution on the location of the prey in time and space to aid capture via a combination of ram and suction feeding (Fig. 9.7).

For dolphins and bats performing trained echolocation tasks in relatively uncluttered environments, a clicking rate matched to the range of the target (the TWTT plus lag time model) may maximize the chance of a reward (Au 1993). In contrast, a beaked whale searching for specific prey in a target-rich environment seems to maintain a broad auditory scene and thus a low clicking rate until it draws close enough to a prey to warrant a high update rate locked to the range of the selected target. For the fast repetition buzzes, the problem of range ambiguity (where the animal may click so fast that echoes from a previous click arrives after a new click has been emitted) is seemingly solved by reducing the SL by more than 20 dB by which the volume of water generating detectable echoes will be reduced by more than an order of magnitude (Madsen et al. 2005a; Johnson et al. 2006), while at the same time reducing the problem of forward masking of echoes arriving with short delays after click emission during prey capture.

Though the overall acoustic behavior pattern of beaked whales is somewhat stereotyped when they target individual prey, they do demonstrate adaptive use of their sonars when targeting schools of prey. A study by Johnson et al. (2008) demonstrated that the clicking rate at the beginning of buzzes is related to the distance between whale and prey, supporting the idea that whales focus on a specific prey target during the buzz. Generalizing from these findings from beaked whales to all toothed whales in the wild would suggest that they use the timing and information in the echo to adapt their acoustic and locomotor behavior for the subsequent capture attempt (Johnson et al. 2008; Madsen et al. 2013b).

9.4.4 Other Species of Toothed Whales

Though Blainville's beaked whales are the best studied in the field, it is clear that other toothed whales may employ biosonar systems that in some ways differ. The largest of the toothed whales, the sperm whales, have been tagged extensively with D-tags in different areas of the world, showing essentially the same overall deep water foraging and biosonar behavior for females (Watwood et al. 2006) whereas males at high latitudes also forage in shallow water (Teloni et al. 2008). Sperm whales, like beaked whales, are deep divers, normally performing 40–60 min foraging dives to depths of 500–2,000 m. However, during descent they start clicking at much shallower depths between 10 and 200 m, using very long click intervals of more than a second. As they descend, the click intervals are gradually reduced in accordance with the reduced distance to either the sea floor or the depth at which they end up foraging (Madsen et al. 2002b). That acoustic behavior supports the contention that sperm whales operate a high-powered, low-frequency long-range sonar system that can likely locate patches of prey at ranges of more than 500 m (Madsen et al. 2002b; Møhl et al. 2003). When sperm whales reach a depth between 400 and 1,000 m, they will switch to shorter ICIs around 400–500 ms, and at times perform fast repetition rate click trains lasting 3–15 s with ICIs around 20 ms. An ICI of around 400–500 ms probably reveals search and approach phases and the fast ICIs around 20 ms are linked to rapid orientation changes in likely conjunction with foraging (Miller et al. 2004), and should hence be coined "buzzes" as they also reflect the terminal phase of prey acquisition (Madsen et al. 2002b; Miller et al. 2004).

At the other end of the size scale the porpoises (*Phocoena phocoena*) and finless porpoises (*Neophocaena phocaenoides*) are found. Their echolocation behavior has been uncovered in the wild by revealing tagging studies of T. Akamatsu and colleagues using A-tags. The porpoises seem to click almost continuously at high rates with ICIs between 10 and 100 ms while searching for prey in shallow water (Akamatsu et al. 2005; Li et al. 2005) using source levels between 160 and 200 dB re 1 μPa (pp) (Villadsgaard et al. 2007). When they detect a prey item they will decrease their ICIs gradually (Akamatsu et al. 2005; Li et al. 2007) and about 1 m away from their prey they initiate a buzz that may reach repetition rates of 500 clicks per second during which the prey is caught some two-thirds into the buzz (DeRuiter et al. 2009; Verfuss et al. 2009; Wisniewska et al. 2012). Both search for and capture of prey involves repeated rolls perhaps to increase the volume of ensonified water, and to maneuver around the prey during captures (Akamatsu et al. 2010).

Despite that delphinids such as bottlenosed dolphins, belugas, and false killer whales having been the favored species for captive studies on toothed whale echolocation, very little is still known about the echolocation behavior of these species in the wild, a fact that future studies hopefully will change to allow for a most interesting comparison with the multitude of captive studies on these species (as reviewed by Au 1993).

Thus, compared to captive animals, toothed whales in the wild generally produce echolocation clicks of higher source level, higher frequency, higher directionality, and longer ICIs, and use buzzes to acquire prey. The available data from free-ranging toothed whales also suggest that there are significant scaling effects in terms of ICIs, source levels, and frequency, so that large, deep diving toothed whales use lower frequencies, longer ICIs, and higher source levels than smaller toothed whales generally living in shallower water (Fig. 9.9). However, in common to them all is their use of click intervals longer than the TWTT, and the lag time (ICI minus TWTT) is longer (sometimes much longer) than TWTT during search and approach phases. It is also clear that the concept of fixed lag times (Au 1993) must be dispensed with as ICIs change both within and between species depending on the echolocation task. The three phases of search, approach, and capture for bats given by ICIs and outputs are less clear in toothed whales. Thus it is harder from the ICIs, at least for some species, to define the detection distance and thus an approach phase as for bats, but when toothed whales are about a body length away from the prey, they all switch to a buzz, where the ICIs and SL both drop by about an order of magnitude (Fig. 9.9), and the prey is captured some two-thirds into the buzz. Buzzes have, to our knowledge, only rarely been recorded or at least rarely been reported in the many years of pen and tank experiments, perhaps because the trained animals are fed and do not normally collect food themselves or simply because most lab experiments in the past often involved stationed animals (Au 1993). Early studies of toothed whale sonar in captivity did report fast clicking rates when the studied animals closely approached targets (Norris et al. 1961; Morozov et al. 1972), but this

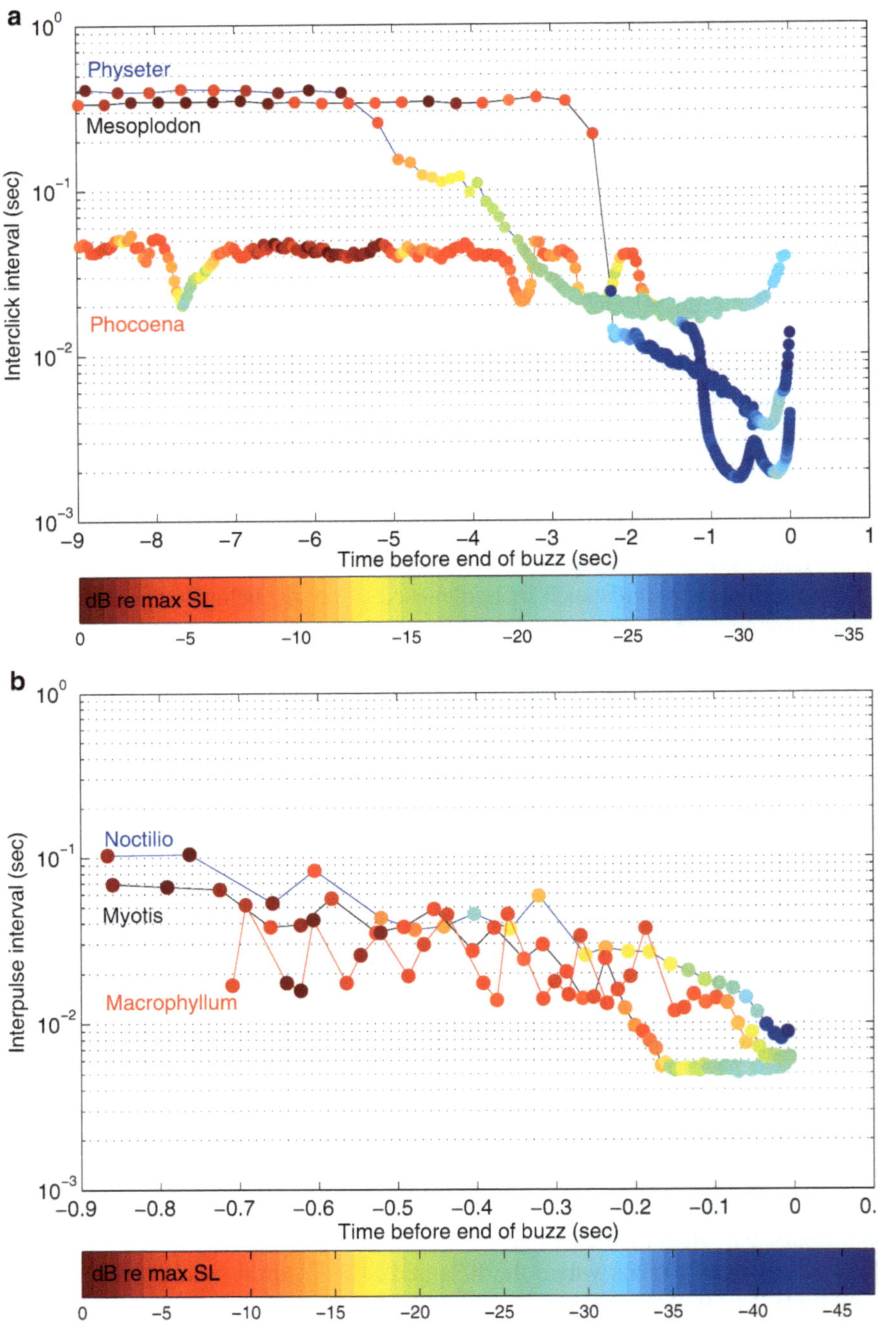

Fig. 9.9 Both toothed whales (**a**) and bats (**b**) decrease click or call intervals as well as source level (emitted energy color coded in decibels) when closing in on prey. (**a**) Three species of toothed whales (*Physeter*, *Mesoplodon*, *Phocoena*) show the same pattern of SL reductions and increased click repetition rates in the buzz where resolution is traded for output. Note the scaling where the *Physeter* buzz operate at ICIs comparable to the ICIs in the search phase of the *Phocoena* with a body mass that is three orders of magnitude smaller. (**b**) A comparable pattern is seen in three species of bats (*Noctilio*, *Myotis*, *Macrophyllum*) but on a time scale that is an order of magnitude smaller. Note the larger changes in ICIs during the approach phase compared to toothed whales. (Modified from Madsen and Surlykke 2013)

buzzing behavior was largely ignored during the next 40 years of research mainly involving animals stationed at a fixed distance from their artificial targets (Au 1993). This critical part of biosonar-based foraging in toothed whales was first rediscovered with the field recordings of Gordon (1987) and Miller et al. (1995), and recent tag studies have all shown that buzzing is indeed involved in prey capture. Based on the currently available data from a growing number of field studies, we therefore hypothesize that all toothed whales use buzzing in the final stages of capture. It seems that the click rates are scaled to the size of the animal so that the buzz rate of a sperm whale is comparable to the clicking rate during searching for a small porpoise (Fig. 9.9). This scaling is at play despite that porpoise and sperm whales swim at comparable speeds when closing in on prey, meaning that sperm whales get about on order of magnitude fewer updates on prey location per distance covered during buzzing than porpoises. It therefore seems that click rates during buzzing are related to absolute maneuverability rather than speed during prey captures. Some species display significant adjustments of ICI and SL in the approach phase (Au and Benoit-Bird 2003), whereas others do not (Madsen et al. 2005a; 2013b). We predict that future studies will uncover that such adjustments or lack thereof are not necessarily species, but rather context specific (Madsen et al. 2013b), and that toothed whales operate a very plastic sonar system in terms of output levels and click intervals to form an acoustic field of view that can be adapted to the task at hand.

9.5 Predator–Prey Interactions

One potential downside for a predator using an active orientation system, such as echolocation, is that it loudly announces its location to prey with sensory means to detect the emitted energy. Thus prey of echolocating bats and toothed whales may be alerted by the sonar signals and take evasive actions on the basis of detecting the powerful sonar pulses, an adaptation that has evolved independently in many families of nocturnal insects, but apparently only very rarely in the prey of toothed whales.

9.5.1 Bats and Their Prey

Roeder and Treat (1961) were the first to study the ultrasound sensitive ears of nocturnal moths. In what has become a classical neuroethological study, Roeder (1967) started the search for bat sensitive ears in other nocturnal insects. The search by Roeder and his successors was successful. At present, ultrasound-sensitive ears have been found in a variety of nocturnal insects of at least seven insect orders, for example, Lepidoptera (moths), Neuroptera (lacewings), and Orthoptera (katydids, crickets, and grasshoppers) (Hoy and Robert 1996; Miller and Surlykke 2001). Many of these insects, but not all, are silent and apparently do not use sound for intraspecific communication, but they all react to bat-like sound pulses with sudden evasive maneuvers, increasing the insect's chance of survival by at least 40 % (Roeder and Treat 1961; Miller and MacLeod 1966).

The hearing thresholds of insects of around 30–50 dB re 20 μPa are much higher than those of bats, which have thresholds of around 0 dB re 20 μPa at their best frequency, as do most other land mammals. However, because insects detect the outgoing sounds of the bats, whereas bats must detect the much fainter returning echo reduced by twice the transmission loss and the target strength of the prey, insects typically have much longer detection ranges for bat calls than bats have for insect echoes (Surlykke et al. 1999; but see Goertlitz et al. 2010). The insects' detection thresholds vary quite a lot, but with a tendency to correlate with size, such that larger insects are also more sensitive, in part due to bigger tympanums. Because bats will be able to detect larger insects further away this correlation ensures the favorable ratio of detection ranges for the insects, while keeping energetically costly false alarms low. Surlykke et al. (1999) estimated from a range of moth sizes that the moth's detection range of a bat was approximately 10 times that of the bat's detection of the moth. The 40 % survival value of bat sensitive ears and acoustically evoked anti-bat defensive flight behaviors were deduced based on interactions where the bats had detected the insects, but insects' long detection ranges may indicate that many more insects fly away long before detected by bats, and hence that survival value of ultrasound sensitive ears may be much higher (Miller 1984; Surlykke et al. 1999).

These adaptations across insect orders and families point to the high selection pressure exerted on nocturnal insects by echolocating bats, and the relative ease with which part of the cuticle with mechano receptors can be turned into an ultrasonic receiver (Surlykke et al. 2003). Thus, there seems to be little doubt that many nocturnal insects have evolved ultrasonic hearing specifically in response to selective pressure by bats and thus to counter bat predation (ter Hofstede et al. 2009). The so-called arms race between bats and their prey implies that bats may also have evolved countermeasures against insect defenses. However, this notion is contentious (Fullard et al. 2005; Ratcliffe and Nydam 2008; ter Hofstede et al. 2008), as bats seem to feed opportunistically; the diets of insectivorous bats reflect the seasonal changes, but with a bias toward non-hearing groups (Fullard 1998; ter Hofstede et al. 2008). Most ultrasound-sensitive insects have best detection thresholds between 20 and 80 kHz, and some bats echolocate below or above this frequency range, but it is not necessarily to avoid being detected by their ultrasound hearing prey. Low frequencies may be employed by large bats going for large prey to increase detection ranges whereas high frequencies may increase resolution and directionality in highly cluttered habitats. However, no matter why bats call at frequencies outside the hearing range of most insects, it is clear that moths and other hearing insects react less or not at all to these allotonic bats (reviewed by Fullard 1998; Miller and Surlykke 2001). For example, Fullard et al. (2008) showed that moths did not react at all to the calls of the little *Hipposiderid ater* echolocating with CF calls at 160 kHz, possibly explaining the over-representation of hearing moths in the diets of allotonic bats (Rydell and Arlettaz 1994; Bogdanowicz et al. 1999).

9.5.2 Toothed Whales, Their Prey, and Predators

For echolocating toothed whales the costs of echolocation may not only be that their prey can hear them, but also that other toothed whales will detect and kill them for food. Thus, the acoustic behavior of echolocating toothed whales may be shaped by the physics of operating biosonars in water, but also by their prey and potential predators. Killer whales prey on all other marine toothed whale species, and other bigger delphinids may harass or kill smaller toothed whales. Different tactics seem to be employed to reduce such partially sound-based predation when echolocating. Many delphinids form large groups and hence use the power of numbers and coordinated defense strategies against killer whales. Others seem to employ acoustic crypsis via frequency range of their sonar pulses or their acoustic behavior. All of the narrow-band, high-frequency species produce echolocation clicks with little or no energy below 100 kHz. That likely relates to the need to achieve high directionality for a small animal, and to exploit the low ambient noise levels at high frequencies (Madsen et al. 2005b; Kyhn et al. 2010), but perhaps also to the fact that killer whales hear very poorly beyond approximately 90 kHz, and that the sharp low-frequency cut-off of narrow-band high-frequency clicks hence may serve to reduce the predation risk for these small toothed whales (Morisaka and Connor 2007). Several species of beaked whale produce clicks, or any sounds for that matter, only at depths of more than 200 m (Aguilar de Soto et al. 2011), and will surface many hundreds of meters away from where they produced the last click at depth. This behavior may be explained as acoustic crypsis to reduce acoustic cues to guide killer whale predation on these deep divers (Tyack et al. 2006). Mammal eating killer whales produce much fever echolocation clicks and sounds in general than their fish eating conspecifics (Barrett-Lennard et al. 1996; Deecke et al. 2005), probably to avoid that their mammalian prey can hear them. Thus, transient killer whales are examples of echolocating toothed whales that rarely echolocate because of the acute hearing capabilities of their prey.

Like for bats and their insect prey, another cost of toothed whale echolocation may be related to prey avoidance reactions mediated by detection of the powerful ultrasonic clicks with source levels often in excess of 210 dB re μPa (pp) (Astrup 1999). The first suggestion that fish may be able to detect intense ultrasound from toothed whales was based on observations that blueback herring could be deterred by ultrasound (Nestler et al. 1992). Those observations were followed by a dedicated experiment in which cod were conditioned to ultrasonic pulses from a commercial echosounder (Astrup and Møhl 1993). The response thresholds for cod were very high and close to the limit of the maximum energy a toothed whale can produce. If such apparent ultrasound detection is of any survival value to the cod, they should also respond when they are not conditioned. That was tested by Schack et al. (2008), who could not detect any responses in cod to simulated echolocation clicks with received levels up to 215 dB re 1 μPa (pp), making it very unlikely that cod respond to the ultrasonic pulses of toothed whales in the wild.

However, as shown by Nestler et al. (1992), members of the fish subfamily Alosinae do actually react to ultrasonic pulses (Mann et al. 1997) in a way that is consistent with predator avoidance in responding more strongly to higher exposure levels (Plachta and Popper 2003; Wilson et al. 2008) and by consistently swimming away from the ultrasonic sound source (Plachta and Popper 2003; Wilson et al. 2011). Both characteristics are apparently restricted to the Alosinae, and in general, it seems that the vast majority of fish species cannot detect ultrasound from toothed whales. That may in part be because toothed whales do not exert a significant predation pressure compared to other predators in water, or relate to the fact that it is much harder to evolve an ultrasonic pressure detector in water than in air (Wilson et al. 2013). Wilson et al. (2009) identified a possible novel ultrasonic pressure detector in allis shad in the form of a small air volume that resonates around 40 kHz by which nearby hair cells or other mechanoreceptors in the lateral line may be stimulated. Thus, contrary to all other studied fish, members of the Alosinae subfamily can detect ultrasound seemingly because of a special coupling between a small gas volume and their lateral line.

Cephalopods are another major food source for toothed whales, but there is no indication that cephalopods have evolved ultrasound detection. Wilson et al. (2007) detected no response in the longfin squid (*Loligo pealeii*) exposed to 215 dB re 1 μPa (pp) at 80 kHz. Subsequent hearing studies on the same species have shown that they hear very poorly above 500 Hz (Mooney et al. 2010), as expected owing to the lack of air volumes that can act as pressure to particle motion transducers in these mollusks. Though it appears that the vast majority of aquatic animals cannot hear ultrasound, all studied species are very sensitive to low-frequency particle motion to which they react strongly with no signs of habituation (Karlsen et al. 2004). The detection of such low-frequency hydrodynamic stimuli seemingly confers a large and general advantage when trying to avoid predation attempts from any predator under water whether it is echolocating or not. So marine prey do in fact likely detect their echolocating predators acoustically, not by using the very intense ultrasonic echolocation pulses at long ranges, but rather by detecting the last second head wave or water flow from suction during predation attempts. Thus biosonars in air and water have had quite dissimilar effects on their prey: In water echolocation by toothed whales may have triggered evolution of dedicated ultrasound pulse detectors in very few prey species (Alosinae), in contrast to the situation in air, where many orders of nocturnal insects have independently evolved ultrasound sensitive ears apparently only or mainly to detect echolocating bats.

Several prey items of toothed whales also show signs of active and passive acoustic crypsis. The call rates of midshipmen have been shown to decrease during times of playbacks of dolphin whistles, suggesting that the fish try to reduce passive sonar detection of dolphins that use the calls to home in on for predation (Gannon et al. 2005; Remage-Healey et al. 2006). Passive acoustic crypsis may perhaps also play a role for some prey of toothed whales. For example, cephalopods have vastly different body compositions from highly muscular fast species to so-called blimps; the ammoniacal cephalopods, which have very little muscle mass. From target strength measurements it has been shown that the muscle mass of the mantle is likely the main backscattering object in squid (Madsen et al. 2007), and it is therefore implicated that blimps will have low target strengths and hence face a lower risk of biosonar detection compared to muscular squid, which, in turn, have more capabilities to evade predators if detected (Madsen et al. 2007). It remains to be tested if the passive acoustic crypsis of blimp cephalopods in the deep sea is the direct result of a selection pressure for passive acoustic crypsis or due to other life history strategies in a low productivity environment. Either way, the result is that they are indeed more difficult to detect for echolocating toothed whales compared to more muscular cephalopods.

The most extreme interaction between toothed whales and their prey involves that the toothed whale via very high sound pressures may be able to debilitate their prey to facilitate capture. That idea was first advanced by Yablokov and Belchowitch (1963, as cited by Berzin 1972) and since significantly elaborated on by Norris and Møhl (1983). Considering the apparent challenges that a large sperm whale faces when trying to subdue small, agile prey, the debilitation theory is appealing in the sense that it provides a convenient way in which a large whale with a powerful sound source (Møhl et al. 2003) can use sound to ease the capture of small, agile prey. Unfortunately, the available data do not support this attractive hypothesis: Though this so-called biological big bang theory gained some initial support from experiments showing that very powerful, low-frequency, fast rise time impulses could affect both cephalopods (Mackay and Pegg 1988) and fish (Zagaeski 1987), later experiments using ultrasound pulses similar to those of toothed whales have failed to show any level of debilitation in neither squid (Wilson et al. 2007) nor fish (Benoit-Bird et al. 2006; Schack et al. 2008). A further observation that is inconsistent with acoustic debilitation of prey is the apparent universal reduction in sound pressure during buzzing when a toothed whale is within about a body length of their prey targets (Fig. 9.9).

9.6 Summary: Comparison of Biosonars in Air and Water

Echolocating animals in the wild probably use an array of senses, including vision, when the conditions support such multimodal sensing. Under circumstances in which only acoustic cues are available, echolocating animals rely to a large degree on active acoustic sampling of the environment to find food. Ultimately, the combination of sensory inputs given by the sound emission rate and the motor outputs (Moss et al. 2011; Madsen et al. 2013b) must result in situations in which enough prey items are captured by the echolocating animal to meet its energy requirements.

Biosonar-based foraging evolved independently in two distant mammalian groups, bats and toothed whales, and led to the successful speciation in both to fill an array of foraging niches under poor light conditions in air and water. Despite very different evolutionary starting points in the very different media of air and water, echolocating bats and toothed whales use surprisingly similar frequencies and sampling rates to search for, approach, and capture prey. Hence though the biggest echolocating bat weighs less than a single tooth of an echolocating sperm whale, both bats and whales echolocate between 15 and 150 kHz, generally, and have buzz rates from 50 to >200 pulses per second, although they try to intercept prey that differ in size by more than four orders of magnitude. This striking evolutionary convergence is based in part on fundamental shared features of the mammalian auditory system, but also by the physics of sound production, propagation, and reflection in air and water (Madsen and Surlykke 2013). A functional biosonar system for locating small prey items calls for high frequencies to provide directional sound beams to increase source levels and reduce clutter, and further to provide geometric backscatter and a spectral basis for target discrimination. Bats are several orders of magnitude smaller than toothed whales and the difference in size of their prey is equally large, which suggests that bats should echolocate at higher frequencies than toothed whales to achieve the same directionality. However, despite that directionality is likely a major driving force for call frequencies both in bats and toothed whales, the severe atmospheric absorption of ultrasound in air compared to water (Fig. 9.3) has had an opposing effect in bats. As a consequence, bats and toothed whales produce echolocation signals in a surprisingly similar frequency range from 10 to 200 kHz. The result is that bat echolocation beams (DIs of 6–18 dB) are two to five times broader than those of echolocating toothed whales (DIs of 22–32 dB). However, owing to the very large impedance differences between insect cuticle and air compared to the much lower low impedance differences between water and fish tissues, echolocating bats ensonify targets with much higher target strengths for their size than is the case for toothed whales and their aquatic prey.

In combination with the physics of production and propagation of ultrasound in air, it means that maximum detection ranges of typical prey for bats are estimated to be on the order of 2–10 m, compared to the 25–500 m for toothed whales. Given that bats move forward at speeds between 3 and 8 m s^{-1}, it means that they typically cover about a prey detection distance per second. Toothed whales, on the other hand, move at speeds around 2 m per s^{-1} and hence only cover fractions of a typical detection distance per second, leaving much more time to gather echo information from the ensonified target compared to bats (Madsen and Surlykke 2013). Consequently, bats go through their approach and capture phases on time scales that are one to two orders of magnitude faster than toothed whales (Fig. 9.9). This means that toothed whales can employ a deliberate mode of sensorimotor operation in which the sensory volume is large compared to the stopping volume (Madsen et al. 2013b). This is very different from bats that operate in a reactive mode, where they normally have less than 1 s between detection and interception of prey (Kalko 1995). These very different ratios between maximum detection ranges and speeds of forward motion and the resulting differences in sensorimotor operation may explain why there is little evidence for prey selection by bats (Barclay and Brigham 1994; Koselj et al. 2011) and considerable evidence for prey selection by toothed whales in the wild (Madsen et al. 2005a), even though bats in the lab in fact can discriminate targets based on infinitesimal spectral echo differences when given the time (Simmons et al. 1974). Second, bats must be able to employ a much faster feedback in their sonar system to guide motor patterns in split second interception attempts (Madsen and Surlykke 2013). They do, on the other hand, have the advantage compared to toothed whales that their capture mechanism involves parts of the tail or wing membrane that make up an area considerably bigger than their mouth, whereas toothed whales must position precisely around their larger prey to engulf it with a relatively much smaller mouth area (Werth 2006; Miller 2010). Nevertheless, bat echolocation must provide higher absolute prey location capabilities in a shorter time than is the case for toothed whales, a problem that is further compounded by their prey's capability to detect and respond evasively to the emitted ultrasonic cries.

Another remarkable convergence is seen in the acoustic behavior of bats and toothed whales (Fig. 9.9), which both wait for the echoes to return before emitting the next sonar pulse. Bats catching insects on the wing emit powerful calls at low repetition rates during search; reduce their the output levels and pulse durations, but increase the repetition rate during approach; and emit short, low-level calls at high repetition rates, the "buzz," just before capture. Toothed whales also use high output clicks at low repetition rates (2–40 Hz) while searching for prey, and at least for some species, the approach phase involves, like for bats, a reduction in click levels and an increase in repetition rates. When about a body length away from their prey, all studied toothed whale species in the wild switch to a buzz consisting of low-output, high repetition rate clicks. Thus, recent field data show that both bats and toothed whales employ dramatic gaze adjustments by using a low-level, fast repetition rate buzz for fine-scale tracking of their prey for capture, trading output intensity for update rate on their actively generated auditory scene (Madsen and Surlykke 2013). The function of the buzz is still not fully understood; neither are the central processing mechanisms at these high sampling rates (pulse/echo to pulse/echo or running average), but the ubiquitous nature of high rate buzzes in both air and water when bats and whales home in on moveable prey is not only striking and interesting, but also highly suggestive of a key function for sonar perception in those last moments of prey capture. All of the above indicate common and general characteristics of biosonar in two very different environments.

Interestingly, there are some differences in the timing of buzzing and capture in bats and toothed whales. Bats, both those emitting sounds through the mouth and through the nostrils, stop buzzing right before the prey is scooped up with their wing or tail membranes, whereas toothed whales seem to capture their prey about two-thirds into the buzz, and hence keep clicking after prey capture (Madsen and Surlykke 2013). It is still unclear why that is, but because bats bend their heads down into their tail membranes to grab their prey, it may make little sense to keep calling. Toothed whales, on the other hand, produce sonar signals with their noses in a way that is largely decoupled from the mouth and gular region used for capture and swallowing prey, allowing whales to click and engulf prey simultaneously. Thus they can continue sampling the water ahead of them to avoid obstacles or recapture the prey, should it escape the first capture attempt.

Though both toothed whales and bats change their acoustic gaze when they approach their prey items, the degree to which they do it differs. Both groups reduce the energy output per sonar pulse during buzzing, but while bats do it both by reducing the peak pressure and the duration, toothed whales generally do not seem to change their click durations, but instead reduce their peak pressures dramatically. This is a part of a broader conclusion that bats have a much more plastic sound production system, where bandwidth, duration, sweep rate, and peak power can be adjusted to produce a plethora of different biosonar signals within a single bat species. This plasticity is based on superfast muscle control of the vocal cords (Elemans et al. 2011) and configuration of the upper airways in bats, allowing for fast changes in vocal outputs (see Au & Suthers, Chap. 3). Toothed whales produce much shorter clicks using the right pair of phonic lips in their nasal complex (Madsen et al. 2010, 2013a), with a mass and configuration that offers less plasticity in terms of duration and frequency of the emitted sonar pulses compared to bats (see Au & Suthers, Chap. 3).

In conclusion, echolocation in bats and toothed whales is an example of striking evolutionary convergence where two very distantly related groups of mammals independently evolved the capability to hunt and navigate in the dark using ultrasound. The last 60 years of research have formed a solid basic understanding of how these animals use echolocation to detect, discriminate, and track targets in noise and clutter. More recently, the advent of microcontroller technology has enabled studies with more ecological validity of how these animals use such capabilities while hunting in the wild where their sixth sense evolved. However, both bats and toothed whales researchers as well as this book chapter face the problem that current field techniques provide only snapshots of the sensory ecology of a limited number of species from which general inferences about biosonar operation in the wild are drawn. We hope that bat and toothed whale researchers can learn more from each other in the future and that technological advances will enable increasingly detailed, long-term studies on a broader range of free ranging echolocators to allow biosonar researchers a deeper look into Griffin's magic well via a strong synergy with captive studies to understand in detail how these intriguing animals evolved to sense actively with sound in the dark.

Acknowledgments We stand on the shoulders of many before us to provide the insights presented here, and we are indebted to our mentors, colleagues, and students for inputs and discussions over the years. Specifically, we wish to extend our sincere gratitude to Bertel Møhl and Lee Miller, who have pioneered technology and research on fundamental aspects of the sensory physiology of bats, toothed whales, and their prey both in the field and in the lab. We thank N. Aguilar Soto, T. Akamatsu, W. Au, K. Beedholm, S. Brinkløv, E. Fordyce, L. Jakobsen, F. Jensen, M. Johnson, L. Miller, P. Nachtigall, A. Popper, J. Ratcliffe, M. Wahlberg, M. Wilson, and D. Wisniewska for helpful discussions and critique. We also wish to thank K. Beedholm, M. de Freitas and M. Ladegaard for help with figures and editing. P. T. Madsen and A. Surlykke were supported by frame grants the Danish National Science Council for Independent Research while writing this chapter.

References

Aguilar de Soto, N., Johnson, M. P., Madsen, P. T., Tyack, P. L., Bocconcelli, A., & Borsani, J. F. (2006). Does intense ship noise disrupt foraging in deep-diving Cuvier's beaked whales (*Ziphius Cavirostris*)? *Marine Mammal Science*, 22(3), 690–699.

Aguilar de Soto N., Johnson, M. P., Madsen, P. T., Díaz, F., Domínguez, I., Brito, A., & Tyack, P. L. (2008). Cheetahs of the deep sea: Deep foraging sprints in short finned pilot whales off Tenerife (Canary Islands). *Journal of Animal Ecology*, 77(5), 936–947.

Aguilar de Soto, N., Madsen, P. T., Tyack, P. L., Arranz, P., Marrero, J., Fais, A., Revelli, E., & Johnson, M. P. (2011). No shallow talk: Cryptic strategy in the vocal communication of Blainville's beaked whales. *Marine Mammal Science*, 28(2), E75–E92.

Akamatsu, T., Wang, D., Nakamura, K., & Wang, K. (1998). Echolocation range of captive and free-ranging baiji (*Lipotes vexillifer*), finless porpoise (*Neophocaena phocaenoides*), and bottlenose dolphin (*Tursiops truncatus*). *Journal of the Acoustical Society of America*, 104(4), 2511–2516.

Akamatsu, T., Wang, D., Wang, K. X., & Naito, Y. (2005). Biosonar behaviour of free-ranging porpoises. *Proceedings of the Royal Society of London B: Biological Sciences*, 272(1565), 797–801.

Akamatsu, T. Wang, D., Wang, K., Li, S., & Dong, S. (2010). Scanning sonar of rolling porpoises during prey capture dives. *Journal of Experimental Biology*, 213(1), 146–152.

Arranz, P., Aguilar de Soto, N., Madsen, P. T., Brito, A., Bordes, F., & Johnson, M. P. (2011). Following a foraging fish-finder: diel habitat use of Blainville's beaked whales revealed by echolocation. *PLoS ONE*, doi:10.1371/journal.pone.0028353.

Astrup, J. (1999). Ultrasound detection in fish—a parallel to the sonar-mediated detection of bats by ultrasound-sensitive insects? *Comparative Biochemistry and Physiology A: Molecular and Integrative Physiology*, 124(1), 19–27.

Astrup, J., & Møhl, B. (1993). Detection of intense ultrasound by the cod *Gadus morhua*. *Journal of Experimental Biology*, 182, 71–80.

Au, W. W. L. (1993). *The sonar of dolphins*. New York: Springer-Verlag.

Au, W. W. L. (2004). A comparison of the sonar capabilities of bats and dolphins. In J. Thomas, C. Moss, & M. Vater (Eds.), *Echolocation in bats and dolphins* (pp. 76–98). Chicago: University of Chicago Press.

Au, W. W. L., & Benoit-Bird, K. J. (2003). Automatic gain control in the echolocation system of dolphins. *Nature*, 423(6942), 861–863.

Au, W.W. L., & Benoit-Bird, K. J. (2008). Broadband backscatter from individual Hawaiian mesopelagic boundary community animals with implications for spinner dolphin foraging. *Journal of the Acoustical Society of America*, 123(5), 2884–2894.

Au, W. W. L., & Moore, P. W. B. (1984). Receiving beam patterns and directivity indices of the Atlantic bottlenose dolphin *Tursiops truncatus*. *Journal of the Acoustical Society of America*, 75, 255–262.

Au, W. W. L., & Simmons, J. A. (2007). Echolocation in dolphins and bats. *Physics Today*, 60(9), 40–45.

Au, W. W. L., & Wursig, B. (2004). Echolocation signals of dusky dolphins (*Lagenorhynchus obscurus*) in Kaikoura, New Zealand. *Journal of the Acoustical Society of America*, 115(5), 2307–2313.

Au, W. W. L., Floyd, R. W., Penner, R. H., & Murchison, A. E. (1974). Measurement of echolocation signals of the Atlantic bottlenose dolphin, *Tursiops truncatus* Montagu, in open waters. *Journal of the Acoustical Society of America*, 56(4), 1280–1290.

Au, W.W., Carder, D. A., Penner, R. H., & Scronce, B. L. (1985). Demonstration of adaptation in beluga whale echolocation signals. *Journal of the Acoustical Society of America*, 77(2), 726–730.

Au, W. W. L., Moore, P. W. B., & Pawloski, D. (1986). Echolocation transmitting beam of the Atlantic bottlenose dolphin. *Journal of the Acoustical Society of America*, 80(2), 688–694.

Au, W. W. L., Pawloski, D., Nachtigall, P. E., Blonz, M., & Gisner, R. G. (1995). Echolocation signals and transmission beam pattern of a false killer whale (*Pseudorca crassidens*). *Journal of the Acoustical Society of America*, 98(1), 51–59.

Au, W. W. L., Kastelein, R. A., Rippe, T., &Schooneman, N. M. (1999). Transmission beam pattern and echolocation signals of a harbor porpoise (*Phocoena phocoena*). *The Journal of the Acoustical Society of America*, 106, 3699–3705.

Au, W. W. L., Ford, J. K. B., Horne, J. K., & Allman, K. A. N. (2004). Echolocation signals of free-ranging killer whales (*Orcinus orca*) and modeling of foraging Chinook salmon (*Oncorhynchus tshawytscha*). *Journal of the Acoustical Society of America*, 115(2), 901–909.
Au, W. W., Benoit-Bird, K. J., & Kastelein, R. A. (2007). Modeling the detection range of fish by echolocating bottlenose dolphins and harbor porpoises. *Journal of the Acoustical Society of America*, 121(6), 3954–3962.
Au, W. W. L., Branstetter, B. K., Benoit-Bird, K. J., & Kastelein, R. A. (2009). Acoustic basis for fish prey discrimination by echolocating dolphins and porpoises. *Journal of the Acoustical Society of America*, 126(1), 460–467.
Barclay, R. M. R., & Brigham, R. M. (1991). Prey detection, dietary niche breadth, and body size in bats: Why are aerial insectivorous bats so small? *American Naturalist*, 137(5), 693–703.
Barclay, R. M. R., & Brigham, R. M. (1994). Constraints on optimal foraging: A field test of prey discrimination by echolocating insectivorous bats. *Animal Behaviour*, 48(5), 1013–1021.
Barrett-Lennard, L. G., Ford, J. K. B., & Heise, K. A. (1996). The mixed blessing of echolocation: Differences in sonar use by fish-eating and mammal-eating killer whales. *Animal Behaviour*, 51(3), 553–565.
Beedholm, K., & Miller, L. (2007). Automatic gain control in harbor porpoises (*Phocoena phocoena*)? Central versus peripheral mechanisms. *Aquatic Mammals*, 33(1), 69–75.
Beedholm, K., & Møhl, B. (2006). Directionality of sperm whale sonar clicks and its relation to piston radiation theory. *Journal of the Acoustical Society of America*, 119(2), EL14–EL19.
Belwood, J. J. & Morris, G. K. (1987). Bat predation and its influence on calling behavior in neotropical katydids. *Science*, 238(4823), 64–67.
Benoit-Bird, K. J., & Au, W. W. L. (2001). Target strength measurements of Hawaiian mesopelagic boundary community animals. *Journal of the Acoustical Society of America*, 110(2), 812–819.
Benoit-Bird, K. J., Au, W. W. L., & Kastelein, R. (2006). Testing the odontocete acoustic prey debilitation hypothesis: No stunning results. *Journal of the Acoustical Society of America*, 120(2), 1118–1123.
Berzin, A. A. (1972). *The sperm whale*. Jerusalem: Israel Program for Scientific Translations.
Bogdanowicz, W., Fenton, M. B., & Daleszczyk, K. (1999). The relationships between echolocation calls, morphology and diet in insectivorous bats. *Journal of Zoology (London)*, 247(3), 381–393.
Brinkløv, S., Kalko, E. K. V., & Surlykke, A. (2009). Intense echolocation calls from two 'whispering' bats, *Artibeus jamaicensis* and *Macrophyllum macrophyllum* (Phyllostomidae). *Journal of Experimental Biology*, 212(1), 11–20.
Brinkløv, S., Kalko, E. K. V., & Surlykke, A. (2010). Dynamic adjustment of biosonar intensity to habitat clutter in the bat *Macrophyllum macrophyllum* (Phyllostomidae). *Behavioral Ecology and Sociobiology*, 64(11), 1867–1874.
Brinkløv, S., Jakobsen, L., Ratcliffe, J. M., Kalko, E. K. V., & Surlykke, A. (2011). Echolocation call intensity and directionality in flying short-tailed fruit bats, *Carollia perspicillata* (Phyllostomidae). *Journal of the Acoustical Society of America*, 129(1), 427–435.
Cranford, T. W., Amundin, M., & Norris, K. S. (1996). Functional morphology and homology in the odontocete nasal complex: Implications for sound generation. *Journal of Morphology*, 228(3), 223–285.
Dawson, S. M. (1991). Clicks and communication—the behavioural and social contexts of Hector's dolphin vocalisations. *Ethology*, 88(4), 265–276.
Deecke, V. B., Ford, J. K. B., & Slater, P. J. B. (2005). The vocal behavior of mammal-eating killer whales: Communicating with costly calls. *Animal Behaviour*, 69(2), 395–405.
DeRuiter, S. L., Bahr, A., Blanchet, M.-A., Hansen, S. F., Kristensen, J. H., Madsen, P. T., Tyack, P. L., & Wahlberg, M. (2009). Acoustic behaviour of echolocating porpoises during prey capture. *Journal of Experimental Biology*, 212(19), 3100–3107.
Elemans, C. P. H., Mead, A. F., Jakobsen, L., & Ratcliffe, J. M. (2011). Superfast muscles set maximum call rate in echolocating bats. *Science*, 333(6051), 1885–1888.
Fahlke, J. M., Gingerich, P. D., Welsh, R. C., & Wood A. R. (2011). Cranial asymmetry in Eocene archaeocete whales and the evolution of directional hearing in water. *Proceedings of the National Academy of Sciences of the USA*, 108(35), 14545–14548.

Fenton, M. B. (1990). The foraging behaviour and ecology of animal-eating bats. *Canadian Journal of Zoology*, 68(3), 411–422.
Fenton, M. B., & Bell, G. P. (1981). Recognition of species of insectivorous bats by their echolocation calls. *Journal of Mammalogy*, 62(2), 233–243.
Fenton, M. B., Portfors, C. V., Rautenback, I. L., & Waterman, J. M. (1998). Compromises: Sound frequencies used in echolocation by aerial-feeding bats. *Canadian Journal of Zoology*, 76(6), 1174–1182.
Fenton, M. B., Faure, P. A., & Ratcliffe, J. M. (2012). Evolution of high duty cycle echolocation in bats. *Journal of Experimental Biology*, 215(Pt 17), 2935–2944.
Foote, K. G. (1980). Importance of the swimbladder in acoustic scattering by fish: A comparison of gadoid and mackerel target strengths. *Journal of the Acoustical Society of America*, 67(6), 2084–2089.
Fordyce, R. E. (2009). Cetacean evolution. In W. F. Perrin, J. G. M. Thewissen, & B. Würsig (eds.), *Encyclopedia of marine mammals* (pp. 201–207). San Diego: Elsevier.
Fullard, J. H. (1998). The sensory coevolution of moths and bats. In R. R. Hoy, A. N. Popper, & R.R. Fay (Eds.) *Comparative hearing: Insects* (pp. 279–326). New York: Springer-Verlag.
Fullard, J. H., Ratcliffe, J. M., & Guignion, C. (2005). Sensory ecology of predator-prey interactions: Responses of the AN2 interneuron in the field cricket, *Teleogryllus oceanicus* to echolocation calls of sympatric bats. *Journal of Comparative Physiology A: Neuroethology, Sensory, Neural, and Behavioral Physiology*, 191(7), 605–618.
Fullard, J. H., Jackson, M. E., Jacobs, D. S., Pavey, C. R., & Burwell, C. J. (2008). Surviving cave bats: Auditory and behavioural defences in the Australian noctuid moth, *Speiredonia spectans*. *Journal of Experimental Biology*, 211(Pt 24), 3808–3815.
Gannon, D. P., Barros, N. B., Nowacek, D. P., Read, A. J., Waples, D. M., & Wells, R. S. (2005). Prey detection by bottlenose dolphins, *Tursiops truncatus*: An experimental test of the passive listening hypothesis. *Animal Behaviour*, 69(3), 709–720.
Geipel, I., Jung, K., & Kalko, E. K. V. (2013). Perception of silent and motionless prey on vegetation by echolocation in the gleaning bat *Micronycteris microtis*. *Proceedings of the Royal Society of London B: Biological Sciences*, 280, 20122830. doi: 10.1098/rspb.2012.2830.
Genzel, D., & Wiegrebe, L. (2008). Time-variant spectral peak and notch detection in echolocation-call sequences in bats. *Journal of Experimental Biology*, 211(Pt 1), 9–14.
Ghose, K., & Moss, C. F. (2006). Steering by hearing: A bat's acoustic gaze is linked to its flight motor output by a delayed, adaptive linear law. *Journal of Neuroscience*, 26, 1704–1710.
Gillam, E. H., & McCracken, G. F. (2007). Variability in the echolocation of *Tadarida brasiliensis*: Effects of geography and local acoustic environment. *Animal Behaviour*, 74(2), 277–286.
Goertlitz, H. R., ter Hofstede, H. M., Zeale, M. R. K., Jones, G., & Holderied, M. W. (2010). An aerial hawking bat uses stealth echolocation to counter moth hearing. *Current Biology*, 20(17), 1568–1572.
Gordon, J. C. (1987). *The behaviour and ecology of sperm whales off Sri Lanka*. PhD thesis, Darwin College, Cambridge.
Griffin, D. R. (1953). Bat sounds under natural conditions, with evidence for echolocation of insect prey. *Journal of Experimental Zoology*, 123(3), 435–465.
Griffin, D. R. (1958). *Listening in the dark*. New Haven, CT: Yale University Press.
Griffin, D. R. (1971). The importance of atmospheric attenuation for the echolocation of bats (Chiroptera). *Animal Behaviour*, 19(1), 55–61.
Griffin, D. R., & Galambos, R. (1941). The sensory basis of obstacle avoidance by flying bats. *Journal of Experimental Zoology*, 86(3), 481–506.
Griffin, D. R., Webster, F. A., & Michael, C. R. (1960). The echolocation of flying insects by bats. *Animal Behaviour*, 8(3), 141–154.
Guillén-Servent, A., & Ibáñez, C. (2007). Unusual echolocation behavior in a small molossid bat, *Molossops temminckii*, that forages near background clutter. *Behavior Ecology and Sociobiology*, 61(10), 1599–1613.
Hiryu, S., Hagino, T., Riquimaroux, H., & Watanabe, Y. (2007). Echo-intensity compensation in echolocating bats (*Pipistrellus abramus*) during flight measured by a telemetry microphone. *Journal of the Acoustical Society of America*, 121(3), 1749–1757.

Hiryu, S., Shiori, Y., Hosokawa, T., Riquimaroux, H., & Watanabe, Y. (2008). On-board telemetry of emitted sounds from free-flying bats: Compensation for velocity and distance stabilizes echo frequency and amplitude. *Journal of Comparative Physiology A: Neuroethology, Sensory, Neural, and Behavioral Physiology*, 194(9), 841–851.

Holderied, M. W., & von Helversen, O. (2003). Echolocation range and wingbeat period match in aerial-hawking bats. *Proceedings of the Royal Society of London B: Biological Sciences*, 270(1530), 2293–2299.

Holderied, M. W., Korin, C., Fenton, M. B., Parsons, S., Robinson, S., & Jones, G. (2005). Echolocation call intensity in the aerial hawking bat *Eptesicus bottae* (Vespertilionidae) studied using stereo videogrammetry. *Journal of Experimental Biology*, 208(Pt 7), 1321–1327.

Holland, R. A., & Wikelski, M. (2009). Studying the migratory behavior of individual bats: Current techniques and future directions. *Journal of Mammalogy*, 90(6), 1324–1329.

Houser, D. S., Helweg, D. A., & Moore, P. W. (1999). Classification of dolphin echolocation clicks by energy and frequency distributions. *Journal of the Acoustical Society of America*, 106, 1579–1591.

Houston, R. D., Boonman, A. M., & Jones, G. (2004). Do echolocation signal parameters restrict bats' choice of prey? In J. A. Thomas, C. F. Moss, & M. Vater (Eds.), *Echolocation in bats and dolphins* (pp. 339–345). Chicago: Chicago University Press.

Hoy, R. R., & Robert, D. (1996). Tympanal hearing in insects. *Annual Review of Entomology*, 41, 433–450.

Hristov, N. I., Betke, M., & Kunz, T. H. (2008). Applications of thermal infrared imaging for research in aeroecology. *Integrative and Comparative Biology*, 48(1), 50–59.

Jakobsen, L., & Surlykke, A. (2010). Vespertilionid bats control the width of their biosonar sound beam dynamically during prey pursuit. *Proceedings of the National Academy of Sciences of the USA*, 107(31), 13930–13935.

Jakobsen, L., Ratcliffe, J. M., & Surlykke, A. (2013). Convergent acoustic field of view in echolocating bats. *Nature*, 493(7430), 93–96.

Jensen, M. E., & Miller, L. A. (1999). Echolocation signals of the bat *Eptesicus serotinus* recorded using a vertical microphone array: Effect of flight altitude on searching signals. *Behavioral Ecology and Sociobiology*, 47(1–2), 60–69.

Jensen, F. H., Bejder, L., Wahlberg, M., & Madsen, P. T. (2009). Biosonar adjustments to target range of echolocating bottlenose dolphins (*Tursiops* sp.) in the wild. *Journal of Experimental Biology*, 212(Pt 8), 1078–1086.

Johnson, M. P., & Tyack, P. L. (2003). A digital acoustic recording tag for measuring the response of wild marine mammals to sound. *IEEE Journal of Oceanic Engineering*, 28(1), 3–12.

Johnson, M. P., Madsen, P. T., Zimmer, W. M. X., Aguilar de Soto, N., & Tyack, P. L. (2004). Beaked whales echolocate on prey. *Proceedings of the Royal Society of London B: Biological Sciences*, 271(Suppl 6), S383–S386.

Johnson, M. P., Madsen, P. T., Zimmer, W. M. X., Aguilar de Soto, N., & Tyack, P. L. (2006). Foraging Blainville's beaked whales (*Mesoplodon densirostris*) produce distinct click types matched to different phases of echolocation. *Journal of Experimental Biology*, 209(Pt 24), 5038–5050.

Johnson, M. P., Hickmott, L. S., Aguilar de Soto, N., & Madsen, P. T. (2008). Echolocation behaviour adapted to prey in foraging Blainville's beaked whale (*Mesoplodon Densirostris*). *Proceedings of the Royal Society of London B: Biological Sciences*, 275(1631), 133–139.

Johnson, M. P., Aguilar de Soto, N., & Madsen, P. T. (2009). Studying the behaviour and sensory ecology of marine mammals using acoustic recording tags: A review. *Marine Ecology Progress Series*, 395, 55–73.

Jones, G. (1999). Scaling of echolocation call parameters in bats. *Journal of Experimental Biology*, 202(Pt 23), 3359–3367.

Jones, G., & Holderied, M. W. (2007). Bat echolocation calls: Adaptation and convergent evolution. *Proceedings of the Royal Society of London B: Biological Sciences*, 274(1612), 905–912.

Jones, G., & Teeling, E. C. (2006). The evolution of echolocation in bats. *Trends in Ecology and Evolution*, 21(3), 149–156.

Jones, B. A., Stanton, T. K., Lavery, A. C., Johnson, M. P., Madsen, P. T., & Tyack, P. L. (2008). Classification of broadband echoes from prey of a foraging Blainville's beaked whale. *Journal of the Acoustical Society of America*, 123(3), 1753–1762.

Jung, K., Kalko, E. K. V., & von Helversen, O. (2007). Echolocation calls in Central American emballonurid bats: Signal design and call frequency alternation. *Journal of Zoology*, 272(2), 125–137.

Kalko, E. K. V. (1995). Insect pursuit, prey capture and echolocation in Pipistrelle bats (Microchiroptera). *Animal Behaviour*, 50(4), 861–880.

Kalko, E. K. V., Friemel, D., Handley, C. O., Jr. & Schnitzler, H. U. (1999). Roosting and foraging behavior of two Neotropical gleaning bats, *Tonatia silvicola* and *Trachops cirrhosus* (Phyllostomidae). *Biotropica*, 31(2), 344–353.

Karlsen, H. E., Piddington, R. W., Enger, P. S., & Sand, O. (2004). Infrasound initiates directional fast-start escape responses in juvenile roach *Rutilus rutilus*. *Journal of Experimental Biology*, 207(24), 4185–4193.

Kellogg, W. N., Kohler, R., & Morris, H. N. (1953). Porpoise sounds as sonar signals. *Science*, 117(3036), 239–243.

Kloepper, L. N., Nachtigall, P. E., Gisiner, R., & Breese, M. (2010). Decreased echolocation performance following high-frequency hearing loss in the false killer whale (*Pseudorca crassidens*). *Journal of Experimental Biology*, 213(Pt 21), 3717–3722.

Koblitz, J. C., Wahlberg, M., Stilz, P., Madsen, P. T., Beedholm, K., & Schnitzler, H. U. (2012). Asymmetry and dynamics of a narrow sonar beam in an echolocating harbor porpoise. *Journal of the Acoustical Society of America*, 131, 2315–2324.

Koselj, K., Schnitzler, H. U., & Siemers, B. M. (2011). Horseshoe bats make adaptive prey-selection decisions, informed by echo cues. *Proceedings of the Royal Society of London B: Biological Sciences*, 278, 3034–3041.

Kössl, M. & Vater, M. (1995). Cochlear structure and function in bats. In A. N. Popper & R. R. Fay (Eds.), *Hearing by bats* (pp. 191–234). New York: Springer-Verlag.

Kyhn, L. A., Tougaard, J., Jensen, F., Wahlberg, M., Stone, G., Yoshinaga, A., Beedholm, K., & Madsen, P. T. (2009). Feeding at a high pitch: Source parameters of narrow band, high frequency clicks from echolocating off-shore hourglass dolphins and coastal Hector's dolphins. *Journal of the Acoustical Society of America*, 125(3), 1783–1791.

Kyhn, L. A., Jensen, F. H., Beedholm, K., Tougaard, J., Hansen M., & Madsen, P. T. (2010). Echolocation in sympatric Peale's dolphins (*Lagenorhynchus australis*) and Commerson's dolphins (*Cephalorhynchus commersonii*) producing narrow-band high-frequency clicks. *Journal of Experimental Biology*, 213(11), 1940–1949.

Lang, A. B., Kalko, E. K. V., Römer, H., Bockholdt, C., & Dechmann, D. K. N. (2006). Activity levels of bats and katydids in relation to the lunar cycle. *Oecologia*, 146(4), 659–666.

Li, S. Wang, K., Wang, D., & Akamatsu, T. (2005). Echolocation signals of the free-ranging Yangtze finless porpoise (*Neophocaena phocaenoides asiaeorientialis*). *Journal of the Acoustical Society of America*, 117(5), 3288–3296.

Li, S., Wang, D., Wang, K., Akamatsu, T., Ma, Z., & Han, J. (2007). Echolocation click sounds from wild inshore finless porpoise (*Neophocaena phocaenoides sunameri*) with comparisons to the sonar of riverine *N. p. asiaeorientalis*. *Journal of the Acoustical Society of America*, 121(6), 3938–3946.

Mackay, R. S., & Pegg, J. (1988). Debilitation of prey by intense sounds. *Marine Mammal Science*, 4, 356–359.

Madsen, P. T., & Surlykke, A. (2013). Functional convergence in bat and toothed whale biosonars. *Physiology*, 28, 276–283.

Madsen, P. T., & Wahlberg, M. (2007). Recording and quantification of ultrasonic echolocation clicks from free-ranging toothed whales. *Deep-Sea Research Part I: Oceanographic Research Papers*, 54(8), 1421–1444.

Madsen, P. T., Payne, R., Kristiansen, N. U., Wahlberg, M., Kerr, I., & Møhl, B. (2002a). Sperm whale sound production studied with ultrasound time/depth-recording tags. *Journal of Experimental Biology*, 205(Pt 13), 1899–1906.

Madsen, P. T., Wahlberg, M., & Møhl, B. (2002b). Male sperm whale (*Physeter macrocephalus*) acoustics in a high-latitude habitat: Implications for echolocation and communication. *Behavioral Ecology and Sociobiology*, 53(1), 31–41.

Madsen, P. T., Kerr, I., & Payne, R. (2004). Echolocation clicks of two free-ranging, oceanic delphinids with different food preferences: False killer whales *Pseudorca crassidens* and Risso's dolphins *Grampus griseus*. *Journal of Experimental Biology*, 207(Pt 11), 1811–1823.

Madsen, P. T., Johnson, M. P., Aguilar De Soto, N., Zimmer, W. M. X., & Tyack, P. L. (2005a). Biosonar performance of foraging beaked whales (*Mesoplodon densirostris*). *Journal of Experimental Biology*, 280 (2), 181–194.

Madsen, P. T., Carder, D. A., Beedholm, K., & Ridgway, S. H. (2005b). Porpoise clicks from a sperm whale nose: Convergent evolution of toothed whale echolocation clicks? *Bioacoustics*, 15(2), 195–206.

Madsen, P. T., Wilson, M., Johnson, M. P., Hanlon, R. T., Bocconcelli, A., Aguilar de Soto, N., & Tyack, P. L. (2007). Clicking for calamari: Toothed whales can echolocate squid *Loligo pealeii*. *Aquatic Biology*, 1(2), 141–150.

Madsen, P. T., Wisniewska, D. M., & Beedholm, K. (2010). Single source sound production and dynamic beam formation in echolocating harbour porpoises (*Phocoena phocoena*). *Journal of Experimental Biology*, 213(Pt 18), 3105–3110.

Madsen, P. T., Lammers, M., Wisniewska, D., & Beedholm, K.(2013a) Nasal sound production in echolocating delphinids (*Tursiops truncatus* and *Pseudorca crassidens*) is dynamic, but unilateral: Clicking on the right side and whistling on the left side. *Journal of Experimental Biology*, 216, 4091–4102.

Madsen, P. T., Aguilar de Soto, N., Arranz, P., & Johnson, M. (2013b). Echolocation in Blainville's beaked whales (*Mesoplodon densirostris*). *Journal of Comparative Physiology A: Neuroethology, Sensory, Neural, and Behavioral Physiology,* 199, 451–469.

Mann, D. A., Lu, Z., & Popper, A. N. (1997). A clupeid fish can detect ultrasound. *Nature*, 389(6649), 341.

Marimuthu, G., & Neuweiler, G. (1987). The use of acoustical cues for prey detection by the Indian false vampire bat, *Megaderma lyra*. *Journal of Comparative Physiology A: Neuroethology, Sensory, Neural, and Behavioral Physiology*, 160(4), 509–515.

Miller, L. A. (1984). Hearing in green lacewings and their responses to the cries of bats. In M. Canard & Y. Sçmçria (Eds.), *Biology of Chrysopidae* (pp. 134–149). The Hague: Dr. W. Junk Publishers.

Miller, L. A. (2010). Prey capture by harbor porpoises (*Phocoena phocoena*); a comparison between echolocators in the field and in captivity. *Journal of the Marine Acoustical Society of Japan*, 37(3), 156–168.

Miller, L. A., & Degn, H. J. (1981). The acoustic behavior of four species of vespertilionid bats studied in the field. *Journal of Comparative Physiology A: Neuroethology, Sensory, Neural, and Behavioral Physiology*, 142(1), 67–74.

Miller, L. A., & MacLeod, E. G. (1966). Ultrasonic sensitivity: A tympanal receptor in the green lace wing *Chrysopa carnea*, *Science*, 154(3751), 891–893.

Miller, L. A., & Surlykke, A. (2001). How some insects detect and avoid being eaten by bats: Tactics and countertactics of prey and predator. *BioScience*, 51(7), 571–582.

Miller, L. A., Pristed, J. Møhl, B., & Surlykke, A. (1995). The click-sounds of narwhals (*Monodon monoceros*) in Inglefiels Bay, Northwest Greenland. *Marine Mammal Science*, 11(4), 491–502.

Miller, P. J. O., Johnson, M. P., & Tyack, P. L. (2004). Sperm whale behaviour indicates the use of rapid echolocation click buzzes 'creaks' in prey capture. *Proceedings of the Royal Society of London B: Biological Sciences*, 271(1554), 2239–2247.

Mogensen, F., & Møhl, B. (1979). Sound radiation patterns in the frequency domain of cries from a vespertilionid bat. *Journal of Comparative Physiology A: Neuroethology, Sensory, Neural, and Behavioral Physiology*, 134(2), 165–171.

Møhl, B. (1988). Target detection by echolocating bats. In P. E. Nachtigall & P. W. Moore (Eds.), *Animal sonar: Processes and performances* (pp. 435–450). New York: Plenum Press.

Møhl, B., & Andersen, S. (1973). Echolocation: High frequency component in the click of the harbour porpoise (*Phocoena phocoena* L.). *Journal of the Acoustical Society of America*, 54(5), 1368–1372.

Møhl, B., Surlykke, A., & Miller, L. A. (1990). High intensity narwhal click. In J. Thomas & R. Kastelein (Eds.), *Sensory abilities of cetaceans* (pp. 295–304). New York: Plenum Press.

Møhl, B., Wahlberg, M, Madsen, P. T., Heerfordt, A., & Lund, A. (2003). The monopulsed nature of sperm whale clicks. *Journal of the Acoustical Society of America*, 114(2), 1143–1154.

Mooney, T. A., Hanlon, R. T., Christensen-Dalsgaard, J., Madsen, P. T., Ketten, D. R. & Nachtigall P. E. (2010). Sound detection by the Longfin Squid (*Loligo pealeii*) studied with auditory evoked potentials: sensitivity to low-frequency particle motion and not pressure. *Journal of Experimental Biology*, 213(21), 3748–3759.

Moore, P. W., & Pawloski, J. (1990). Investigations on the control of echolcation pulses in the dolphin (*Tursiops truncatus*)." In J. Thomas & R. Kastelein (Eds.), *Sensory abilities of cetaceans* (pp. 305–316). New York: Plenum Press.

Moore, P. W., Dankiewicz, L. A., & Houser, D. S. (2008). Beamwidth control and angular target detection in an echolocating bottlenose dolphin (*Tursiops truncatus*). *Journal of the Acoustical Society of America*, 124(5), 3324–3332.

Morisaka, T., & Connor, R. C. (2007). Predation by killer whales (*Orcinus orca*) and the evolution of whistle loss and narrow-band high frequency clicks in odontocetes. *Journal of Evolutionary Biology*, 20(4), 1439–1458.

Morozov, V. P., Akopian, A. I., Burdin, V. I., Zaitseva, K. A., & Sokovykh, Y.A. (1972). Tracking frequency of the location signals of dolphins as a function of distance to the target. *Biofizika*, 17, 139–145.

Moss, C. F., Bohn, K., Gilkenson, H., & Surlykke, A. (2006). Active listening for spatial orientation in a complex auditory scene. *PLoS Biology*, 4(4), 615–626.

Moss, C. F., Chiu, C., & Surlykke, A. (2011). Adaptive vocal behavior drives perception by echolocation in bats. *Current Opinion in Neurobiology*, 21(4), 645–652.

Nestler, J. M., Ploskey, G. R., & Pickens, J. (1992). Responses of blueback herring to high-frequency sound and implications for reducing entrainment at hydropower dams. *North American Journal of Fisheries Management*, 12(4), 667–683.

Neuweiler, G. (1989). Foraging ecology and audition in echolocating bats. *Trends in Ecology and Evolution*, 4(6), 160–166.

Norris, K.S., & Møhl, B. (1983). Can odontocetes debilitate prey with sound? *American Naturalist*, 122(1), 85–104.

Norris, K. S., Prescott, J. H., Asa-Dorian, P. V., & Perkins, P. (1961). An experimental demonstration of echo-location behavior in the porpoise, *Tursiops truncatus* (Montagu). *Biological Bulletin*, 120(2), 163–176.

Nørum, U., Brinkløv, S., & Surlykke, A. (2012). New model for gain control of signal intensity to object distance in echolocating bats. *Journal of Experimental Biology*, 215(Pt 17), 3045–3054.

Nummela, S., Thewissen, J. G. M., Bajpal, S., Hussain, S. T., & Kumar, K. (2004). Eocene evolution of whale hearing. *Nature*, 430(7001), 776–778.

Pierce, G. W., & Griffin, D. R. (1938). Experimental determination of supersonic notes emitted by bats. *Journal of Mammalogy*, 19(4), 454–455.

Plachta, D. T., & Popper, A. N. (2003). Evasive responses of American shad (*Alosa sapidissima*). *Acoustic Research Letters Online*, 4(2), 25–30.

Pye, J. D. (1993). Is fidelity futile? The 'true' signal is illusory, especially with ultrasound. *Bioacoustic*, 4(4), 271–286.

Ratcliffe, J. M., & Nydam, M. L. (2008). Multimodal warning signals for a multiple predator world. *Nature*, 455(7209), 96–99.

Ratcliffe, J. M., Elemans, C. P. H., Jakobsen, L., & Surlykke, A. (2013). How the bat got its buzz. *Biology Letters*, 9, 20121031. doi: 10.1098/rsbl.2012.1031.

Remage-Healey, L., Nowacek, D. P., & Bass, A. H. (2006). Dolphin foraging sounds suppress calling and elevate stress hormone levels in a prey species, the Gulf toadfish. *Journal of Experimental Biology*, 209(Pt 22), 4444–4451.

Roeder, K. D. (1967). *Nerve cells and insect behavior*. Cambridge, MA: Harvard University Press.
Roeder, K. D., & Treat, A. E. (1961). The detection and evasion of bats by moths. *American Scientist*, 49, 135–148.
Rydell, J., & Arlettaz, R. (1994). Low-frequency echolocation enables the bat *Tadarida teniotis* to feed on tympanate insects. *Proceedings of the Royal Society of London B: Biological Sciences*, 257(1349), 175–178.
Safi, K., & Siemers, B. M. (2010). Implications of sensory ecology for species coexistence: Biased perception links predator diversity to prey size distribution. *Evolutionary Ecology*, 24(4), 703–713.
Schack, H. B., Malte, H., & Madsen, P. T. (2008). The response of Atlantic cod (*Gadus morhua*) to ultrasound-emitting predators: stress, behavioural changes or debilitation? *Journal of Experimental Biology*, 211(Pt 13), 2079–2086.
Schaub, A., Ostwald, J., & Siemers, B. M. (2008). Foraging bats avoid noise. *Journal of experimental Biology*, 211(Pt 19), 3174–3180.
Schmidt, S., & Thaller, J. (1994). Temporal auditory summation in the echolocating bat, *Tadarida brasiliensis*. *Hearing Research*, 77(1–2), 125–134.
Schnitzler, H.-U., & Henson, O. W. J. (1980). Performance of airborne animal sonar systems: I. Microchiroptera. In R. G. Busnel & J. F. Fish. (Eds.), *Animal sonar systems* (pp. 109–181). New York: Plenum Press.
Schnitzler, H.-U., & Kalko, E. K. V. (2001). Echolocation by insect-eating bats. *BioScience*, 51(7), 557–569.
Schuchmann, M., & Siemers, B. M. (2010). Variability in echolocation call intensity in a community of horseshoe bats: A role for resource partitioning or communication? *PLoS ONE*, 5(9), e12842.
Siemers, B. M., & Schnitzler, H.-U. (2000). Natterer's bat (*Myotis nattereri* Kuhl, 1818) hawks for prey close to vegetation using echolocation signals of very broad bandwidth. *Behavioral Ecology and Sociobiology*, 47(6), 400–412.
Siemers, B. M., & Schnitzler, H.-U. (2004). Echolocation signals reflect niche differentiation in five sympatric congeneric bat species. *Nature*, 429(6992), 657–661.
Simmons, J. A., Lavender, W. A., Lavender, B. A., Doroshow, C. A., Kiefer, S. W., Livingston, R., Scallet, A. C., & Rowley, D. E. (1974). Target structure and echo spectral discrimination by echolocating bats. *Science*, 186(4169), 1130–1132.
Simmons, J. A., Lavender, W. A., Lavender, B. A., Childs, J. E., Hulebak, K., Rigden, M. R., Sherman, J., & Woolman, B. (1978). Echolocation by free-tailed bats (*Tadarida*). *Journal of Comparative Physiology A: Neuroethology, Sensory, Neural, and Behavioral Physiology*, 125(4), 291–299.
Simmons, J. A., Freedman, E. G., Stevenson, S. B., Chen, L., & Wohlgenant, T. J. (1989). Clutter interference and the integration time of echoes in the echolocating bat, *Eptesicus fuscus*. *Journal of the Acoustical Society of America*, 86(4), 1318–1332.
Simmons, N. B., Seymour, K. L., Habersetzer, J., & Gunnell, G. F. (2008). Primitive early Eocene bat from Wyoming and the evolution of flight and echolocation. *Nature*, 451(7180), 818–821.
Strother, G. K., & Mogus, M. (1970). Acoustical beam patterns for bats: Some theoretical considerations. *Journal of the Acoustical Society of America*, 48(6), 1430–1432.
Surlykke, A., & Bojesen, O. (1996). Integration time for short broad band clicks in echolocating FM-bats (*Eptesicus fuscus*). *Journal of Comparative Physiology A: Neuroethology, Sensory, Neural, and Behavioral Physiology*, 178(2), 235–241.
Surlykke, A., & Kalko, E. K. V. (2008). Echolocating bats cry out loud to detect their prey. *PLoS ONE*, 3(4), e2036.
Surlykke, A., & Moss, C. F. (2000). Echolocation behavior of big brown bats, *Eptesicus fuscus*, in the field and the laboratory. *Journal of the Acoustical Society of America*, 108(5 Pt 1), 2419–2429.
Surlykke, A. Miller, L. A., Møhl, B., Andersen, B. B., Christensen-Dalsgaard, J., & Jørgensen, M. B. (1993). Echolocation in two very small bats from Thailand: *Craseonycteris thonglongyai* and *Myotis siligorensis*. *Behavioral Ecology and Sociobiology*, 33(1), 1–12.

Surlykke, A., Filskov, M., Fullard, J. H., & Forrest, E. (1999). Auditory relationships to size in noctuid moths: Bigger is better. *Naturwissenschaften*, 86(5), 238–241.

Surlykke, A., Yack, J. E., Spence, A. J., & Hasenfuss, I. (2003). Hearing in hooktip moths (Drepanidae: Lepidoptera). *Journal of Experimental Biology*, 206, 2653–2663.

Surlykke, A., Ghose, K., & Moss, C. F. (2009). Acoustic scanning of natural scenes by echolocation in the big brown bat, *Eptesicus fuscus*. *Journal of Experimental Biology*, 212(Pt 7), 1011–1020.

Surlykke A., Jakobsen, L., Kalko, E. K. V., & Page, R. A. (2013). Echolocation intensity and directionality of perching and flying fringe-lipped bats, *Trachops cirrhosus* (Phyllostomidae) *Frontiers in Integrative Physiology*, 4, Article 143.

Teloni, V., Johnson, M. P., Miller, P. J. O., & Madsen, P. T. (2008). Shallow food for deep divers: Dynamic foraging behavior of male sperm whales in a high latitude habitat. *Journal of Experimental Marine Biology and Ecology*, 354(1), 119–131.

ter Hofstede, H. M., & Fullard, J. H. (2008). The neuroethology of song cessation in response to gleaning bat calls in two species of katydids, *Neoconocephalus ensiger* and *Amblycorypha oblongifolia*. *Journal of Experimental Biology*, 211(Pt 15), 2431–2441.

ter Hofstede, H. M., Ratcliffe, J. M., & Fullard, J. H. (2008). Nocturnal activity positively correlated with auditory sensitivity in noctuoid moths. *Biology Letters*, 4(3), 262–265.

ter Hofstede, H. M., Killow, J., & Fullard, J. H. (2009). Gleaning bat echolocation calls do not elicit antipredator behaviour in the Pacific Weld cricket, *Teleogryllus oceanicus* (Orthoptera: Gryllidae). *Journal of Comparative Physiology A: Neuroethology, Sensory, Neural, and Behavioral Physiology*, 195(8), 769–776.

Tyack, P. L., Johnson, M., Aguilar Soto, N., Sturlese, A., & Madsen, P. T. (2006). Extreme diving of beaked whales. *Journal of Experimental Biology*, 209, 4238–4253.

Ulanovsky, N., & Moss, C. F. (2008). What the bat's voice tells the bat's brain. *Proceedings of the National Academy of Sciences of the USA*, 105(25), 8491–8498.

Urick, R. J. (1995). *Principles of underwater sound*, 3rd ed. Los Altos Hills: Peninsula Publishing.

Verfuss, U. K., Miller, L. A., Pilz, P. K. D., & Schnitzler, H. U. (2009). Echolocation by two foraging harbor porpoises. *Journal of Experimental Biology*, 212(Pt 6), 823–834.

Veselka, N., McErlain, D. D., Holdsworth, D. W., Eger, J. L., Chherm, R. K., Mason, M. J., Brain, K. L., Faure, P. A., & Fenton, M. B. (2010). A bony connection signals laryngeal echolocation in bats. *Nature*, 463(7283), 939–942.

Villadsgaard, A., Wahlberg, M., & Tougaard, J. (2007). Echolocation signals of wild harbour porpoises, *Phocoena phocoena*. *Journal of Experimental Biology*, 210(Pt 1), 56–64.

Wahlberg, M. (2002). The acoustic behaviour of diving sperm whales observed with a hydrophone array. *Journal of Experimental Marine Biology and Ecology*, 281(1–2), 53– 62.

Wahlberg, M., Jensen, F. H., Aguilar de Soto, N., Beedholm, K., Bejder, L., Oliveira, C., Rasmussen, M., Simon, M., Villadsgaard, A., & Madsen, P. T. (2011a). Source parameters of echolocation clicks from wild bottlenose dolphins (*Tursiops aduncus* and *Tursiops truncatus*). *Journal of the Acoustical Society of America*, 130(4), 2263–2274.

Wahlberg, M., Beedholm, K., Heerfordt, A., & Møhl, B. (2011b). Characteristics of biosonar signals from the northern bottlenose whale, *Hyperoodon ampullatus*. *Journal of the Acoustical Society of America*, 130(5), 3077–3084.

Watwood, S. L., Miller, P. J. O., Johnson, M. P., Madsen, P. T., & Tyack, P. L. (2006). Deep-diving foraging behavior of sperm whales (*Physeter Macrocephalus*). *Journal of Animal Ecology*, 75(3), 814–825.

Weinberg, M., Meyer, C. F. J., & Kalko, E. K. V. (2006). Activity pattern of the trawling phyllostomid bat, *Macrophyllum macrophyllum*, in Panama. *Biotropica*, 38(1), 69–76.

Werth, A. J. (2006). Mandibular and dental variation and the evolution of suction feeding in odontoceti. *Journal of Mammalogy*, 87(3), 579–588.

Wilson, W. W., & Moss, C. F. (2004). Sensory-motor behavior of free-flying Fm bats during target capture. In J. Thomas, C. F. Moss, & M. Vater (Eds.), *Echolocation in bats and dolphins* (pp. 22–26). Chicago: University of Chicago Press.

Wilson, M., Hanlon, R. T., Tyack, P. L., & Madsen, P. T. (2007). Intense ultrasonic clicks from echolocating toothed whales do not elicit anti-predator responses or debilitate the squid *Loligo pealeii*. *Biology Letters*, 3(3), 225–227.

Wilson, M., Acolas, M. L., Bégout, M. L., Madsen, P. T., & Wahlberg, M. (2008). Allis shad (*Alosa alosa*) exhibit an intensity-graded behavioral response when exposed to ultrasound. *Journal of the Acoustical Society of America*, 124 (4), EL243–EL247.

Wilson, M., Montie, E. W., Mann, K. A., & Mann, D. A. (2009). Ultrasound detection in the gulf menhaden requires gas-filled bullae and an intact lateral line. *Journal of Experimental Biology*, 212(Pt 21), 3422–3427.

Wilson, M., Schack H. B., Madsen, P. T., Surlykke, A., & Wahlberg, M. (2011). Directional escape behavior in allis shad (*Alosa alosa*) exposed to ultrasonic clicks mimicking an approaching toothed whale. *Journal of Experimental Biology*, 214(Pt 1), 22–29.

Wilson, M., Wahlberg, M., Surlykke, A. M., & Madsen, P. T. (2013). Ultrasonic predator-prey interactions in water—a parallel to the bat-moth interaction in air? *Frontiers in Physiology*, 4, | Article 137.

Wisniewska, D. M., Johnson, M., Beedholm, K., Wahlberg, M., & Madsen, P. T. (2012). Acoustic gaze adjustments during active target selection in echolocating porpoises. *Journal of Experimental Biology*, 215, 4358–4373.

Zagaeski, M. (1987). Some observations on the prey stunning hypothesis. *Marine Mammal Science*, 3(3), 275–279.

Zimmer, W. M. X., Johnson, M. P., Madsen, P. T., & Tyack, P. L. (2005). Echolocation clicks of Cuvier's beaked whales (*Ziphius cavirostris*). *Journal of the Acoustical Society of America*, 117(6), 3919–3927.

MIX
Papier aus verantwortungsvollen Quellen
Paper from responsible sources
FSC® C105338

If you have any concerns about our products,
you can contact us on
ProductSafety@springernature.com

In case Publisher is established outside the EU,
the EU authorized representative is:
Springer Nature Customer Service Center GmbH
Europaplatz 3, 69115 Heidelberg, Germany

Printed by Libri Plureos GmbH
in Hamburg, Germany